Stong

Chemical biology ...

09621

514 398-3864

Current Topics in Microbiology and Immunology

Volume 322

Tony Romeo

Editor

Bacterial Biofilms

🜨 Springer

T. Romeo
Department of Microbiology and Immunology
Emory University School of Medicine
3105 Rollins Research Center
1510 Clifton Rd., N.E., Atlanta
GA 30322, USA
romeo@microbio.emory.edu

ISBN 978-3-540-75417-6   e-ISBN 978-3-540-75418-3
DOI 10.1007/978-3-540-75418-3

Current Topics in Microbiology and Immunology ISSN 007-0217x

Library of Congress Catalog Number: 2007942367

Cover Design: WMXDesign GmbH, Heidelberg, Germany

Printed on acid-free paper

9 8 7 6 5 4 3 2 1

springer.com

Preface

Throughout the biological world, bacteria thrive predominantly in surface-attached, matrix-enclosed, multicellular communities or biofilms, as opposed to isolated planktonic cells. This choice of lifestyle is not trivial, as it involves major shifts in the use of genetic information and cellular energy, and has profound consequences for bacterial physiology and survival. Growth within a biofilm can thwart immune function and antibiotic therapy and thereby complicate the treatment of infectious diseases, especially chronic and foreign device-associated infections. Modern studies of many important biofilms have advanced well beyond the descriptive stage, and have begun to provide molecular details of the structural, biochemical, and genetic processes that drive biofilm formation and its dispersion. There is much diversity in the details of biofilm development among various species, but there are also commonalities. In most species, environmental and nutritional conditions greatly influence biofilm development. Similar kinds of adhesive molecules often promote biofilm formation in diverse species. Signaling and regulatory processes that drive biofilm development are often conserved, especially among related bacteria. Knowledge of such processes holds great promise for efforts to control biofilm growth and combat biofilm-associated infections.

This volume focuses on the biology of biofilms that affect human disease, although it is by no means comprehensive. It opens with chapters that provide the reader with current perspectives on biofilm development, physiology, environmental, and regulatory effects, the role of quorum sensing, and resistance/phenotypic persistence to antimicrobial agents during biofilm growth. The next chapters are devoted to common problematic biofilms, those that colonize venous and urinary catheters. The final series of chapters examines biofilm formation by four species that are important pathogens and well-studied models, one of which, *Yersinia pestis,* cleverly adopts a biofilm state of growth within its insect vector to promote disease transmission to mammalian hosts.

Thanks are due to the authors who devoted their time and energy to write these chapters, the anonymous reviewers for their thoughtful comments, and Anne Clauss at Springer for excellent editorial assistance. I hope that you find the chapters to be interesting and informative, and that this volume provides both a snapshot of this burgeoning field and a stimulus for further investigation.

Tony Romeo

Contents

Contributors

C. Aguilar
Department of Microbiology and Molecular Genetics,
Harvard Medical School, Boston, MA 02115, USA

G.G. Anderson
Department of Microbiology and Immunology, Dartmouth Medical School,
Hanover, NH 03755, USA

R.M. Donlan
Division of Healthcare Quality Promotion, Centers for Disease Control and
Prevention, 1600 Clifton Road, N.E., Mail Stop C16, Atlanta, GA 30333, USA,
rld8@cdc.gov

A.M. Earl
Department of Microbiology and Molecular Genetics, Harvard Medical School,
Boston, MA 02115, USA

D.L. Erickson
Laboratory of Zoonotic Pathogens, Rocky Mountain Laboratories, NIH, NIAID,
Hamilton, MT 59840, USA

J.-M. Ghigo
Groupe de Génétique des Biofilms, Institut Pasteur, CNRS URA 2172, 25 rue du
Dr. Roux, 75724 Paris Cedex 15, France,
jmghigo@pasteur.fr

C.C. Goller
Department of Microbiology and Immunology, Emory University School of
Medicine, 3105 Rollins Research Center,
1510 Clifton Rd., N.E., Atlanta, GA 30322, USA

J.K. Hatt
Research Service, Veterans Affairs Medical Center, Emory University School
of Medicine, Atlanta, GA, USA

B.J. Hinnebusch
Laboratory of Zoonotic Pathogens, Rocky Mountain Laboratories, NIH, NIAID,
Hamilton, MT 59840, USA, jhinnebusch@niaid.nih.gov

Y. Irie
Department of Microbiology, University of Washington, Box 357242, HSB
Room K-343B, 1959 NE Pacific St., Seattle, WA 98195-7242, USA

R. Kolter
Department of Microbiology and Molecular Genetics, Harvard Medical School,
Boston, MA 02115, USA, rkolter@hms.harvard.edu

K.P. Lemon
Department of Microbiology and Molecular Genetics, Harvard Medical School,
Boston, MA, 02115, USA

K. Lewis
Antimicrobial Discovery Center and Department of Biology, Northeastern
University, 360 Huntington Avenue, Boston, MA 02459, USA, k.lewis@neu.edu

G.A. O'Toole
Department of Microbiology and Immunology, Dartmouth Medical School,
Hanover, NH 03755, USA, George.O'Toole@Dartmouth.edu

M. Otto
Laboratory of Human Bacterial Pathogenesis, National Institute of Allergy and
Infectious Diseases, The National Institutes of Health, Rocky Mountain
Laboratories, Hamilton, MT, USA, motto@niaid.nih.gov

M.R. Parsek
Department of Microbiology, University of Washington, Box 357242, HSB Room
K-343B, 1959 NE Pacific St., Seattle, WA 98195-7242, USA,
parsem@u.washingon.edu

P.N. Rather
Department of Microbiology and Immunology, Emory University School of
Medicine 3001 Rollins Research Center, Atlanta, GA, USA, prather@emory.edu

T. Romeo
Department of Microbiology and Immunology, Emory University School
of Medicine, 3105 Rollins Research Center , 1510 Clifton Rd., N.E., Atlanta,
GA, 30322 USA, romeo@microbio.emory.edu

Beloin A. Roux
Groupe de Génétique des Biofilms, Institut Pasteur, CNRS URA 2172, 25 rue du
Dr. Roux, 75724 Paris Cedex 15, France

A. M. Spormann
Departments of Chemical Engineering of Civil and Environmental Engineering,
and of Biological Sciences, Clark Center E250, Stanford University, Stanford,
CA 94305-5429, USA,
spormann@stanford.edu

A.H. Tart
Department of Microbiology and Immunology, Wake Forest University School
Medicine, Medical Center Blvd., Winston Salem, NC 27157, USA

H.C. Vlamakis
Department of Microbiology and Molecular Genetics, Harvard Medical School,
Boston, MA 02115, USA

D.J. Wozniak
Department of Microbiology and Immunology, Wake Forest University School
Medicine, Medical Center Blvd., Winston Salem, NC 27157, USA

Biofilm Development with an Emphasis on *Bacillus subtilis*

K. P. Lemon, A. M. Earl, H. C. Vlamakis, C. Aguilar, and R. Kolter(✉)

Abstract Our understanding of the molecular mechanisms involved in biofilm formation has increased tremendously in recent years. From research on diverse bacteria, a general model of bacterial biofilm development has emerged. This model can be adjusted to fit either of two common modes of unicellular existence: nonmotile and motile. Here we provide a detailed review of what is currently known about biofilm formation by the motile bacterium *Bacillus subtilis*. While the ability of bacteria to form a biofilm appears to be almost universal and overarching themes apply, the combination of molecular events necessary varies widely, and this is reflected in the other chapters of this book.

In most natural settings, bacteria are found predominantly in biofilms (Henrici 1933; Costerton et al. 1999; Hall-Stoodley et al. 2004). The widespread recognition that biofilms impact myriad environments, from water pipes to indwelling devices in hospital patients, led to an increased interest in investigating the molecular mechanisms underlying the formation and maintenance of these communities. As a consequence, we have recently witnessed much growth in our knowledge of biofilms. The ability to form biofilms, once considered the domain of a few species, is now seen as a nearly universal attribute of microorganisms. It has also become evident that the pathways utilized by bacteria to build biofilms are extremely

R. Kolter
Department of Microbiology and Molecular Genetics, Harvard Medical School,
Boston, MA, 02115 USA
rkolter@hms.harvard.edu

T. Romeo (ed.), *Bacterial Biofilms.*
Current Topics in Microbiology and Immunology 322.
© Springer-Verlag Berlin Heidelberg 2008

se, varying enormously among different species and under different
onmental conditions. There are, however, several common features among all
biofilms examined to date:

1. Constituent cells are held together by an extracellular matrix composed of
   exopolysaccharides (EPS), proteins, and sometimes nucleic acids (Whitchurch
   et al. 2002; Branda et al. 2005; Lasa 2006).
2. Biofilm development occurs in response to extracellular signals, both environmental and self-produced (Kolter and Greenberg 2006; Spoering and Gilmore 2006).
3. Biofilms afford bacteria with protection from a wide array of environmental
   insults, as diverse as antibiotics (Mah and O'Toole 2001), predators (Kadouri
   et al. 2007), and the human immune system (Singh et al. 2000; Fedtke et al.
   2004; Leid et al. 2005).

Initial studies on bacterial biofilms were predominantly descriptive. By applying
novel microscopic approaches, most notably laser scanning confocal microscopy, a
whole new universe of biofilm architecture became apparent. These approaches,
coupled with time-lapse video microscopy and microsensors, have given us a more
complete view of the complex structure of biofilms. These descriptive advances
have been followed in the last decade by an outburst in the number of molecular
genetic analyses carried out in biofilms. Today, investigators are applying a wide
range of molecular biological approaches to the study of the regulatory processes
that underlie biofilms. As is made clear in the chapters of this book, there are many
examples where molecular genetics has had great impact on biofilm research. For
almost every organism that has been investigated, we can now draw working
genetic models for the steps in the pathways of biofilm development. There are
genes expressed during each step of development that can serve as reporters of that
stage and there are genes whose functions are essential for each particular step to
be completed successfully. In spite of these advances, one can argue that biofilm
genetics is still in its infancy. Investigators continue to identify novel genes that are
either essential for, or expressed during, biofilm formation in many different organisms. Even among genes that have been previously identified, we have only recently
begun to elucidate exactly how some of these genes' products contribute to biofilm
development, maintenance, and dissolution.

1 How Do We Study Biofilms in the Laboratory?

Pipelines, catheters, teeth, plant roots, and the lungs of cystic fibrosis patients
are but a few of the most widely recognized surfaces where the effects of biofilms
are readily apparent. The biofilms that form on such surfaces almost invariably
house a complex mixture of species, rendering them not particularly amenable to
molecular genetic studies. To be able to address questions regarding the molecular
basis of biofilm formation, investigators have developed artificial biofilm model
systems that are easy to control and reproducible from laboratory to laboratory.

While there are numerous laboratory conditions that favor biofilm formation, investigators have routinely utilized four general systems for the study of biofilms. First among these systems is the flow cell (Christensen et al. 1999; Branda et al. 2005). Flow cells are small chambers with transparent surfaces where submerged biofilms can form and be continually fed fresh nutrients. The submerged biofilms that form on flow cells are particularly amenable to observation through confocal scanning laser microscopy. This allows for the capture of images of biofilm development in real time. The results obtained using flow cells have provided us with the familiar images of submerged biofilms consisting of mushroom-like structures separated by water-filled channels (Christensen et al. 1999; Branda et al. 2005). However, flow cells can be cumbersome and are not easily adapted for high throughput mutant screens. Submerged biofilms can also be studied in batch culture under conditions of no flow in microtiter dishes (O'Toole and Kolter 1998b; O'Toole et al. 1999). In this system, large numbers of samples can be quickly analyzed. Using the microtiter dish assay system, many investigators have carried out high-throughput screens and identified genes involved in biofilm formation and maintenance in numerous bacterial species (O'Toole and Kolter 1998a, 1998b; Pratt and Kolter 1998; Watnick and Kolter 1999; Watnick et al. 2001; Valle et al. 2003). The floating pellicles that form at the liquid-air interface of standing cultures represent another form of biofilm that is easily studied and adaptable for mutant screens (Guvener and McCarter 2003; Friedman and Kolter 2004; Enos-Berlage et al. 2005). Finally, the colonies that grow on the surface of agar dishes and demonstrate macroscopically complex architecture are now widely recognized as a form of biofilm (reviewed in Branda et al. 2005). This complex colony morphology correlates with production of extracellular matrix and the morphological variation observed in colonies often correlates with cells' ability to form robust biofilms in other assays. Like pellicles and the biofilms that form on the walls of microtiter dish wells, colonies are amenable to high-throughput screens to identify genes involved in biofilm formation and maintenance.

While all four systems for studying biofilm formation have been successful in broadening our understanding of biofilm development among diverse microorganisms, it is important to note that there can be variation among the phenotypes observed as one moves between systems. For example, mutants that exhibit a biofilm defect in one system may have imperceptible or no phenotype in another (O'Toole and Kolter 1998b). The converse also holds true; there are classes of biofilm mutants that do have a reproducible phenotype across all systems, for example, mutants defective for extracellular matrix production (Friedman and Kolter 2004). Ultimately, of these four general systems, no single one stands out as clearly superior; rather, the methods complement each other. Analyses of the phenotypic changes expressed by different mutants using combinations of several, or all four, of these systems can greatly aid our understanding of the role that different gene products play in biofilm development.

Individual species of bacteria vary greatly with regards to the environmental conditions under which they will produce maximal amounts of biofilm. These optimal conditions may, in fact, be telling us something about the biology and/or ecology of the organism. We have also noted that many commonly used laboratory strains produce only frail or weak biofilms when compared to wild strains of the same species. In a number of instances, it has been possible to show that this stems from laboratory strains having accumulated numerous mutations over years of passaging through liquid cultures in a process we refer to as domestication (Branda et al. 2001; Valle et al. 2003). In working with liquid cultures of dispersed populations of cells, we appear to have unwittingly enriched for strains that have lost some of their potential to form structured multicellular communities while growing rapidly in liquid culture.

2   A General Model for Biofilm Development

Biofilm formation is a developmental process in which bacteria undergo a regulated lifestyle switch from a nomadic unicellular state to a sedentary multicellular state where subsequent growth results in structured communities and cellular differentiation. Results of prior work by many groups allow the construction of a hypothetical developmental model for biofilm formation that can be generalized for many different bacterial species. This model can be adjusted to fit either of two general modes of unicellular lifestyle: nonmotile and motile.

In the case of non-motile species (Fig. 1), when conditions are propitious for biofilm formation, individual bacteria appear to increase the expression of adhesins on their outer surface, i.e., they increase their "stickiness". This increased stickiness promotes both cell-cell adherence and cell-surface adherence when these bacteria encounter a surface (Gotz 2002). For example, in the case of some strains of staphylococcal species, surface-expressed proteins, including Bap, promote cell-cell interaction and contribute to the extracellular matrix (Lasa and Penades 2006). Many other species, both nonmotile and motile, harbor homologs of Bap. A unifying feature of these large extracellular proteins is the presence of repeated domains. At the level of the *bap* gene, these repeats have been shown to be recombinogenic, resulting in the production of proteins of variable length within a biofilm population (Latasa et al. 2006). Yet, the significance of this variability in the size of Bap proteins within a biofilm remains unknown. Nonmotile species also produce exopolysaccharides (EPS) that form an integral part of the extracellular matrix. One example of this is the PIA or PNAG EPS, produced by the gene products of the *ica* operon of staphylococcal species. Thus, in nonmotile bacteria, changes in cell surface proteins, along with the production of EPS, play a critical role in the initiation of biofilm formation (Gotz 2002; Latasa et al. 2005).

In the case of motile species (Fig. 2), when conditions favor biofilm formation, individual bacteria localize to a surface and initiate a dramatic lifestyle switch. Motility is lost and bacteria begin to produce an extracellular matrix that holds the cells together. For a number of motile organisms, the dominant role for flagella in initiation of biofilm formation is to provide motility as flagella-minus and paralyzed flagella mutants are comparably defective in biofilm formation (Pratt and Kolter

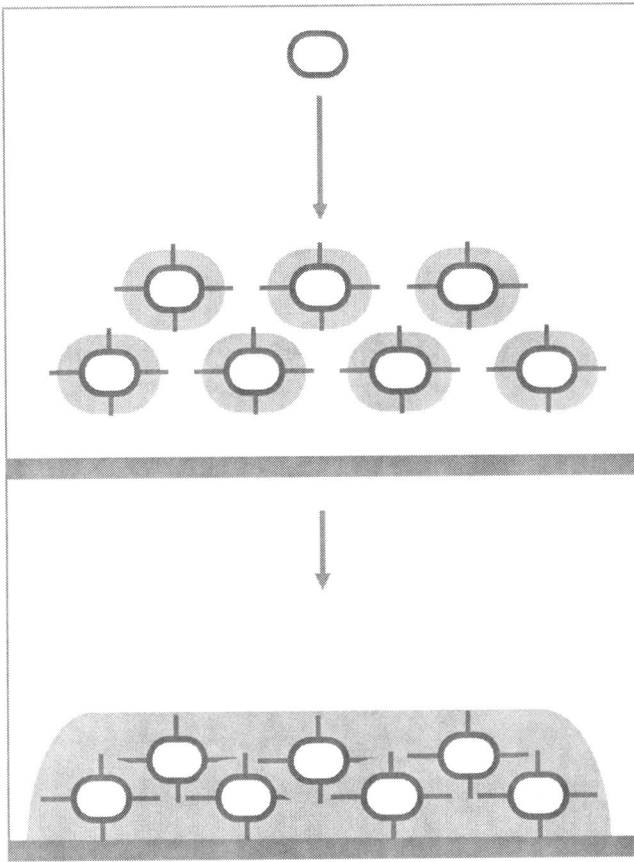

Fig. 1 General model for biofilm formation by nonmotile bacteria. Thick gray lines represent surfaces. (Top panel) To initiate biofilm formation nonmotile cells increase expression of adhesins. (Bottom panel) This results in surface-adhered, matrix-enclosed cells in a biofilm

1998; Watnick and Kolter 1999; Lemon et al. 2007). In fact, in *Listeria monocytogenes* supplying exogenous cell movement directed toward the surface via centrifugation restores wild type levels of initial surface adhesion to nonmotile mutants (Lemon et al. 2007). In these cases, it appears that motility is the driving force that overcomes repulsive forces between the bacteria and the surface. Initial encounters with a surface usually lead to transient adherence. This transient adherence can result in either a stable surface association, and a subsequent switch to biofilm development, or in a return to planktonic existence.

The first strides in understanding the molecular mechanisms of biofilm formation were made in Gram-negative Proteobacteria, especially *Vibrio cholerae*, *Escherichia coli*, *Pseudomonas fluorescens*, and *Pseudomonas aeruginosa*, using surface-adhered biofilm assays (O'Toole and Kolter 1998a, 1998b; Pratt and Kolter 1998; Watnick and Kolter 1999; Watnick et al. 2001). Based on mutant phenotypes from these organisms, biofilm formation can be divided into five genetically distinct stages:

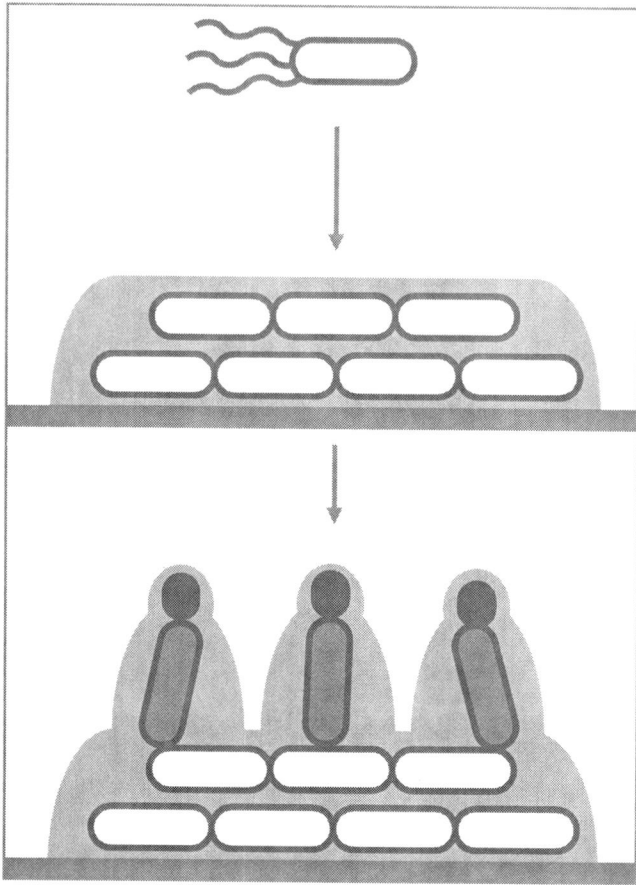

Fig. 2 General model for biofilm formation by motile bacteria. Thick gray lines represent surfaces. (Top panel) Motile, planktonic cells transition to nonmotile, matrix-producing, surface-adhered cells in a biofilm. (Bottom panel) Subsequently, cells differentiate within the biofilm

1. Initial surface attachment
2. Monolayer formation
3. Migration to form multilayered microcolonies
4. Production of extracellular matrix
5. Biofilm maturation with characteristic three-dimensional architecture (O'Toole et al. 2000)

These general stages provide a paradigm for studying biofilms formed by motile bacteria, although the precise details vis-à-vis regulation of this process do vary greatly from species to species. While initial surface attachment is dependent on flagella-mediated motility in a wide variety of motile bacteria, in some Gram-negative bacteria, microcolony formation and final three-dimensional architecture are also dependent on type IV pili-associated surface motility, which is notably absent in

Gram-positive bacteria (with the exception of Clostridia ssp.; Varga et al. 20_ O'Toole et al. 2000). After cells have adhered, matrix production begins; the extracellular matrix serves as an organizing principle that permits the building of structured communities within which there can be extensive cellular differentiation, shown as cells of different shapes and shades in Fig. 2. While these genetically defined stages of biofilm formation correlate with the temporal progression of biofilm development, it should be noted that the production of extracellular matrix appears to overlap with all stages that occur after initial surface adhesion. Also, while cellular differentiation within a biofilm is illustrated in our model for biofilm formation by a motile bacterium (see Figs. 1 and 2), we postulate that many bacteria, both motile and nonmotile, undergo comparable processes during biofilm formation.

While it is generally agreed upon that matrix production and motility are mutually exclusive, much remains to be learned regarding the molecular mechanisms that underlie this lifestyle switch for most bacterial species. Through a combination of genetic and biochemical approaches, our group, in close collaboration with the group of Richard Losick, has begun to identify the molecular regulatory circuitry that governs the transition from motile cells to matrix-producing cells of the Gram-positive soil bacterium *Bacillus subtilis*.

3 *Bacillus subtilis* as a Model System for Studying Biofilm Formation

B. subtilis, a Gram-positive motile rod-shaped bacterium, is best known for its ability to become competent and undergo sporulation in response to starvation and high population densities (Grossman 1995). The regulatory processes controlling *B. subtilis* sporulation and competence have been extensively characterized (Sonenshein et al. 2002; Piggot and Hilbert 2004). At the molecular level, the regulation of *B. subtilis* endospore formation is probably the best understood microbial developmental process; however, until very recently sporulation has been analyzed almost exclusively from the perspective of a single cell and not as a process occurring within a spatially organized community.

Most of the *B. subtilis* biofilm data come from studies on the development of complex, wrinkled colonies and from the development of pellicles at an air-liquid interface, although some studies have focused on solid surface-associated biofilms (Hamon and Lazazzera 2001; Stanley et al. 2003; Hamon et al. 2004). Biofilm formation by *B. subtilis* follows a distinct developmental pathway (Branda et al. 2001) (Fig. 3). After inoculation of standing cultures in a defined minimal medium containing glycerol as the major carbon source (MSgg; Branda et al. 2001), motile cells proliferate throughout the liquid as planktonic cells until they reach a density of approximately 5×10^7 cfu/ml after 1 day at room temperature (Fig. 3a). At that point, the vast majority of the cells begin to migrate to the air-liquid interface, where they form a floating biofilm or pellicle on the surface of the medium. The

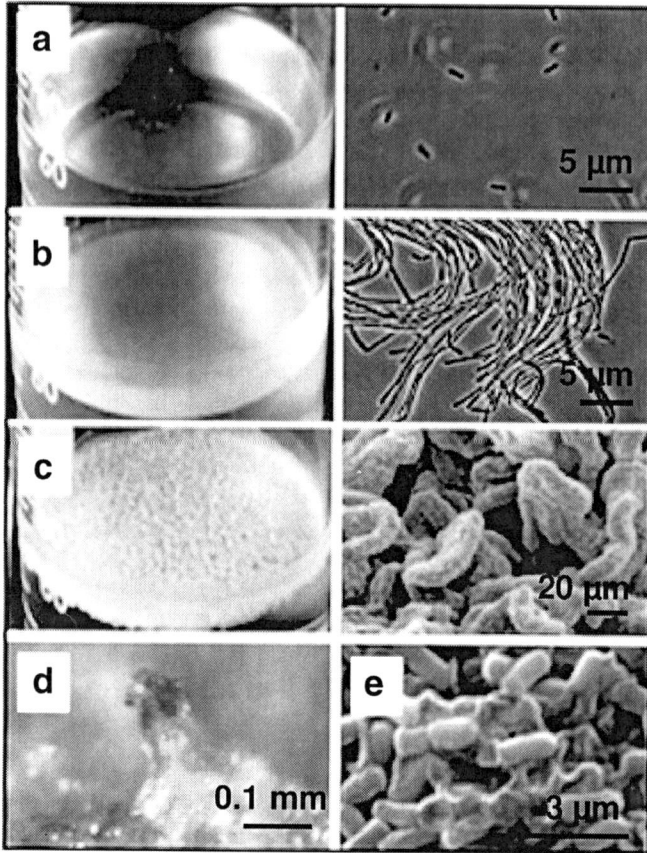

Fig. 3 Development of a *B. subtilis* biofilm. Panels **a-c** show pellicles on the *left* and microscopic images on the *right*. **a** and **b** are phase-contrast images, **c** and **e** are SEM images. **d** is from a dissecting microscope

pellicle is readily apparent but flat after 3 days (Fig. 3b). At this and subsequent times, the few remaining planktonic cells (<10^5 cfu/ml) retain their motility and do not sporulate. In contrast, cells within the pellicle undergo dramatic differentiation as they continue to proliferate. Cells become nonmotile and form long chains that are aligned in parallel (Fig. 3b). After 5 days of incubation, and as the cell mass increases, the pellicle begins to wrinkle, and, within the wrinkles, some groups of cells begin to grow as aerial projections (Fig. 3c). The tips of these projections serve as preferential sites of sporulation, as evidenced by the localized expression of the sporulation-specific gene *sspE* fused to *lacZ* (Branda et al. 2001) (Fig. 3d). Because a similar spatial organization of sporulation is characteristic of myxobacterial fruiting bodies, we refer to the *B. subtilis* aerial structures as fruiting body-like structures. Aerial structures indistinguishable from those observed in pellicles form at the edges of colonies grown on agar plates. To achieve such spatiotemporal organization, the cells rely on an extracellular matrix to hold them together; scanning electron microscopy of cells from a 5-day-old colony reveals that they are indeed enclosed in such a matrix (Fig. 3e).

It is important to note that these structured communities and their high degree of cellular differentiation are only apparent when wild isolates are analyzed. Most of the standard laboratory strains, derivatives of strain *B. subtilis* 168, do not display such robust community structure, presumably as a result of domestication. Thus, most studies have focused primarily on the wild strain *B. subtilis* NCIB3610 (henceforth referred to as wild or 3610).

4   The Genetic Circuitry of *Bacillus subtilis* Biofilm Formation

During biofilm development, *B. subtilis* switches from being flagellated, motile single cells to growing in long chains of nonmotile cells that form parallel bundles (Branda et al. 2001). Figure 4 is a simplified view of the key players in this lifestyle switch. The transcriptional regulator SinR serves as the master regulator governing this switch (Kearns et al. 2005). In motile cells, SinR represses the transcription of genes responsible for matrix production and indirectly promotes cell separation and motility (Branda et al. 2006). SinR is constitutively produced, and when conditions become favorable for biofilm formation, SinR activity is antagonized. SinI and two newly identified proteins, YlbF and YmcA, all serve to directly and/or indirectly

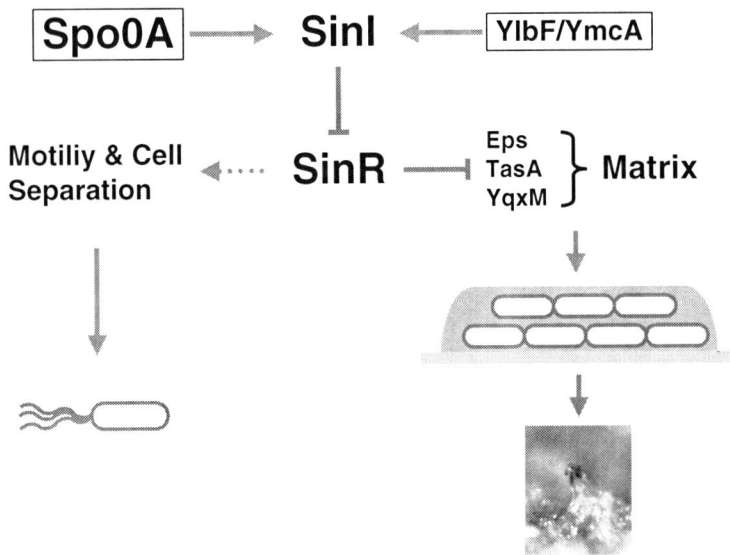

Fig. 4 Simplified view of the genetic circuitry governing *B. subtilis*'s lifestyle switch from nomadic to a sedentary existence

antagonize SinR activity. Lowered SinR activity results in loss of motility, cell chain formation, and matrix production. The extracellular matrix responsible for proper biofilm development in 3610 consists primarily of an exopolysaccharide (EPS) and a protein, TasA (Branda et al. 2006). Once this matrix is produced, the community develops a high degree of spatiotemporal organization culminating with sporulation occurring preferentially at the tips of aerial structures.

Prior to the discovery of SinR as the master regulator of biofilm formation in *B. subtilis*, SpoOA and σ^H were identified as transcriptional factors involved in biofilm development (Branda et al. 2001; Hamon and Lazazzera 2001). Two transcriptional profiling studies had identified members of the SpoOA and σ^H regulons (Fawcett et al. 2000; Britton et al. 2002). One fifteen-gene operon designated as *yveK-T yvfA-F*, later renamed *epsA-O*, under control of both SpoOA and σ^H, was predicted to encode products likely to be involved in EPS synthesis and export (Branda et al. 2001). EpsA and B are similar to enzymes that regulate EPS chain length, EpsC is similar to nucleotide sugar synthesizing enzymes, EpsD, E, F, H, J, L, and M are all predicted to be glycosyl transferases, EpsK is similar to proteins involved in saccharide export, and EpsG is similar to proteins involved in polymerization of EPS repeating units. Mutants lacking EpsG and EpsH, as well a mutant lacking the entire *eps* operon, all produce flat colonies and extremely fragile pellicles. Microscopic examination of these mutants revealed that the product(s) of these genes is important for structuring the community. Phase-contrast microscopic analyses made it clear that *eps* mutants still proliferate as long chains, but these chains no longer align, nor they are bound together (Fig. 5) (Branda et al. 2001). Scanning electron microscopy (SEM) also revealed bare cells with only small amounts of extracellular material remaining.

In addition to the EPS component of the matrix, three proteins, encoded in the three-gene operon *yqxM-sipW-tasA*, were identified as involved in matrix assembly in a transposon mutant screen for genes involved in biofilm formation (Branda et al. 2004). In-frame deletion mutations in any of the genes of the three-gene operon *yqxM-sipW-tasA* result in defective pellicle formation and defective colony architecture. Microscopic analyses demonstrate that, like the *eps* mutants, *tasA* and *yqxM* mutants produce cell chains that are not held together and are defective for extracellular matrix production. The *tasA* and *yqxM* mutants alone or in combination, as well as a mutant deleted for the entire operon, have similar phenotypes, suggesting that TasA and YqxM act via the same mechanism. The *yqxM* and *tasA* genes encode preproteins that are converted to their mature, secreted forms by the product of *sipW*, a dedicated signal peptidase (Stover and Driks 1999a, 1999b). Previous to these findings, relatively little was known about the function of YqxM and TasA. YqxM was detected in culture supernatants, but only in the presence of high salt, suggesting that it is a cell-surface-associated protein (Stover and Driks 1999a). TasA was detected in the supernatant as well as associated with both cells and spores, and has been reported to have a poorly characterized antimicrobial activity (Serrano et al. 1999; Stover and Driks 1999b).

TasA is present in the biofilm's extracellular matrix. When pellicles were separated from the culture medium, no TasA was detected in the medium (Branda

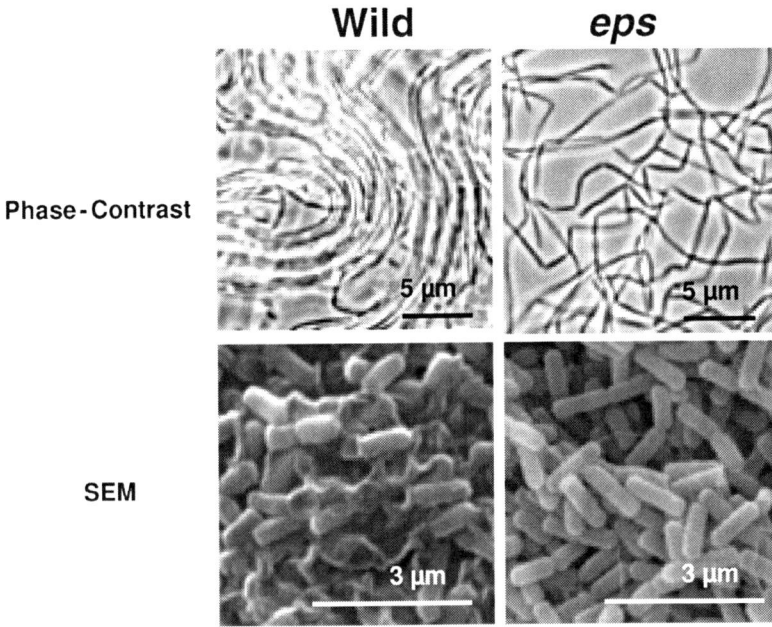

Fig. 5 Phenotype of *eps* mutant

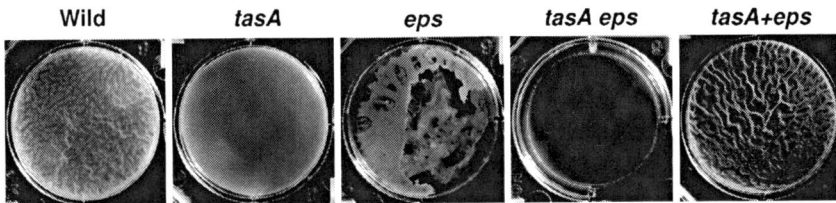

Fig. 6 Phenotype of *tasA*, *eps*, and *tasAeps* mutants and extracellular complementation in *tasA+eps* co-culture

et al. 2006). When mild sonication of the pellicle was used to separate cells from the matrix material, most of the TasA was shown to be present in the matrix fraction. Quite interestingly, TasA remains cell-associated and is not delivered to the matrix fraction when cells lack YqxM, leading to the hypothesis that YqxM is involved in delivering TasA to the matrix (Branda et al. 2006).

While single *eps* or *tasA* mutants still produce weak, unstructured pellicles, an *eps tasA* double mutant produces no pellicle whatsoever, suggesting that the products of these two operons represent the major structural components of the matrix (Fig. 6). Quite strikingly, when an *eps* mutant is co-cultured with a *tasA* mutant, there is restoration of the wild pellicle phenotype, suggesting that these components exert their function outside of the cell. In contrast, it was not possible to restore the wild pellicle phenotype by co-culturing *tasA* and *yqxM* mutants, consistent with the idea that YqxM is needed to deliver TasA to the matrix. Poly-γ-glutamate has also been shown to be an extracellular polymer important for biofilm formation in a different wild strain of *B. subtilis* (Stanley and Lazazzera 2005). However, mutants unable to produce poly-γ-glutamate display a wild type biofilm phenotype in *B. subtilis* 3610 (Branda et al. 2006).

Mutants lacking SinR or SinI greatly affect biofilm development (Kearns et al. 2005). In the absence of SinI, no pellicle forms and colonies are flat, while the lack of SinR results in extremely wrinkled pellicles and colonies (Fig. 7). *eps* mutations are epistatic to *sinR*, i.e., the *eps* flat colony phenotype is retained in a *sinR eps* double mutant. DNA footprinting and gel shift analyses using purified SinR revealed that SinR binds directly to the promoter regions of both the *eps* (Kearns et al. 2005) and *yqxM-sipW-tasA* operons (Chu et al. 2006). Also, SinR binding to the *eps* regulatory region is inhibited if purified SinR protein is complexed with purified SinI prior to mixing with DNA (Kearns et al. 2005). Thus, SinR acts as a transcriptional repressor of the genes involved in producing the extracellular matrix, and SinI can antagonize its action.

The involvement of SinR and SinI in the regulation of *epsA-O* and *yqxM-sipW-tasA* explains the indirect effects of Spo0A and σ^H on extracellular matrix synthesis. The *sinI* and *sinR* genes are adjacent to each other, with *sinI* lying upstream. The *sinR* gene is transcribed primarily from a constitutive promoter dependent on the major housekeeping sigma factor σ^A, while *sinI* is transcribed from two σ^A-dependent promoters, the major one also being dependent on Spo0A~P (Shafikhani et al. 2002). The σ^H effect is probably due to the fact that *spo0A* itself contains a σ^H-dependent promoter (Predich et al. 1992). Therefore, mutants lacking Spo0A or σ^H will express *sinI* at a lower level, so that the negative effects of SinR on matrix synthesis will not be antagonized, resulting in defects in biofilm development (Fig. 3). Another regulatory protein known to control *B. subtilis* biofilm formation is AbrB (Hamon and Lazazzera 2001). However, just exactly how AbrB acts is not yet known.

Spo0A is not the only signal transducer feeding into the pathway regulating extracellular matrix synthesis. Two genes, *ylbF* and *ymcA*, when mutated lead to

Wild   *sinI*   *sinR*

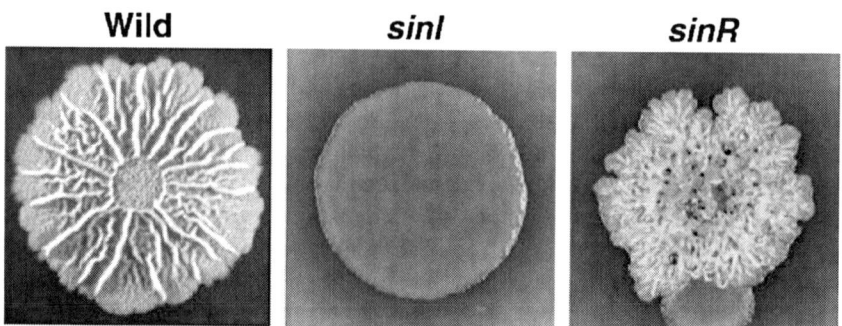

Fig. 7 Colony phenotype of *sinI* and *sinR*

flat colonies and no pellicles (Branda et al. 2004). In mutants lacking YlbF or YmcA, suppressor mutants take over the surface of the culture and form late-arising pellicles (Kearns et al. 2005). These suppressors that produce hyperwrinkled colonies do, indeed, harbor suppressor mutations in their *sinR* genes (Kearns et al. 2005). Thus, it appears that YlbF and YmcA function upstream of SinR. Because the expression of *ylbF* and *ymcA* does not appear to be regulated by either Spo0A or σ^H (Britton et al. 2002), we posit that YlbF and YmcA feed into the SinI-SinR circuitry via a different pathway (Fig. 3).

SinR functions as a master regulator of the lifestyle switch in *B. subtilis* (Fig. 4). In the model, SinR acts as a direct repressor of the genes involved in extracellular matrix production (*epsA-O* and *yqxM-sipW-tasA*). At the same time and through a mechanism that remains largely unknown, SinR acts positively to influence motility and cell separation. During vegetative growth, cells swim, are unit length, and do not produce extracellular matrix. When nutrient limitation is sensed, presumably through both the Spo0A/σH and the YlbF/YmcA pathways, SinI activity increases and SinR is antagonized. In the absence of SinR the expression of matrix components is de-repressed and cell separation and the assembly of motility machinery ceases. As a result, the cells switch to a mode of life where they form chains, become enclosed in a self-produced extracellular matrix, and stop making flagella. Synthesis of the matrix renders the cells able to attain a high degree of spatiotemporal organization, culminating in the production of spores at the tips of aerial projections.

5　The Future of Biofilm Development Research

Elucidation of the genes, proteins, and molecular mechanisms involved in *B. subtilis* biofilm formation continues and, though much progress has been made in the past 5 years, much remains to be done. Among Gram-positive bacteria, the molecular mechanisms of biofilm formation appear to be species-specific. For example, the master regulators of biofilm formation in *B. subtilis* (the transcriptional repressor SinR; Kearns et al. 2005), *Staphylococcus* (the transcriptional activator SarA; Beenken et al. 2003; Valle et al. 2003; Tormo et al. 2005) and *Enterococcus* (the response-regulator FsrA; Hancock and Perego 2004) are not homologs of each other. In the future, we can expect the combination of genetics, biochemistry, and microscopy to yield an ever-increasing understanding of the molecular mechanisms of biofilm formation unique to many bacteria. Invariably, microbes carry out fascinating, and often unexpected, processes when presented with the greater organizing potential afforded by a surface. Once on a surface, microbial cells can begin long-term relationships with each other; therein lies the transition from unicellularity to multicellularity. Analyses of microbial activities on surfaces will continue to provide new insights into the marvelous and astounding diversity of the microbial world.

Acknowledgements Biofilm work in our laboratory is funded by a grant from the NIH to R.K. (GM58213). K.P.L. was the recipient of an NIH Mentored Clinical Scientist Development Award (K08 AI070561) and A.M.E was the recipient of an NIH postdoctoral fellowship (GM072393).

References

Beenken KE, Blevins JS, Smeltzer MS (2003) Mutation of *sarA* in *Staphylococcus aureus* limits biofilm formation. Infect Immun 71:4206-4211

Branda SS, Gonzalez-Pastor JE, Ben-Yehuda S, Losick R, Kolter R (2001) Fruiting body formation by *Bacillus subtilis*. Proc Natl Acad Sci U S A 98:11621-11626

Branda SS, Gonzalez-Pastor JE, Dervyn E, Ehrlich SD, Losick R, Kolter R (2004) Genes involved in formation of structured multicellular communities by *Bacillus subtilis*. J Bacteriol 186:3970-3979

Branda SS, Vik S, Friedman L, Kolter R (2005) Biofilms: the matrix revisited. Trends Microbiol 13:20-26

Branda SS, Chu F, Kearns DB, Losick R, Kolter R (2006) A major protein component of the *Bacillus subtilis* biofilm matrix. Mol Microbiol 59:1229-1238

Britton RA, Eichenberger P, Gonzalez-Pastor JE, Fawcett P, Monson R, Losick R, Grossman AD (2002) Genome-wide analysis of the stationary-phase sigma factor (sigma-H) regulon of *Bacillus subtilis*. J Bacteriol 184:4881-4890

Christensen BB, Sternberg C, Andersen JB, Palmer RJ Jr, Nielsen AT, Givskov M, Molin S (1999) Molecular tools for study of biofilm physiology. Methods Enzymol 310:20-42

Chu F, Kearns DB, Branda SS, Kolter R, Losick R (2006) Targets of the master regulator of biofilm formation in *Bacillus subtilis*. Mol Microbiol 59:1216-1228

Costerton JW, Stewart PS, Greenberg EP (1999) Bacterial biofilms: a common cause of persistent infections. Science 284:1318-1322

Enos-Berlage JL, Guvener ZT, Keenan CE, McCarter LL (2005) Genetic determinants of biofilm development of opaque and translucent *Vibrio parahaemolyticus*. Mol Microbiol 55:1160-1182

Fawcett P, Eichenberger P, Losick R, Youngman P (2000) The transcriptional profile of early to middle sporulation in *Bacillus subtilis*. Proc Natl Acad Sci U S A 97:8063-8068

Fedtke I, Gotz F, Peschel A (2004) Bacterial evasion of innate host defenses - the *Staphylococcus aureus* lesson. Int J Med Microbiol 294:189-194

Friedman L, Kolter R (2004) Genes involved in matrix formation in *Pseudomonas aeruginosa* PA14 biofilms. Mol Microbiol 51:675-690

Gotz F (2002) *Staphylococcus* and biofilms. Mol Microbiol 43:1367-1378

Grossman AD (1995) Genetic networks controlling the initiation of sporulation and the development of genetic competence in *Bacillus subtilis*. Annu Rev Genet 29:477-508

Guvener ZT, McCarter LL (2003) Multiple regulators control capsular polysaccharide production in *Vibrio parahaemolyticus*. J Bacteriol 185:5431-5441

Hall-Stoodley L, Costerton JW, Stoodley P (2004) Bacterial biofilms: from the natural environment to infectious diseases. Nat Rev Microbiol 2:95-108

Hamon MA, Lazazzera BA (2001) The sporulation transcription factor Spo0A is required for biofilm development in *Bacillus subtilis*. Mol Microbiol 42:1199-1209

Hamon MA, Stanley NR, Britton RA, Grossman AD, Lazazzera BA (2004) Identification of AbrB-regulated genes involved in biofilm formation by *Bacillus subtilis*. Mol Microbiol 52:847-860

Hancock LE, Perego M (2004) The *Enterococcus faecalis fsr* two-component system controls biofilm development through production of gelatinase. J Bacteriol 186:5629-5639

Henrici AT (1933) Studies of freshwater bacteria. I. A direct microscopic technique. J Bacteriol 25:277-287

Kadouri D, Venzon NC, O'Toole GA (2007) Vulnerability of pathogenic biofilms to *Micavibrio aeruginosavorus*. Appl Environ Microbiol 73:605-614

Kearns DB, Chu F, Branda SS, Kolter R, Losick R (2005) A master regulator for biofilm formation by *Bacillus subtilis*. Mol Microbiol 55:739-749

Kolter R, Greenberg EP (2006) Microbial sciences: the superficial life of microbes. Nature 441:300-302

Lasa I (2006) Towards the identification of the common features of bacterial biofilm development. Int Microbiol 9:21-28

Lasa I, Penades JR (2006) Bap: a family of surface proteins involved in biofilm formation. Res Microbiol 157:99-107

Latasa C, Roux A, Toledo-Arana A, Ghigo JM, Gamazo C, Penades JR, Lasa I (2005) BapA, a large secreted protein required for biofilm formation and host colonization of *Salmonella enterica* serovar *Enteritidis*. Mol Microbiol 58:1322-1339

Latasa C, Solano C, Penades JR, Lasa I (2006) Biofilm-associated proteins. C R Biol 329:849-857

Leid JG, Willson CJ, Shirtliff ME, Hassett DJ, Parsek MR, Jeffers AK (2005) The exopolysaccharide alginate protects *Pseudomonas aeruginosa* biofilm bacteria from IFN-gamma-mediated macrophage killing. J Immunol 175:7512-7518

Lemon KP, Higgins DE, Kolter R (2007) Flagella-mediated motility is critical for *Listeria monocytogenes* biofilm formation. J Bacteriol 189:4418-4424

Mah TF, O'Toole GA (2001) Mechanisms of biofilm resistance to antimicrobial agents. Trends Microbiol 9:34-39

O'Toole GA, Kolter R (1998a) Flagellar and twitching motility are necessary for *Pseudomonas aeruginosa* biofilm development. Mol Microbiol 30:295-304

O'Toole GA, Kolter R (1998b) Initiation of biofilm formation in *Pseudomonas fluorescens* WCS365 proceeds via multiple, convergent signalling pathways: a genetic analysis. Mol Microbiol 28:449-461

O'Toole GA, Pratt LA, Watnick PI, Newman DK, Weaver VB, Kolter R (1999) Genetic approaches to study of biofilms. Methods Enzymol 310:91-109

O'Toole G, Kaplan HB, Kolter R (2000) Biofilm formation as microbial development. Annu Rev Microbiol 54:49-79

Piggot PJ, Hilbert DW (2004) Sporulation of *Bacillus subtilis*. Curr Opin Microbiol 7:579-586

Pratt LA, Kolter R (1998) Genetic analysis of *Escherichia coli* biofilm formation: roles of flagella, motility, chemotaxis and type I pili. Mol Microbiol 30:285-293

Predich M, Nair G, Smith I (1992) *Bacillus subtilis* early sporulation genes *kinA*, *spo0F*, and *spo0A* are transcribed by the RNA polymerase containing sigma H. J Bacteriol 174:2771-2778

Serrano M, Zilhao R, Ricca E, Ozin AJ, Moran CP Jr, Henriques AO (1999) A *Bacillus subtilis* secreted protein with a role in endospore coat assembly and function. J Bacteriol 181:3632-3643

Shafikhani SH, Mandic-Mulec I, Strauch MA, Smith I, Leighton T (2002) Postexponential regulation of *sin* operon expression in *Bacillus subtilis*. J Bacteriol 184:564-571

Singh PK, Schaefer AL, Parsek MR, Moninger TO, Welsh MJ, Greenberg EP (2000) Quorum-sensing signals indicate that cystic fibrosis lungs are infected with bacterial biofilms. Nature 407:762-764

Sonenshein AL, Hoch JA, Losick R (eds) (2002) *Bacillus subtilis* and its closest relatives: from genes to cells. ASM Press, Washington DC

Spoering AL, Gilmore MS (2006) Quorum sensing and DNA release in bacterial biofilms. Curr Opin Microbiol 9:133-137

Stanley NR, Lazazzera BA (2005) Defining the genetic differences between wild and domestic strains of *Bacillus subtilis* that affect poly-gamma-dl-glutamic acid production and biofilm formation. Mol Microbiol 57:1143-1158

Stanley NR, Britton RA, Grossman AD, Lazazzera BA (2003) Identification of catabolite repression as a physiological regulator of biofilm formation by *Bacillus subtilis* by use of DNA microarrays. J Bacteriol 185:1951-1957

Stover AG, Driks A (1999a) Control of synthesis and secretion of the *Bacillus subtilis* protein YqxM. J Bacteriol 181:7065-7069

Stover AG, Driks A (1999b) Secretion, localization, and antibacterial activity of TasA, a *Bacillus subtilis* spore-associated protein. J Bacteriol 181:1664-1672

Tormo MA, Marti M, Valle J, Manna AC, Cheung AL, Lasa I, Penades JR (2005) SarA is an essential positive regulator of *Staphylococcus epidermidis* biofilm development. J Bacteriol 187:2348-2356

Valle J, Toledo-Arana A, Berasain C, Ghigo JM, Amorena B, Penades JR, Lasa I (2003) SarA and not sigmaB is essential for biofilm development by *Staphylococcus aureus*. Mol Microbiol 48:1075-1087

Varga JJ, Nguyen V, O'Brien DK, Rodgers K, Walker RA, Melville SB (2006) Type IV pili-dependent gliding motility in the Gram-positive pathogen *Clostridium perfringens* and other *Clostridia*. Mol Microbiol 62:680-694

Watnick PI, Kolter R (1999) Steps in the development of a Vibrio cholerae El Tor biofilm. Mol Microbiol 34:586-595

Watnick PI, Lauriano CM, Klose KE, Croal L, Kolter R (2001) The absence of a flagellum leads to altered colony morphology, biofilm development and virulence in *Vibrio cholerae* O139. Mol Microbiol 39:223-235

Whitchurch CB, Tolker-Nielsen T, Ragas PC, Mattick JS (2002) Extracellular DNA required for bacterial biofilm formation. Science 295:1487

Physiology of Microbes in Biofilms

A. M. Spormann

Abstract Microbial biofilms are governed by an intricate interplay between physical-chemical factors and the physiological and genetic properties of the inhabiting microbes. Many of the physiological traits that are exhibited in a biofilm environment have been observed and studied previously in detail in planktonic cultures. However, their differential and combinatorial phenotypic expression in distinct subpopulations localized to different regions in a biofilm is the cause for the overall biofilm heterogeneity. In this chapter, the causes and consequences of this interplay are elaborated with a special focus on processes controlling biofilm stability and dispersal.

1   Introduction

Research on microbial biofilms has been motivated mainly by a need to understand the mechanisms leading to the physical persistence of microbes on surfaces and the resistance of microbes to antimicrobial agents in biofilm environments (for reviews,

A. M. Spormann
Departments of Chemical Engineering, of Civil and Environmental Engineering, and of Biological Sciences, Clark Center E250, Stanford University, Stanford, CA 94305-5429, USA
spormann@stanford.edu

see Tolker-Nielsen and Molin 2000; Costerton 1999; Costerton et al. 1999; Watnick and Kolter 2000; O'Toole et al. 2000). The persistence of microbes in biofilms, on one hand, provides a reservoir for these microbes, and, on the other hand, is the cause for the build-up of biomass, which by itself is of great medical and industrial concern (clogging of catheters and pipes, creating drag in ships, etc.). In the past, research on biofilm-forming microbes has been focused largely on identifying the molecular processes that define the initial phases of biofilm formation, such as adhesion via pili, flagella, exopolysaccharide (EPS) production, and perhaps a role for quorum sensing (see the chapter by Y. Irie and M.R. Parsek, this volume). However, the physiological and genetic responses of biofilm microbes to external and self-induced stresses, including the competition for resources, determine the fate of a biofilm and its diverse subpopulations to a large extent. Thus, one of the most consequential challenges for microbes in a biofilm is how to deal with these conditions. One strategy is to simply reduce the growth rate or exhibit a behavior similar to that found in sporulating microbes. However, another one may be to leave and exit a biofilm. This review provides a physiological view of micro-bial life in a biofilm and attempts to reveal unifying principles in the physiology of cell populations in a biofilm environment. First, the current status on the physiolog-ical states of microbes in a biofilm environment and then the physiological and molecular mechanism(s) involved in cellular detachment will be reviewed.

2 Physical-Chemical Parameters Control the Physiology of Biofilm Cells

In most environmentally, medically, and industrially relevant systems, biofilms form at the interface of an aqueous phase and a substratum surface or a gaseous (air) phase. Common to all biofilm systems is that metabolic substrates leading to growth need to be available to the cells. These compounds can be present exclu-sively in the aqueous phase, such as in drinking water pipes or medical catheters, or be partitioned between the aqueous and a solid phase, such as in minerals or insoluble organic matter (e.g., cellulose, chitin, or protein). In addition, the absolute amount as well as the ratio of the different nutrient components in the bulk liquid are important, as these parameters determine which nutrient, for example electron donor, electron acceptor, phosphorous, etc., may become growth limiting (for review, see van Loosdrecht et al. 2002). Many environmental and laboratory bio-film systems are often limited by molecular oxygen (Huang et al. 1998; Xu et al. 1998; Barraud et al. 2006). Which compound may become growth-limiting matters, as this identifies the kind of stress biofilm cells might be experiencing. In any case, there is a net transport of at least some essential nutrient(s) from the aqueous phase to the immobilized cells in a biofilm.

Net transport of substrates into, as well as of metabolites out of, a biofilm is determined by the flow rate of the bulk liquid and by molecular diffusion (Picioreanu et al. 2001; van Loosdrecht et al. 2002) (Fig. 1). Depending on the laboratory or

A

B

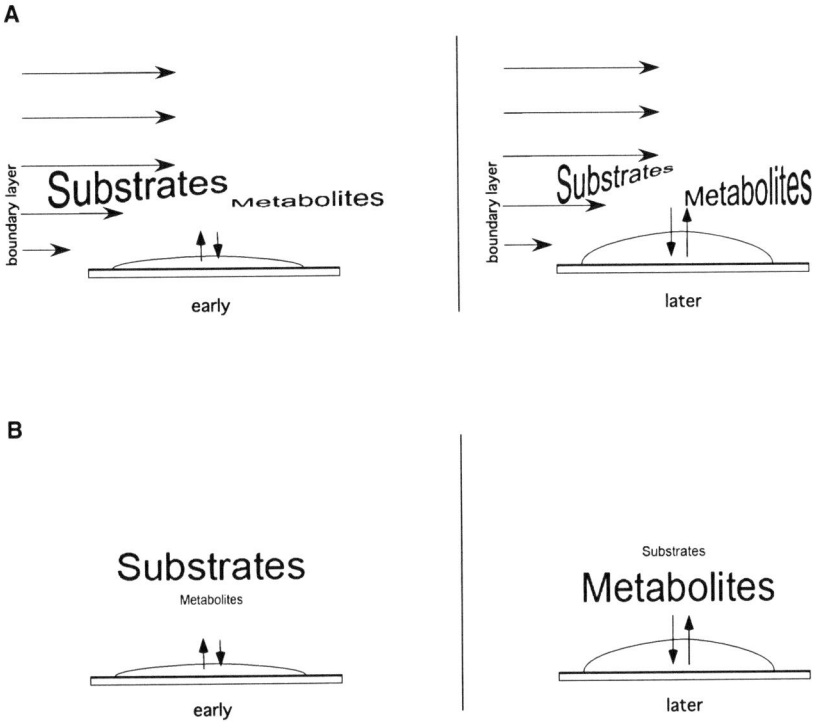

Fig. 1 Schematic representation of physical factors important for physiology of biofilm cells. Role of substrate and metabolite transport in hydrodynamic (**a**) and static (**b**) biofilm systems. Laminar flow is indicated by *horizontal, pointed arrows*. The length indicates the velocity at the specific position in the system. The boundary layer is indicated. Diffusion from the laminar layer into and out of the biofilm are indicated with the *vertical, flat-headed arrows*. The length of the arrow indicates mass transport by diffusion based on the generated concentration gradients. **A** In the early stages (*left panel*) of a forming hydrodynamic biofilm, laminar bulk liquid flow (*from left to right*) carries substrates across the biofilm. The flow rate next to the biofilm is slower than farther out in the bulk liquid, as indicated by the different arrow lengths. Substrates diffuse from the boundary layer into the biofilm. Substrate consumption by the metabolic activity of the cells is indicated by the tapered substrate label. Note that the more pronounced decrease in concentration is in the layers next to the biofilm and not in the bulk liquid. Conversely, the metabolites generated by the biofilm cells diffuse into the boundary layer and from there into the deeper laminar layers, establishing a gradient with higher concentration in the layers closer to the biofilm. At later stages (*right panel*), substantial biomass has accumulated that catalyzes a more rapid (but at an individual cell level, at a more reduced rate) removal of substrates, thus creating steeper gradients. **B** In static systems with no laminar flow or mixing, the dominant transport of nutrients into a biofilm occurs by diffusion. Another notable and physiologically consequential difference between the shear flow (**A**) and the zero flow (**B**) system is the concentration of substrates and metabolites. While the shear flow system has properties of a chemostat system, i.e., very low but constant steady-state concentrations, metabolites accumulate in the static biofilm systems and affect the physiology of cells similarly as the initial high-substrate concentrations

real-world settings, the flow rate of bulk liquid can range from zero (e.g., a fungal biofilm forming at the air-aqueous interface of a standing, half empty coffee cup or static laboratory biofilm systems) to high (e.g., inside of a drinking water pipe or a hydrodynamic laboratory flow chamber). Consequently, the flow rate determines to which extent a biofilm is in a chemostat-like setting (flow rate>0, e.g., in a flow chamber) with a constant rate of substrate flux into and metabolite flux out of a biofilm or in a batch culture type condition (flow rate=0), such as in laboratory 96-well plates, where initial high substrate concentrations are converted into accumulating metabolite concentration during the biofilm development (Fig. 1A, B). A laminar flow also exerts shear stress onto a biofilm, and thereby controls the activity of the biofilm cells (van Loosdrecht et al. 2002).

While laminar flow transports nutrients to the boundary layer (i.e., the layer of fluid in the immediate vicinity of a bounding surface; see Fig. 1), molecular diffusion transports nutrients between the laminar layers in the bulk liquid and between the boundary layer and the biofilm (Fig. 1A). There is evidence that bulk liquid flow through a biofilm is negligible, if present at all, and that net transport processes within biofilms are basically limited by diffusion from the surrounding boundary layer to the interior of developed biofilm. Because of the chemostat-like conditions, diffusion of a continuous supply of substrates at low concentrations, which are found in many natural environments, can support a growing biofilm very well. On the other hand, in the absence of a bulk liquid flow or mixing, i.e., in a static system, high substrate concentrations are required to be initially present in order to promote growth of a biofilm (Fig. 1B). Also here, molecular diffusion will control cellular physiology but to a quite different extent than in hydrodynamic systems. As a consequence, the concentration of chemicals, their gradients, and the distribution of such gradients across a biofilm are different in static biofilms systems relative to hydrodynamic systems, and with that the physiology of biofilm cells.

Figure 2 illustrates the dramatic effect of transport limitation on the extent of formation, the structure, and on the relative position of subpopulations within a biofilm. After identical initial conditions, transport-limited biofilm populations are on different trajectories relative to unlimited biofilm environments. Besides the overall dramatically reduced rate of increase in biofilm biomass, transport-limited biofilms exhibit significantly more structure, indicative of local chemical gradients and associated heterogeneous single-cell physiology (Picioreanu et al. 2001; van Loosdrecht et al. 2002). Notably, there is less mixing of phenotypically identical subpopulations in transport-limited biofilms (Fig. 2).

How do physical conditions such as transport by diffusion, including its limitation, determine the physiology of biofilm cells? Under initial conditions in a hydrodynamic systems (Figs. 1A, 2), when only few cells adhere to the substratum surface as an interspersed monolayer, molecular diffusion along the concentration gradient between the boundary layer and the cells causes mass transfer, thereby enabling cells to grow at high rate (Fig. 3). It should be noted that not all cells grow equally rapidly, and very few do not grow at all, thereby setting

No transport limitation

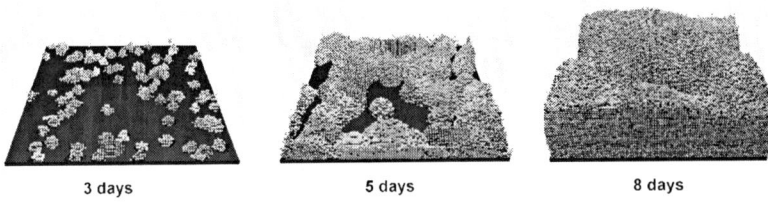

3 days   5 days   8 days

Transport limitation

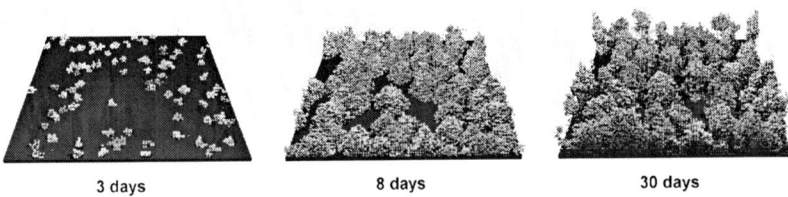

3 days   8 days   30 days

Fig. 2 Simulation of clonal growth of biofilm populations in non-transport-limited (*upper panel*) and in transport-limited (*lower panel*) biofilms (images kindly provided by Cristian Picioreanu after Picioreanu et al. 1998). The two genetically and physiologically identical subpopulations are indicated in *green* and *yellow*, respectively. Note the difference in biofilm biomass at day 8 in both types of biofilm, as well as the distinct architecture of biofilms. Note also the enhanced retention of areas of clonal populations in transport-limited biofilms. See text for details.

subpopulations on to different physiological trajectories (Fig. 3b). Using fusions of *gfp*, encoding unstable GFP, to the ribosomal RNA promoter *rrnB*, studies of *Pseudomonas putida* biofilms developing in flow chambers revealed heterogeneous expression of growth activity (Sternberg et al. 1999; Christensen et al. 1999). While isolated cells exhibited similar Gfp fluorescence, once these cells had grown into flat clusters, cells at the periphery and exposed to the bulk liquid flow exhibited higher growth activity compared to cells in the center of clusters. Stimulation of growth of those inactive cells upon addition of a more readily metabolizable substrate indicates the importance of specific catabolic substrates in the physiology of a microorganism (Sternberg et al. 1999).

As the substrate concentration in the bulk liquid remains constant and the biomass in the biofilm increases, the rate of substrate consumption increases in the biofilm and with that the concentration gradient between the bulk and the biofilm. The increased concentration gradient, in turn, leads to an increased substrate flux into the biofilm (Fig. 3). When the rate of substrate (or more precisely, of the growth-limiting compound) consumption in the biofilm exceeds the net rate of flux into the biofilm, the biofilm becomes transport-limited (Fig. 3c) and growth slows down according to the cellular Monod growth kinetics (Fig. 3d). There is still a net growth of the biofilm population, although no longer at high rate

(Fig. 3d). Because of the increase in cell mass within the biofilm, the entire biofilm becomes metabolically and physiologically stratified: as a consequence of microbial growth under diffusion-limited conditions and of unequal diffusion in the biofilm, the growth rate becomes highly variable between cells, with a few bulk

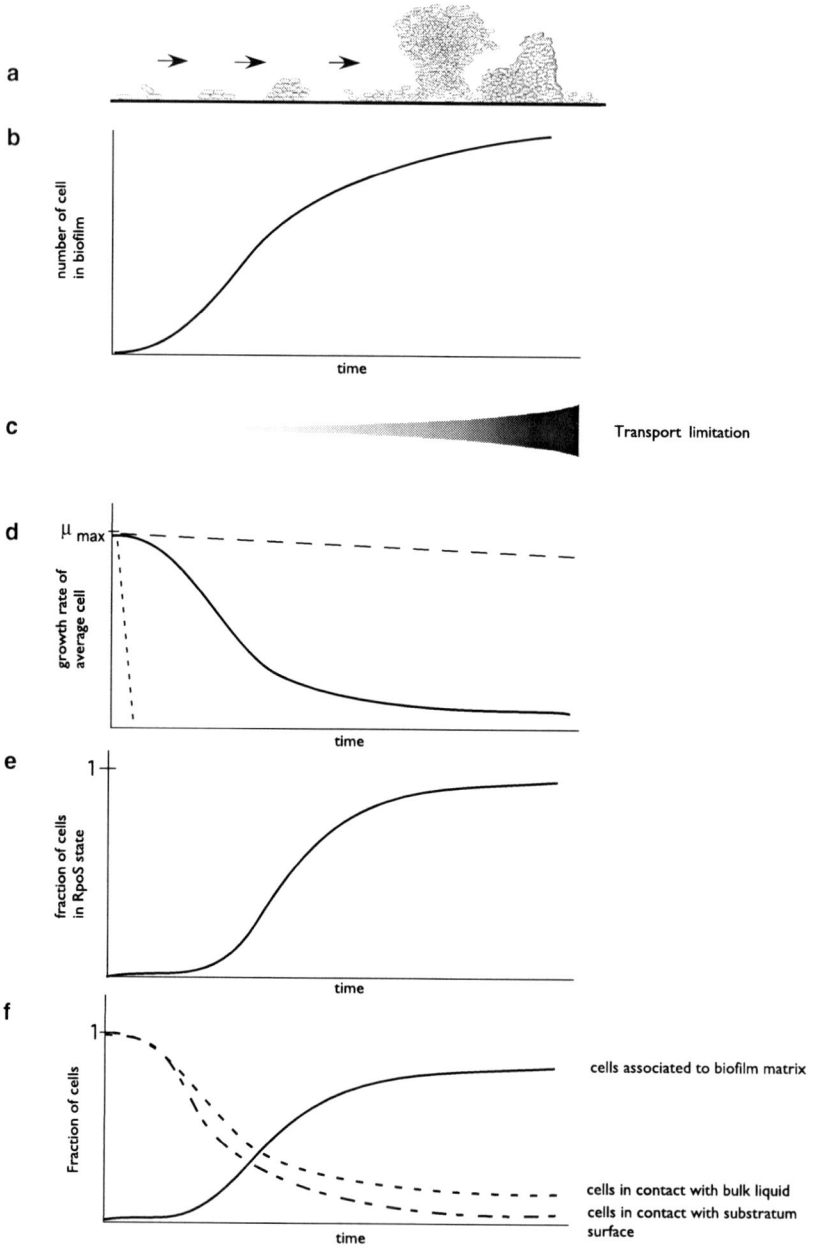

liquid-exposed cells growing rapidly and others not growing at all (Fig. 3d). Consequently, pronounced, self-induced concentration gradients develop within different sections of the biofilm. The extent of such gradients is determined by the local biomass, its growth state and activity, and the net nutrient flux from the bulk liquid (Fig. 3e). This intricate interplay between the physical chemistry and biology is one major cause for the observed biofilm heterogeneity, consisting of steep chemical gradients and subpopulations in different physiological states. This heterogeneity in populations is also reflected in the composite profile of the transcriptome of biofilm cells, which provides evidence for subpopulations being in a logarithmic and stationary phase in the same biofilm (Whiteley et al. 2001; Waite et al. 2005).

◄──

Fig. 3 Conceptional representation of the interplay between physical and biological causes and consequences leading to physiological diversification of subpopulations of nondetaching biofilm cells of γ proteobacteria. Scales were intentionally omitted to focus on general trends rather than specific mathematical relationships. Accordingly, the shape of the graphs is not as important as the general trend and the onset of changes. **a** Schematic representation of a biofilm developing on a surface in a hydrodynamic system with media flowing tangentially to the substratum surface. The four biofilms pictured at different stages indicate increased surface coverage and build-up of biomass. **b** The number of cells is increasing over time in a biofilm. Initially, the increase in biofilm biomass is high following the cellular Monod growth kinetics at saturating substrate concentration. With a constant convective flux, the steadily increasing net rate of substrate consumption exceeds the rate of net nutrient transport into the biofilm, which in turn reduces the actual in situ growth rate and the overall growth slows down. **c** Biofilms become increasingly diffusion limited. Due to the increasing metabolic demand, metabolic processes in biofilms become increasingly limited by substrate transport, and thereby nutrient-limited, which reduces the growth rate. **d** Growth rate of the average biofilm cell. Note that this graph displays only the growth rate of the average cell (theoretically counting the rates exhibited by all biofilm cells divided by the number of cells). It is known that after surface colonization, cells in monolayers grow at different growth rates (see text). The growth rate of the average cell is decreasing because the entire biofilm is becoming increasingly diffusion-limited. Cells exposed directly to the bulk liquid may grow at a higher rate because of more access to nutrients, whereas cells in deeper parts near the substratum surface might have entered stationary phase. The overall growth rate is decreasing with time because the fraction of mass transport-limited cells is increasing and outnumbering the cells at the bulk liquid interface. Dotted lines indicate a spectrum of growth rates displayed by a limited number of individual cells (*short dots* indicate cells rapidly entering a no-growth state (see text), *long dots* indicate cells growing rapidly, e.g., at the bulk liquid surface). **e** The fraction of cells where the physiology is governed by the activity of RpoS. During the initial growth of isolated monolayer cells, the physiology of biofilm cells is relatively independent of RpoS (see text). Because of the reduced growth rate and environmental stresses being generated in the biofilm environment, cellular physiology becomes increasingly dominated by RpoS. It is assumed that only the fast growing, bulk liquid exposed cells in developed biofilms are relatively RpoS-independent. **f** Distribution of localization of cells in a biofilm. Initially, all cells are associated with the substratum surface and also at the bulk liquid interface. As biomass increases, surface coverage increases and saturates, and more cells are localized within the biofilm, i.e., attaching to the biofilm matrix. The fraction of cells exposed to the bulk liquid is also declining due to the increase in matrix-attached cells. However, their fraction is higher than that of the surface contacting cells because of structural heterogeneity being generated in the biofilm

3 Global Regulators Determine the Physiology of Subpopulations of Biofilm Cells

How is cellular physiology controlled by fluctuating growth rates and changing chemical gradients? The most commonly found molecular control of gene expression in γ proteobacteria in response to diverse environmental conditions is mediated by the sigma factor RpoS (σ^{32}, σ^S) (for review, see Hengge-Aronis 2002), although recently CRP was also recognized to play a critical role (Liang et al. 2007). Broadly, RpoS mediates responses to physical, chemical, or biological environmental conditions that result in a switch from maximal to reduced growth or even maintenance metabolism. Such changes can be induced by environmental conditions stressful for cells, such as oxidative stress, near-UV irradiation, heat shock, hyperosmolarity, acidic pH, and ethanol treatment (Hengge-Aronis 2002). More importantly for the biofilm environment, limitation in growth substrates, primarily in electron donor and acceptor, but also in other nutrients such as phosphorous and ammonia, accumulation of salts and metabolites, and high cell density strongly induce the RpoS-dependent general stress response (Hengge-Aronis 2002; Liu et al. 2000) (Fig. 4). RpoS is considered to function as a master regulator, mediating not just a specific physiological response to a single specific stress, but rendering cells

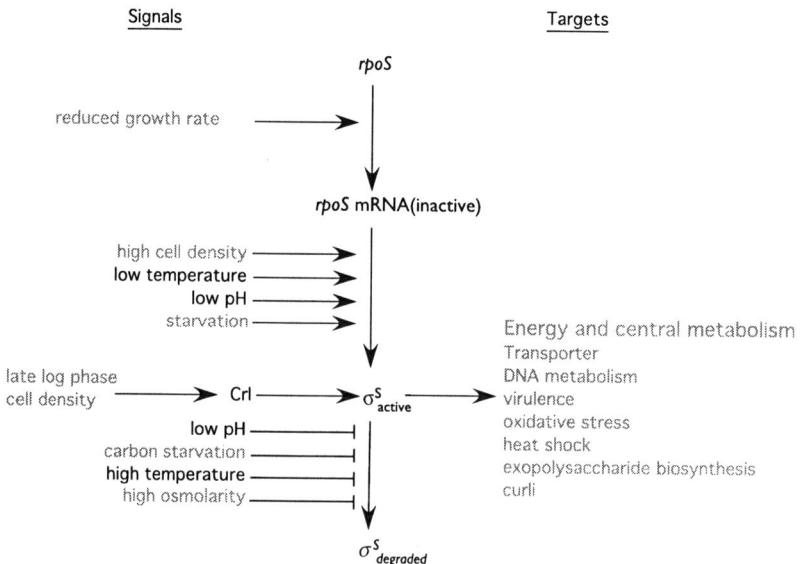

Fig. 4 RpoS regulation of physiological traits in γ proteobacteria important in biofilms (modified after Hengee-Arronis 2002). Transcriptional, translational, and posttranslational control of *rpoS/* RpoS activity by environmental factors (*left panel*) and physiological traits (*right panel*) important for biofilm physiology. See text and Hengee-Arronis 2002 for details.

broadly stress-resistant regardless of whether a specific stress is present. The redundancies, nonlinearities, and internal feed-back of the RpoS circuitry result in a complex network coordinating numerous aspects of cellular physiology and provide a general robustness in the stress response. This preventive property of the general stress response, its almost ubiquitous involvement, and its complex regulatory interconnectedness emphasizes the central role of RpoS in controlling the physiology of microbes. One important consequence of this property of the RpoS circuitry is the increased resistance of biofilm cells to antibiotics and other biocides (reviewed in the chapter by G.G. Anderson and G.A. O'Toole, this volume).

Generally, it is assumed that changes in cellular concentration of RpoS protein determine the extent of the general stress response. RpoS concentration is controlled at the transcriptional, translational, and posttranslational level, and diverse environmental conditions have been shown to act directly or indirectly at one or more of these control levels (Hengge-Aronis 2002; Fig. 4). However, RpoS may not exclusively function as a rheostat, where the general stress response is directly proportional to the cellular RpoS concentration, but may also be subject to direct activity control. The essential role of RpoS in non-logarithmic growth physiology is also underlined by the finding that genetic variants adapted to stationary phase (GASP mutants) carry mutations in *rpoS* (Zinser and Kolter 1999; Farrell and Finkel 2003).

Nutrient limitation, metabolite accumulation, changes in osmolarity and high cell density are conditions experienced by biofilm cells, and a hypothetical course of *rpoS*-dependent physiology of biofilm cells is depicted in Fig. 3e, based on the observations described above and below. The general importance of RpoS in biofilm architecture and the physiology of biofilm cells was revealed by several studies in *Escherichia coli*, *Vibrio cholerae*, and *Pseudomonas aeruginosa*. While *rpoS* generally affects cell density in biofilms, positive and negative regulation was observed. Both in *E. coli* and *P. aeruginosa*, Δ*rpoS* cells grew to either higher (Heydorn et al. 2002) or lower (Adams and McLean 1999; Schembri et al. 2003) cell densities compared to wild type. Although it is unclear whether different experimental conditions are the basis for these apparently contradicting observations, a dual role of *rpoS* in controlling both the expression of adhesion determinants required for the initial and effective adhesion of cells to the substratum and the repression of exopolysaccharide biosynthesis could explain this apparent contradiction. On the other hand, the control of adhesin and EPS biosynthesis *per se* might be more diverse, and some genes encoding enzymes, such as the *E. coli pgaABCD* genes, might be regulated to a large extent independently of RpoS (Goller et al. 2006). In any case, these findings emphasize the importance of *rpoS* in the physiology of biofilm cells as well as the experimental set-up, which controls the physical-chemical factors determining biofilm physiology.

An interesting case for the differential involvement of *rpoS* in biofilms is found in *V. cholerae*. RpoS plays a key role in the late phase of *V. cholerae* infection of the intestine and for the dispersal of biofilms. RpoS is required for detachment of cells from biofilms formed on mucosal surfaces and on glass surfaces in flow chambers (Nielsen et al. 2006; Muller et al. 2007). Interestingly, this RpoS control of a phenotypically similar behavior occurs via two different molecular mechanisms and targets. One target of RpoS is Vibrio polysaccharide (VPS) biosynthesis, encoded by the *vps* genes. RpoS positively controls synthesis of HapR, a negative regulator of *vps* gene expression (Yildiz and Schoolnik 1999; Yildiz et al. 2004; Nielsen et al. 2006; Muller et al. 2007). Notably, *V. cholerae* biofilms form *in vivo* (i.e., rabbit ileal loop) independent of the *vps* genes (Nielsen et al. 2006) but Δ*rpoS* mutants are still defective in the mucosal escape response (Nielsen et al. 2006). The dependence of the mucosal escape response on RpoS in those biofilms is, thus, not due to differential regulation of *vps* genes but to the upregulation of motility and chemotaxis genes and possibly to the repression of other factors, such as TCP, involved in attachment (Nielsen et al. 2006). *V. cholerae* biofilms developing in flow chambers (*in vitro*) also form largely independent of the *vps* genes, where only the extent of biofilm formation is rpoS-controlled (Muller et al. 2007). However, biofilms of Δ*rpoS* mutants forming under those conditions are dramatically thicker and do not detach, due to overproduction of VPS mediated by the *vps* gene products (Muller et al. 2007). Thus, in *in vitro* wild type biofilms, cells can detach because a reduced growth rate in those cells triggers activation of RpoS, which represses *vps* gene expression, enabling cells to detach. Thus, detachment of *V. cholerae* cells from biofilms that form *in vitro* seems incompatible with *vps* gene expression, since it is also the case with *mxdA-D* expression in *Shewanella oneidensis* MR-1 (Thormann et al. 2006). The above examples demonstrate the essential function of RpoS in controlling basic cellular physiology of biofilm cells in response to environmental factors including growth rate and environment. They also show, perhaps more enlightening, that although RpoS controls *vps* expression *in vitro* in biofilms developing in the intestine this regulation is by-passed and muted by other environmental factors. Therefore, observations on the genetics and physiology of biofilm formation that have been made under one condition cannot be simply extrapolated to biofilms forming by the same organism under different conditions.

Another physiological consequence of slow cellular growth and RpoS activity is a dramatically increased resistance to antibiotics and biocides, as observed in biofilm microbes (see Mah and O'Toole 2001, and reviewed in the chapter by G.G. Anderson and G.A. O'Toole, this volume). While several mechanisms, including limitation of diffusion of antibiotics through, and binding to, the biofilm matrix have been discussed as contributing factors, strong experimental evidence suggests that the large fraction of slow or non-growing biofilm cells and their associated physiology might account for most of the observed resistance. As studies with planktonic cells have revealed, slow growth *per se* is sufficient to induce resistance to antibiotics, although differential sensitivities to different antibiotics has been observed (Evans et al. 1991; Duguid et al. 1992a, 1992b). However, biofilm cells are not only growing slowly (Fig. 3) but are also subjected to other stresses, such as high osmolarity, cell density, and accumulating metabolites. Although other mechanisms cannot be ruled out and might be important in individual cases, it is tempting to speculate that most, if not all, of the greatly enhanced phenotypic antibiotic resistance exhibited by biofilm cells is simply a consequence of the nonlinear signal integration of multiple stresses by the RpoS circuitry.

4 Biofilm Stability and Cellular Detachment

The molecular basis of how cells bind to the biofilm matrix and what the molecular mechanisms are that determine biofilm stability are unclear and most likely complex and multifaceted. This complexity is due to the fact that any number of cellular components, for example, EPS, DNA, protein, pili, flagella, phage, and thus far unidentified components can contribute to the attachment of cells to a biofilm matrix (see for a review Sutherland 2001). In this context, *attachment* describes the binding of cells to a biofilm matrix or to other cells, whereas *adhesion* is used to describe binding of cells to a substratum surface. Some of the binding may be easily reversed (e.g., pili-mediated cell-cell contact), while other may require a more elaborate cellular mechanism for breaking (e.g., enzymatic hydrolysis of exopolymers). Moreover, cell attachment can involve more than one component, depending on whether and how many of such components are activated as a function of a cell's physiological state. One might predict that if more mechanistically independent components are expressed and activated at a given time point, a tighter, and maybe more irreversible attachment may dominate a biofilm.

However, some biofilms do dissolve under certain conditions, and give rise to viable detached cells, while others, or even different areas of the same biofilm, do not dissolve (Gjermansen et al. 2005; Thormann et al. 2005; Sauer et al. 2004; Sawyer and Hermanowicz 2000). This observation indicates that biofilms can, operationally defined, exist in either a dissolvable or undissolvable state. This dichotomy has important implications for both the underlying biology but also for treatment strategies for biofilm-related diseases and industrial fouling processes. How the transitioning between these two states is controlled is a central issue of biofilm research (Gjermansen et al. 2005; Thormann et al. 2005; Sauer et al. 2004). Attachment and detachment are mutually exclusive processes and inherently mechanistically linked: those molecular factors that mediate attachment need to be deactivated (or broken) in order to lead to detachment. Consequently, the molecular mechanisms of cell attachment are essential for rationalizing detachment. On the other hand, attachment *per se* is not reversible but depends on the molecular properties of the cell-cell/matrix contact. For example, EPS-based attachment might be irreversible and no detachment other than a sloughing off can be observed. However, other molecular mechanisms, such as a pilus-based attachment, might promote reversible attachment, and, therefore, allow for controlled cell detachment. As will be discussed below, the emerging, common theme for understanding the regulation of detachment is a change in the physiological state of individual biofilm cells. Early research on biofilms revealed that dissolution can be induced upon a change in nutrient conditions (Delaquis et al. 1989; Marshall et al. 1989; Sawyer and Hermanowicz 2000; James et al. 1995). Biofilm dissolution in *Pseudomonas putida* and *S. oneidensis* can be due to oxygen limitation (Applegate and Bryers 1991; Gjermansen et al. 2005; Thormann et al. 2005; Hansen et al. 2007). In this type of dissolution, more cells are affected compared to cell death of phage-induced detachment. In addition, the dissolution occurs rapidly after the environmental perturbation.

4.1   Detachment of S. oneidensis Cells from Biofilms

Next to using well-studied electron acceptors, such as O_2 or nitrate, *S. oneidensis* MR-1 uses insoluble electron acceptors that are present in mineral surfaces, such as Fe(III) and Mn(IV) oxides, which can be catabolically reduced under anoxic conditions (Myers and Nealson 1988; Nealson and Saffarini 1994; Heidelberg et al. 2002). This metabolic interaction of microbes with mineral surfaces leads to reductive mineral dissolutions that are consequential for abiotic redox reactions in soil and sediment environments. Biofilm studies with *S. oneidensis* have been conducted mainly with flow chamber system under oxic conditions, although the essential features are similar under anoxic conditions (Thormann et al. 2004, 2005, 2006; C. Cordova and A.M. Spormann, unpublished observations). *S. oneidensis* cells form relatively homogeneous three-dimensional biofilms without pronounced mushroom-like structures. Stopping the flow of an O_2-limited lactate medium 14-18 h post inoculation induces an immediate and massive cell detachment, wherein between 50% and 80% of the cell mass detaches from the biofilm within 15 min. The detached cells are viable. This detachment is not due to a cell loss induced by shear stress or the removal of the electron donor lactate, but due to oxygen limitation: the removal of oxygen from the medium under flow conditions is sufficient to induce the same extent of biofilm dissolution. These observations suggest that metabolic conditions such as oxygen limitation trigger biofilm dissolution in *S. oneidensis*. Interestingly, this type of detachment decreases with the age and/or thickness of a biofilm, and after 48 h, the biofilm has phenotypically converted to a nondetaching biofilm. Preliminary data suggest that, under these experimental conditions, this switch to an irreversibly attached biofilm is associated with the expression of the *mxd* genes, encoding for the synthesis of EPS in this microorganism (Thormann et al. 2006; S. Shukla and A.M. Spormann, unpublished observations).

Physiological studies provided some insights into potential mechanisms involved in the detachment response. Using IPTG-inducible P_{lac}-*gfp* fusions as constructs to report *in situ* gene induction in these biofilms, it was found that these biofilms were sensitive to transcriptional and translational inhibitors (R. Saville and A.M. Spormann, unpublished observations). However, neither transcription nor translation was required for the rapid detachment response as rifampicin- and tetracycline-inhibited biofilms exhibited the same detachment response upon a stop-of-flow as wild type cells. Thus, the detachment response upon O_2-limitation involves a posttranslational mechanism. Subsequent bioenergetic experiments revealed an important connection between detachment and attachment: addition of the uncoupler CCCP (carbonyl cyanide m-chloro phenyl hydrazone), which acts as a protonophore and depletes a cell of metabolic energy by collapsing the chemiosmotic proton potential, resulted in massive cell detachment even in the absence of a stop-of-flow (R. Saville and A.M. Spormann, unpublished observations). This finding strongly suggests that metabolic energy is required for cell attachment, and a simple dissipation of such energy is sufficient to induce cell detachment. From these observations on *S. oneidensis* biofilm, a direct link between cellular physiology and the attachment/detachment of detachable biofilms can be envisioned: the limitation for oxygen leads to a low respiratory activity and, consequently, to a low energetic state of the cell. This state may not provide sufficient energy to maintain cell attachment and, consequently, trigger detachment. This hypothesis is consistent with the stop-of-flow-induced, O_2-limitation-induced, and CCCP-induced detachment. An alternative explanation is that a decrease in cellular energy *per se*, as induced by a rapid drop in molecular O_2 or by CCCP treatment, is the signal input into a molecular signaling transduction cascade, which then negatively controls cell attachment.

4.2 Detachment of Pseudomonas Biofilms

Biofilm dissolution and cell detachment has been observed and studied in *P. aeruginosa* and *P. putida* strains (Sawyer and Hermanowicz 2000; Sauer et al. 2004; Gjermansen et al. 2005). Interestingly, an increase in carbon (electron donor) concentration can induce biofilm dissolution in *P. aeruginosa* PAO1. A tenfold increase in the carbon substrate concentration, i.e., from 1.8 mM to 18 mM in succinate, citrate, glutamate or glucose, resulted in a rapid loss (~80%) of biofilms grown on a silicone tubing surface (Sauer et al. 2004). However, the extent of induced detachment varied with the carbon source (succinate the most efficient, glucose the least efficient), and also an increase in ammonium chloride concentration as the N-source induced a similar effect. Biofilm dissolution was accompanied with cells gaining swimming motility, but as shown in *S. oneidensis*, motility is not required for the detachment. It is thus conceivable that detachment induced by a carbon source up-shift may be similar to the down-shift observed by Gjermansen et al. Interestingly in *P. aeruginosa*, the addition of protein phosphatase inhibitors prevented up-shift-induced detachment, which led the authors to speculate that protein dephosphorylation may play an important role in the metabolically controlled detachment process (Sauer et al. 2004). A recent finding by Barraud et al. showed that sublethal concentrations of NO, which is a well-known signaling molecule in eukaryotes and a byproduct of anaerobic NO_2^- reduction, can induce *P. aeruginosa* biofilm dissolution (Barraud et al. 2006). In the context of the physiological perspective provided here, it is interesting to note that genes encoding enzymes mediating denitrification were expressed in biofilm populations in these aerobic flow chamber experiments, and deletion mutants in the genes encoding nitrite reductase (*nirS*), but not in NO-consuming NO reductase (*norCB*) were defective in the NO-induced dissolution (Barraud et al. 2006).

Biofilms formed by *P. putida* OUS82, when grown in an oxygen-limited minimal citrate medium in a hydrodynamic flow chamber, develop a substantial three-dimensional structure and undergo massive dissolution (Gjermansen et al. 2005). Biofilm dissolution can also be induced, to a similar extent, by a simple stop of medium flow or by the removal of the carbon source from the medium, even in the absence of a stop of medium flow. Similar to the stop-of-flow response in *S. onei-densis*, the dissolution occurred within a few minutes rather than in tens of minutes or hours. A genetic selection for detachment-deficient mutants identified one class with an interesting phenotype. These mutants formed compact biofilms that were defective in the natural as well as stop-of-flow-induced dissolution. The inactivation of PP0164 encoding a periplasmic protein of an unknown function, conferred that phenotype (Gjermansen et al. 2005). This gene is located in the *lap* genomic region, which was identified previously in *P. fluorescens* and contains the *lapA* gene.

The *P. fluorescens* LapA is a greater than 500-kDa, outer membrane-associated, adhesion protein (Hinsa et al. 2003) that is essential for biofilm formation in this microorganism. Properly localized LapA is required not for initial adhesion but for forming irreversible, stable biofilms. LapA homologs have been identified in numerous other microbes, including *P. putida* and *S. oneidensis* (Hinsa et al. 2003). Activation of the *pho* regulon by low external P_i concentration leads to expression of *rapA*, encoding a c-di-GMP-hydrolyzing phosphodiesterase (PDE), whose PDE activity inhibits LapA secretion, and consequently, the formation of stable biofilms (Monds et al. 2007). PP0164 of *P. putida* encodes a putative periplasmic protein, and the chromosomal gene deletion renders cells sticky and unable to reduce the adhesiveness. Based on this genetic finding, it was speculated that PP0164 and the neighboring PP0165 are involved in signal transduction and c-di-GMP signaling, respectively, and may control cellular adhesiveness, potentially by controlling the secretion and/or localization of LapA.

4.3   Detachment of Vibrio cholerae *Biofilms*

Detachment in *V. cholerae* biofilms has been studied in some detail in strain A1552 (Muller et al. 2007). *V. cholerae* A1552 forms biofilms on a variety of biological and inorganic surfaces. When grown in a flow chamber in defined glycerol minimal medium, rapid surface colonization is followed by the development of pronounced three-dimensional biofilms, that undergo natural dissolution (Muller et al. 2007): at around 72 h after inoculation, up to 80% of the biofilm biomass is synchronously detaching. This detachment can be induced in 60-h-old biofilms simply by stopping the media flow, as was observed for *S. oneidensis* biofilms. In contrast to that system, the environmental or physiological cue(s) triggering detachment in *V. cholerae* are unknown at this point. Interestingly, these three-dimensional bio-films form, although to a reduced extent, in Δ*vps* mutants, which are deficient in production of the major exopolysaccharide VPS (Yildiz and Schoolnik 1999). Such biofilms are fully competent in detaching upon subjecting to a stop-of-flow (Muller et al. 2007). Expression of *vps* genes, which are controlled by *rpoS*, leads to nondetaching biofilms (Casper-Lindley and Yildiz 2004; Muller et al. 2007).

From the observations discussed above, a theme for biofilm progression and detachment in some microbes is emerging: after initial seeding, formation of a substantial three-dimensional biofilm structure occurs largely independent of exopolysaccharides. Initially, cells of these biofilms are not irreversibly attached to the biofilm matrix, but can undergo physiologically induced detachment. Only in later phases, exopolysaccharide production is activated, which, presumably, leads to nondetaching biofilms. On the other hand in some microbes, some EPSs, such as PGA in *E. coli*, mediates both initial adhesion as well as cell attachment, and might, therefore, lead to immediate irreversible attachment. This could also be an explanation for why detachment has not been observed so far in *P. aeruginosa*. This microbe contains at least three gene clusters encoding for the synthesis of distinct exopolysaccharides, and *psl* genes are expressed already upon contact with the substratum surface (Ma et al. 2007). In this context, it is of interest that the stationary-phase-induced carbon storage regulator (CsrA) is a negative regulator of biofilm formation and a positive regulator of biofilm dispersal (Jackson et al. 2002). CsrA controls glycogen metabolism and poly-β-1,6-*N*-acetyl-D-glucosamine synthesis (Jackson et al. 2002). This polymer promotes attachment of cells to surfaces, to other cells, and stabilization of biofilms (Itoh et al. 2005; Wang et al. 2004), suggesting that it may act as an exopolysaccharide-based cell adhesin. However, and very interestingly, overexpression of *csrA* leads to biofilm dispersal, but the underlying mechanisms are so far unknown (Wand et al. 2005).

4.4   c-di-GMP in Cellular Attachment and Detachment

Numerous studies have shown that c-di-GMP signaling regulates the transition of cells between the planktonic to the biofilm state (see also Jenal and Malone 2006). c-di-GMP is formed by diguanylate cyclases (DGC), which commonly contain a GGDEF domain containing this conserved amino acid sequence. Hydrolysis of c-di-GMP proceeds by the activity of phosphodiesterases (PDE) often associated with an EAL or HD-GYP domain, containing these conserved amino acid sequences (Galperin et al. 2001; Dow et al. 2006). In general, elevated cellular levels of c-di-GMP can lead to increased and nondetachable biofilm by EPS- and/or a LapA-dependent attachment/adhesion mechanism(s).

For EPS-dependent attachment, c-di-GMP may be acting as an allosteric activator of an exopolysaccharide synthase, similar as in cellulose synthase, leading to production of extracellular polysaccharides, including cellulose, and concomitant cell attachment (Ross et al. 1987, 1990). Biofilm dissolution in *P. putida* has been shown to be promoted by addition of external cellulase (Gjermansen et al. 2005). In *P. putida* as well as in *S. oneidensis* and other microbes, it was shown that overexpression of DGCs leads to formation of significant biofilm formation and biofilm thickness (Gjermansen et al. 2005; Thormann et al. 2006). On the other hand, overexpression of a PDE, which lowers the internal c-di-GMP pool, leads to decreased biofilm formation, and the dissolution of established biofilms (Thormann et al. 2006). However, c-di-GMP signaling is involved in a variety of cellular functions other than cellulose (or EPS) synthesis, including pathogenesis, pili, and flagella-based motility, and PDE overexpression may be toxic to cells (J. Yu and A.M. Spormann unpublished observations; Jenal and Malone 2006).

In the LapA-mediated attachment in *P. fluorescens*, elevated levels of c-di-GMP are required for proper secretion of LapA and concomitant formation of irreversible cellular attachment (Monds et al. 2007). Low external concentrations of P_i act through a transcriptional mechanism as an environmental signal to reduce secretion of LapA via expression of PDE-encoding *rapA*, resulting in a less stable, dissolving biofilm. It is interesting to note that there is an overall correlation between the enhanced retention of biomass in a biofilm and an elevated cellular c-di-GMP level even though the mode of action of c-di-GMP by allosterically controlling exopoly-saccharide synthase activity or secretion of LapA is mechanistically quite different. It will be interesting to see whether the mechanism of LapA-mediated cell attachment is dependent on metabolic energy.

4.5 *Detachment Induced by Cell Lysis or Death*

Death of biofilm cells can also lead to biofilm dispersion. In *P. aeruginosa*, it was found that phage-induced cell death causes biofilm dispersal (Webb et al. 2003, 2004). The phage-induced cell death was hypothesized to be caused by a mutation in the phage leading to a superinfection and consequent killing of cells. As in lysogenic planktonic cells, physiological conditions that induce the SOS response (DNA damage and ROS) were postulated to be involved in phage induction in biofilm cells, and addition of mitomycin C and hydrogen peroxide to planktonic cultures released phages. Moreover, reactive oxygen species accumulated inside biofilm microcolonies, and induced the Pf1 phage. Interestingly, cells decorated with the Pf1 phage become more adhesive and appear to form better biofilms.

Killing of biofilm cells by EDTA treatment rather than by phage induction also leads to biofilm dispersal (Banin et al. 2006). Similarly, various treatments of biofilms with chemicals, for example NaCl, $CaCl_2$, chelating agents, surfactants, and antimicrobials (hypochlorite, monochloramine) have resulted in partial loss of biofilm biomass (Chen and Stewart 2000). It therefore seems that killing biofilm cells can also lead to biofilm dissolution, but the underlying mechanism(s) might be different from those operating in the above-discussed dissolution.

5 Conclusion

Identifying universal molecular themes in physiology of biofilm cells and biofilm formation is challenging for at least two reasons: (1) the diversity of physiological and molecular wiring in microbes and (2) there are many molecular ways to from a biofilm. Comparative studies on the physiology of planktonic microbes have revealed truly unifying principles but, more interestingly, also a fascinating diversity of molecular mechanisms to generate environmentally fit and adaptive microbes. For example, the classical role of CRP as a catabolite repression protein controlling solely carbon degradation pathways was challenged when in *S. oneidensis* it was found that CRP does not control carbon metabolism but expression of electron acceptor pathways (Saffarini et al. 2003). Similarly, RpoS as a master regulator, as shown in *V. cholerae*, can act quite differently depending on the environmental context, yet leading to a similar phenotypic output. These examples simply reflect the diversity of biology of microorganisms. On the other hand, biofilms are macroscopic, operationally defined entities, where *a priori* it is not obvious, that they form by the same molecular mechanisms in different microbes. Regardless, what is invariant and crucial for microbial life in a biofilm is the interplay between physical-chemical and biological factors that control the physiology of the inhabiting microbes. Major changes in cellular physiology include a reduction in growth rate and the induction of *rpoS* but probably also the activation of other global regulators. Also, some microbes form biofilms that are initially in a dissolvable state where changes in metabolic conditions can lead to rapid dispersal. Dynamic and localized changes of the often self-induced changes in the physical-chemical and biological environment determine the physiological heterogeneity of biofilm cells. Future work with new technologies will provide a better understanding of the physiology of individual cells within the actual biofilm environments.

Acknowledgements I thank Alex T. Nielsen, Soeren Molin, Cristian Picioreanu, and the members of the Spormann laboratory for numerous helpful discussions. Work in this laboratory and for this review was supported through funding from NSF and DoE.

References

Adams JL, McLean RJ (1999) Impact of *rpoS* deletion on *Escherichia coli* biofilms. Appl Environ Microbiol 65(9):4285-4287

Applegate DH, Bryers JD (1991) Effects of carbon and oxygen limitations and calcium concentrations on biofilm removal processes. Biotechnol Bioeng 37(1):17-25

Banin E, Brady KM, Greenberg EP (2006) Chelator-induced dispersal and killing of *Pseudomonas aeruginosa* cells in a biofilm. Appl Environ Microbiol 72(3):2064-2069

Barraud N, Hassett DJ, Hwang SH, Rice SA, Kjelleberg S, Webb JS (2006) Involvement of nitric oxide in biofilm dispersal of *Pseudomonas aeruginosa*. J Bacteriol 188(21):7344-7353

Casper-Lindley C, Yildiz FH (2004) VpsT is a transcriptional regulator required for expression of *vps* biosynthesis genes and the development of rugose colonial morphology in *Vibrio cholerae* O1 El Tor. J Bacteriol 186(5):1574-1578

Chen X, Stewart PS (2000) Biofilm removal caused by chemical treatments. Water Res 34:4229-4233

Christensen BB, Sternberg C, Andersen JB, Palmer RJ Jr, Nielsen AT, Givskov M, Molin S (1999) Molecular tools for study of biofilm physiology. Methods Enzymol 310:20-42

Costerton JW (1999) Introduction to biofilm. Int J Antimicrob Agents 11(3-4):217-221; discussion 237-239

Costerton JW, Stewart PS, Greenberg EP (1999) Bacterial biofilms: a common cause of persistent infections. Science 284(5418):1318-1322

Delaquis PJ, Caldwell DE, Lawrence JR, McCurdy AR (1989) Detachment of *Pseudomonas fluorescens* from biofilms on glass surfaces in response to nutrient stress. Microb Ecol 18(3):199-210

Dow JM, Fouhy Y, Lucey JF, Ryan RP (2006) The HD-GYP domain, cyclic di-GMP signaling, and bacterial virulence to plants. Mol Plant Microbe Interact 19(12):1378-1384

Duguid IG, Evans E, Brown MR, and Gilbert P (1992a) Effect of biofilm culture upon the susceptibility of *Staphylococcus epidermidis* to tobramycin. J Antimicrob Chemother 30(6):803-810

Duguid IG, Evans E, Brown MR, and Gilbert P (1992b) Growth-rate-independent killing by ciprofloxacin of biofilm-derived *Staphylococcus epidermidis*; evidence for cell-cycle dependency. J Antimicrob Chemother 30(6):791-802

Evans DJ, Allison DG, Brown MR, Gilbert P (1991) Susceptibility of *Pseudomonas aeruginosa* and *Escherichia coli* biofilms towards ciprofloxacin: effect of specific growth rate. J Antimicrob Chemother 27(2):177-84

Farrell MJ, Finkel SE (2003) The growth advantage in stationary-phase phenotype conferred by *rpoS* mutations is dependent on the pH and nutrient environment. J Bacteriol 185(24):7044-7052

Galperin MY, Nikolskaya AN, Koonin EV (2001) Novel domains of the prokaryotic two-component signal transduction systems. FEMS Microbiol Lett 203(1):11-21

Gjermansen M, Ragas P, Sternberg C, Molin S, Tolker-Nielsen T (2005) Characterization of starvation-induced dispersion in *Pseudomonas putida* biofilms. Environ Microbiol 7(6):894-906

Goller C, Wang X, Itoh Y, Romeo T (2006) The cation-responsive protein NhaR of *Escherichia coli* activates *pgaABCD* transcription, required for production of the biofilm adhesin poly-beta-1,6-N-acetyl-D-glucosamine. J Bacteriol. 188(23):8022-8032

Hansen SK, Haagensen JA, Gjermansen M, Jorgensen TM, Tolker-Nielsen T, Molin S (2007a) Characterization of a *Pseudomonas putida* rough variant evolved in a mixed-species biofilm with *Acinetobacter* sp. strain C6. J Bacteriol 189(13):4932-4943

Hansen SK, Rainey PB, Haagensen JA, Molin S (2007b) Evolution of species interactions in a biofilm community. Nature 445(7127):533-536

Heidelberg JF, Paulsen IT, Nelson KE et al (2002) Genome sequence of the dissimilatory metal ion-reducing bacterium *Shewanella oneidensis*. Nat Biotechnol 20(11):1118-1123

Hengge-Aronis R (2002) Signal transduction and regulatory mechanisms involved in control of the sigma(S) (RpoS) subunit of RNA polymerase. Microbiol Mol Biol Rev 66(3):373-395

Heydorn A, Ersboll B, Kato J et al (2002) Statistical analysis of *Pseudomonas aeruginosa* biofilm development: impact of mutations in genes involved in twitching motility, cell-to-cell signaling, and stationary-phase sigma factor expression. Appl Environ Microbiol 68(4):2008-2017

Hinsa SM, Espinosa-Urgel M, Ramos JL, O'Toole GA (2003) Transition from reversible to irreversible attachment during biofilm formation by *Pseudomonas fluorescens* WCS365 requires an ABC transporter and a large secreted protein. Mol Microbiol 49(4):905-918

Huang CT, Xu KD, McFeters GA, Stewart PS (1998) Spatial patterns of alkaline phosphatase expression within bacterial colonies and biofilms in response to phosphate starvation. Appl Environ Microbiol 64(4):1526-1531

Itoh Y, Wang X, Hinnebusch BJ, Preston JF 3rd, Romeo T (2005) Depolymerization of beta-1,6-N-acetyl-D-glucosamine disrupts the integrity of diverse bacterial biofilms. J Bacteriol 187(1):382-387

Jackson DW, Simecka JW, Romeo T (2002a) Catabolite repression of *Escherichia coli* biofilm formation. J Bacteriol 184(12):3406-3410

Jackson DW, Suzuki K, Oakford L, Simecka JW, Hart ME, Romeo T (2002b) Biofilm formation and dispersal under the influence of the global regulator CsrA of *Escherichia coli*. J Bacteriol 184(1):290-301

James GA, Korber DR, Caldwell DE, Costerton JW (1995) Digital image analysis of growth and starvation responses of a surface-colonizing *Acinetobacter* sp. J Bacteriol 177(4):907-915

Jenal U, Malone J (2006) Mechanisms of cyclic-di-GMP signaling in bacteria. Annu Rev Genet 40:385-407

Liang W, Pascual-Montano A, Silva AJ, Benitez JA (2007) The cyclic AMP receptor protein modulates quorum sensing, motility and multiple genes that affect intestinal colonization in *Vibrio cholerae*. Microbiology 153(9):2964-2975

Liu X, Ng C, Ferenci T (2000) Global adaptations resulting from high population densities in *Escherichia coli* cultures. J Bacteriol 182(15):4158-4164

Ma L, Lu H, Sprinkle A, Parsek MR, Wozniak D (2007) *Pseudomonas aeruginosa* Psl is a galactose- and mannose-rich exopolysaccharide. J Bacteriol July 13, E-pub ahead of print

Mah TF, O'Toole GA (2001) Mechanisms of biofilm resistance to antimicrobial agents. Trends Microbiol 9(1):34-39

Marshall PA, Loeb GI, Cowan MM, Fletcher M (1989) Response of microbial adhesives and biofilm matrix polymers to chemical treatments as determined by interference reflection microscopy and light section microscopy. App Environ Microbiol 55(11):2827-2831

Monds RD, Newell PD, Gross RH, O'Toole GA (2007) Phosphate-dependent modulation of c-di-GMP levels regulates *Pseudomonas fluorescens* Pf0-1 biofilm formation by controlling secretion of the adhesin LapA. Mol Microbiol 63(3):656-679

Muller J, Miller MC, Nielsen AT, Schoolnik GK, Spormann AM (2007) *vpsA*- and *luxO*-independent biofilms of *Vibrio cholerae*. FEMS Microbiol Lett 275:199-206

Myers CR, Nealson KH (1988) Bacterial manganese reduction and growth with manganese oxide as the sole electron acceptor. Science 240:1319-1321

Nealson KH, Saffarini D (1994) Iron and manganese in anaerobic respiration: environmental significance, physiology, and regulation. Annu Rev Microbiol 48:311-343

Nielsen AT, Dolganov NA, Otto G, Miller MC, Wu CY, Schoolnik GK (2006) RpoS controls the *Vibrio cholerae* mucosal escape response. PLoS Pathog 2(10):e109

O'Toole G, Kaplan HB, Kolter R (2000) Biofilm formation as microbial development. Annu Rev Microbiol 54:49-79

Picioreanu C, van Loosdrecht MC, Heijnen JJ (1998) Mathematical modeling of biofilm structure with a hybrid differential-discrete cellular automaton approach. Biotech Bioeng 58(1):101-116

Picioreanu C, van Loosdrecht MC, Heijnen JJ (2001) Two-dimensional model of biofilm detachment caused by internal stress from liquid flow. Biotechnol Bioeng 72(2):205-218

Ross P, Mayer R, Weinhouse H et al (1987) Regulation of cellulose synthesis in *Acetobacter xylinum* by cyclic diguanylic acid. Nature 325:279-281

Ross P, Mayer R, Weinhouse H et al (1990) The cyclic diguanylic acid regulatory system of cellulose synthesis in *Acetobacter xylinum*. Chemical synthesis and biological activity of cyclic nucleotide dimer, trimer, and phosphothioate derivatives. J Biol Chem 265(31):18933-18943

Saffarini DA, Schultz R, Beliaev A (2003) Involvement of cyclic AMP (cAMP) and cAMP receptor protein in anaerobic respiration of *Shewanella oneidensis* J Bacteriol 185(12):3668-3671

Sauer K, Cullen MC, Rickard AH, Zeef LA, Davies DG, Gilbert P (2004) Characterization of nutrient-induced dispersion in *Pseudomonas aeruginosa* PAO1 biofilm. J Bacteriol 186(21):7312-7326

Sawyer LK, Hermanowicz SW (2000) Detachment of *Aeromonas hydrophila* and *Pseudomonas aeruginosa* due to variations in nutrient supply. Water Sci Technol 41:139-145

Schembri MA, Kjaergaard K, Klemm P (2003) Global gene expression in *Escherichia coli* biofilms. Mol Microbiol 48(1):253-267

Sternberg C, Christensen BB, Johansen T et al (1999) Distribution of bacterial growth activity in flow-chamber biofilms. Appl Environ Microbiol 65(9):4108-4117

Sutherland IW (2001) The biofilm matrix - an immobilized but dynamic microbial environment. Trends Microbiol 9(5):222-227

Thormann KM, Saville R, Shukla S et al (2004) Initial phases of biofilm formation in *Shewanella oneidensis* MR-1. J Bacteriol 186(23):8096-8104

Thormann KM, Saville R, Shukla S, Spormann AM (2005) Induction of rapid detachment in *Shewanella oneidensis* MR1 biofilms. J Bacteriol 187(3):1014-1021

Thormann KM, Duttler SA, Saville R et al (2006) Control of formation and cellular detachment from *Shewanella oneidensis* MR-1 biofilms by cyclic-di-GMP. J Bacteriol 188(7):2681-2691

Tolker-Nielsen T, Molin S (2000) Spatial organization of microbial biofilm communities. Microb Ecol 40(2):75-84

van Loosdrecht MC, Heijnen JJ, Eberl H, Kreft J, Picioreanu C (2002) Mathematical modelling of biofilm structures. Antonie Van Leeuwenhoek 81(1-4):245-256

Waite RD, Papakonstantinopoulou A, Littler E, Curtis MA (2005) Transcriptome analysis of *Pseudomonas aeruginosa* growth: comparison of gene expression in planktonic cultures and developing and mature biofilms. J Bacteriol 187(18):6571-6576

Wang X, Preston JF 3rd, Romeo T (2004) The *pgaABCD* locus of *Escherichia coli* promotes the synthesis of a polysaccharide adhesin required for biofilm formation. J Bacteriol 186(9):2724-2734

Wang X, Dubey AK, Suzuki K, Baker CS, Babitzke P, Romeo T (2005) CsrA post-transcriptionally represses *pgaABCD*, responsible for synthesis of a biofilm polysaccharide adhesin of *Escherichia coli*. Mol Microbiol 56(6):1648-1663

Watnick P, Kolter R (2000) Biofilm city of microbes. J Bacteriol 182(10):2675-2679

Webb JS, Thompson LS, James S et al (2003) Cell death in *Pseudomonas aeruginosa* biofilm development. J Bacteriol 185(15):4585-4592

Webb JS, Lau M, Kjelleberg S (2004) Bacteriophage and phenotypic variation in *Pseudomonas aeruginosa* biofilm development. J Bacteriol 186(23):8066-8073

Whiteley M, Bangera MG, Bumgarner RE et al (2001) Gene expression in *Pseudomonas aeruginosa* biofilms. Nature 413(6858):860-864

Xu KD, Stewart PS, Xia F, Huang CT, McFeters GA (1998) Spatial physiological heterogeneity in *Pseudomonas aeruginosa* biofilm is determined by oxygen availability. Appl Environ Microbiol 64(10):4035-4039

Yildiz FH, Schoolnik GK (1999) *Vibrio cholerae* O1 El Tor: identification of a gene cluster required for the rugose colony type, exopolysaccharide production, chlorine resistance, and biofilm formation. Proc Natl Acad Sci U S A 30(96):4028-4033

Yildiz FH, Liu XS, Heydorn A, Schoolnik GK (2004) Molecular analysis of rugosity in a *Vibrio cholerae* O1 El Tor phase variant. Mol Microbiol 53(2):497-515

Zinser ER, Kolter R (1999) Mutations enhancing amino acid catabolism confer a growth advantage in stationary phase. J Bacteriol 181(18):5800-5807

Environmental Influences on Biofilm Development

C. C. Goller and T. Romeo(✉)

Abstract Bacterial biofilms are found under diverse environmental conditions, from sheltered and specialized environments found within mammalian hosts to the extremes of biological survival. The process of forming a biofilm and the eventual return of cells to the planktonic state involve the coordination of vast amounts of genetic information. Nevertheless, the prevailing evidence suggests that the overall progression of this cycle within a given species or strain of bacteria responds to environmental conditions via a finite number of key regulatory factors and pathways, which affect enzymatic and structural elements that are needed for biofilm formation and dispersal. Among the conditions that affect biofilm development are temperature, pH, O_2 levels, hydrodynamics, osmolarity, the presence of specific ions, nutrients, and factors derived from the biotic environment. The integration of these influences ultimately determines the pattern of behavior of a given bacterium with respect to biofilm development. This chapter will present examples of how environmental conditions affect biofilm development, most of which come from studies of species that have mammalian hosts.

T. Romeo
Department of Microbiology and Immunology, Emory University School of Medicine, 3105 Rollins Research Center, 1510 Clifton Rd., N.E., Atlanta, GA, 30322 USA
romeo@microbio.emory.edu

T. Romeo (ed.), *Bacterial Biofilms.*
Current Topics in Microbiology and Immunology 322.
© Springer-Verlag Berlin Heidelberg 2008

1   Introduction

In the past decade, substantial advances in the understanding of the genetic and physiological bases of biofilm formation have been made. Dramatic differences in gene expression patterns exist between planktonic and sessile cells, and indeed even between different stages of biofilm development (e.g., Sauer et al. 2003). Nevertheless, the environmental and genetic factors that promote the transition from planktonic to sessile communities are only beginning to be understood in a few model organisms (reviewed by Stanley and Lazazzera 2004). It is clear that different species and even strains of bacteria can exhibit unique patterns of response to the environment. What environmental conditions predispose various species of bacteria to initiate a given biofilm? How are the molecular genetic, biochemical, and structural elements that mediate biofilm development regulated in response to environmental conditions? The following sections describe some of the environmental influences on biofilm development in the context of the molecular genetics and biochemistry of the biofilm development cycle (Fig. 1).

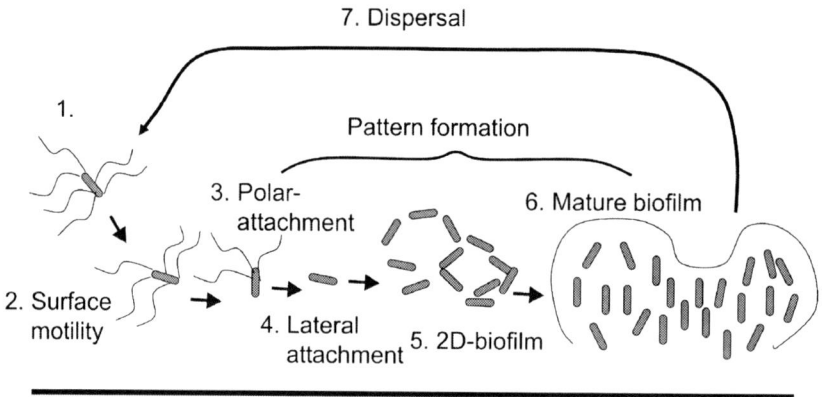

Fig. 1 A model for biofilm development. Planktonic cells (*1*) use motility to approach and swim on a surface (*2*). Upon interacting with the substratum by a pole, cells can become reversibly attached, which may allow for sampling of the environment before committing to a sessile lifestyle (*3*). Next, cells become laterally attached to the surface, involving adhesins such as PGA or LapA (*4*). During this time, the attachment of cells begins to create a two-dimensional biofilm, which in *E. coli*, exhibits distinct periodicity in cellular distribution (*5*). The biofilm grows in thickness as more cells are incorporated into its structure. Extracellular polysaccharides and other substances are produced, resulting in more firmly attached cells within an extracellular matrix. The architecture of the biofilm may be modified by production of surfactant and release of attached cells (*6*). In response to environmental or physiological clues, cells may be released from the matrix and return to a planktonic state, thus completing the developmental cycle (*7*). The entire process of biofilm development is dynamic and is influenced by numerous environmental factors

2   Surface Factors and Hydrodynamic Effects

Virtually any material that comes into contact with fluids containing bacteria is a substrate for biofilm formation. The roughness, chemistry, and presence of conditioning films affect attachment of bacterial cells to a surface. While rough surfaces are readily colonized because shear forces are diminished and surface area is increased in rougher surfaces (Donlan 2002), studies have indicated that nondomesticated strains of at least some species seem to colonize smooth surfaces equally as well (Donlan and Costerton 2002). Studies have also demonstrated that microorganisms typically attach more rapidly to hydrophobic surfaces such as plastics than to hydrophilic glass or metals (reviewed by Donlan 2002). For instance, hydrophobic substrata promote biofilm formation by most clinical isolates of *Staphylococcus epidermidis* (Cerca et al. 2005). Hydrophobic interactions between the cell surface and the substratum may enable the cell to overcome repulsive forces and attach irreversibly (Donlan 2002). A notable exception is that *Listeria monocytogenes* forms biofilms more rapidly on hydrophilic than on hydrophobic surfaces (Chavant et al. 2002).

Submerged surfaces adsorb solutes and small particles, including bacteria (Geesey 2001). Studies dating back to the 1940s showed that glass surfaces adsorb nutrients from sea water, with consequent effects on metabolic activity associated with bacterial attachment (e.g., ZoBell 1943). Furthermore, the metabolic activities of bacteria associated with a surface cause temporal and spatial changes in the three-dimensional chemical gradients at the liquid-solid interface (Geesey 2001; Rani et al. 2007). When surfaces exposed to fluid environments adsorb proteins, coatings or conditioning films are formed that alter the surface properties and affect attachment of bacteria (Dunne 2002; Murga et al. 2001; Tieszer et al. 1998). For example, the proteinaceous conditioning films called acquired pellicles that develop on tooth enamel within the oral cavity are colonized within hours by Gram-positive cocci (Donlan and Costerton 2002; Rickard et al. 2003). The surface of a central venous catheter is in direct contact with the bloodstream and becomes coated with platelets, plasma, and tissue proteins including albumin, fibrinogen, fibronectin, and laminin (see the chapter by R.M. Donlan, this volume). This coating acts as a conditioning film that is colonized by organisms such as *Staphylococcus aureus*, which adheres to fibronectin, fibrinogen, and laminin via large surface proteins known as MSCRAMMs (microbial surface components recognizing adhesive matrix molecules) (see the chapter by M. Otto, this volume; Mack et al. 2007; Patti et al. 1994).

Fluid flow or hydrodynamics influences biofilm structure and can have dramatic effects on the type of biofilm that is formed. Physical properties of biofilms such as cell density and strength of attachment can be affected by fluid sheer (reviewed by Stoodley et al. 2002 a, 2002b; van Loosdrecht et al. 2002). Furthermore, biofilms grown under low flow conditions may form isotropic structures, whereas higher unidirectional flow may produce filamentous cells or groupings of cells with evidence of directionality (Stoodley et al. 1999, 2002a). *Pseudomonas aeruginosa* biofilms grown under high shear were more strongly attached than those grown under lower shear (Stoodley et al. 2002b). Others speculate that turbulent flow may enhance bacterial adhesion and biofilm formation by impinging cells on the surface (Donlan and Costerton 2002). In contrast, rolling of entire staphylococcal microcolonies over surfaces has been observed in biofilms grown under turbulent flow, perhaps allowing mature biofilms to colonize new surfaces downstream (Hall-Stoodley and Stoodley 2005; Rupp et al. 2005). Similarly, *Escherichia coli* attachment to mannose-coated surfaces via the type 1 fimbrial adhesive subunit, FimH, is shear-dependent. At low shear, the cells tended to roll over the surface; however, as shear was increased, they became more firmly attached (Anderson et al. 2007; Thomas et al. 2004). Weak rolling adhesion at low shear force allows for cells to spread out and colonize more surface area than under high shear stress, where cells remain in tight microcolonies. Thus, preferred sites of colonization may be those with the necessary flow to maintain a stable interaction between the bacteria and host proteins (Isberg and Barnes 2002). In a study of *E. coli* biofilm formation under flow, fluid flow altered the spatial organization of cell attachment patterns (Agladze et al. 2003). While these and other studies document the important role of hydrodynamics in biofilm development and structure, little is known about the possible molecular genetic responses to fluid flow.

3   Approach and Initial Attachment to the Surface

3.1   Motility and Chemotaxis

Although both motile and nonmotile species form biofilms, in motile species, the ability to move using flagella or pili is generally required for efficient cell-to-surface attachment. Microscopic observations indicate that motility promotes both initial interaction with the surface and movement along it (O'Toole and Kolter 1998; Pratt and Kolter 1998). However, there are reports suggesting that motility may only be important for biofilm formation under certain conditions (McClaine and Ford 2002). Motility may be needed to overcome the repulsive forces generated between cellular and abiotic surfaces and to permit favorable cell-surface interactions required for attachment (Geesey 2001). However, flagellar motility is not essential for initial adhesion and biofilm formation when the cell is equipped with an efficient adhesin (Jackson et al. 2002b; Prigent-Combaret et al. 2000; Wang et al. 2004). Furthermore, steric hindrance and/or movement caused by a flagellum can destabilize cellular attachments. Accordingly, motility genes are repressed after the bacterium attaches to the surface (Prigent-Combaret et al. 1999). Another example of the complex influence of environmental conditions on motility and biofilm development is the finding that while twitching motility via type IV pili appears to be needed for *P. aeruginosa* biofilm formation (O'Toole and Kolter 1998), overstimulation of twitching by the chelation of iron with lactoferrin, a component of innate immunity, prevents this bacterium from establishing productive surface contacts and forming biofilm (Singh et al. 2002).

Surface motility is widespread among flagellated Gram-negative bacteria. When it involves groups of long, hyperflagellated cells, moving as an organized mass, it is referred to as swarming motility. In *P. aeruginosa*, swarming motility is regulated through Rhl quorum sensing, while swimming is not. In recent studies, Rhl-dependent quorum sensing and nutritional conditions determined whether a flat, uniform biofilm or a structured biofilm was formed (Shrout et al. 2006). In contrast to motility, chemotaxis is not required for *E. coli* biofilm development in batch cultures (Pratt and Kolter 1998). However, in topologically constrained environments, chemotaxis may be important for assembling a quorum of cells that can initiate biofilm development (Park et al. 2003).

Expression of the genes involved in flagellum synthesis, motility, and chemotaxis in *E. coli* occurs in a hierarchical fashion, permitting ordered synthesis and assembly of the flagellum components (e.g., Macnab 2003; Soutourina and Bertin 2003). The master regulator $FlhD_2C_2$ is a DNA-binding protein that is directly or indirectly required for expression of all other motility and chemotaxis genes, over 50 in total. These are expressed from at least 15 operons, clustered at several regions on the chromosome. Expression of the *flhDC* operon serves as a pivotal point for integrating environmental signals (Fig. 2). Its expression is controlled by numerous regulators including H-NS, Crp, EnvZ-OmpR, CsrA, QseBC, LrhA, and RcsCDB, which sense environmental conditions such as osmolarity (H-NS, EnvZ-OmpR), envelope stress (RcsCDB), nutritional conditions (Crp), or quorum sensing (QseBC).

In *E. coli*, high osmolarity and acetyl-phosphate levels inhibit *flhDC* expression and motility through the phosphorylation and subsequent binding of OmpR to the *flhDC* promoter region (Shin and Park 1995). The synthesis of flagella is also controlled by growth temperature: cells are not flagellated at 42°C, perhaps because of competition for the heat shock chaperones DnaK, DnaJ, and GrpE, which are needed for flagellum gene expression (Shi et al. 1992). Furthermore, *flhDC* and flagellum biosynthesis are regulated by catabolite repression, i.e., activated by the cyclic AMP-Crp complex, and are repressed by the nucleoid-associated protein H-NS (Silverman and Simon 1974; Soutourina et al. 1999). Overall, stressful conditions such as high concentrations of salts, sugars, or alcohols, high temperature, both low and high pH, or conditions of blocked DNA replication inhibit flagellum biosynthesis (Maurer et al. 2005; Shin and Park 1995; Soutourina et al. 2002).

The Csr (carbon storage regulator) system of *E. coli* also controls motility and flagellum biosynthesis. The RNA binding protein CsrA positively regulates *flhDC* expression by binding to the untranslated leader and stabilizing this mRNA (Wei et al. 2001). Although much information has been obtained concerning the regulatory circuitry and mechanisms of this complex system (e.g., Romeo 1998; Suzuki et al. 2002, 2006; Weilbacher et al. 2003), the environmental signals are still somewhat obscure. At the present time, it is evident that quorum sensing via SdiA and environmental pH affect the expression of noncoding RNA antagonists that sequester CsrA (Babitzke and Romeo 2007; Suzuki et al. 2002; Mondragon et al. 2006). Importantly, while CsrA activates motility, its dominant role in biofilm formation

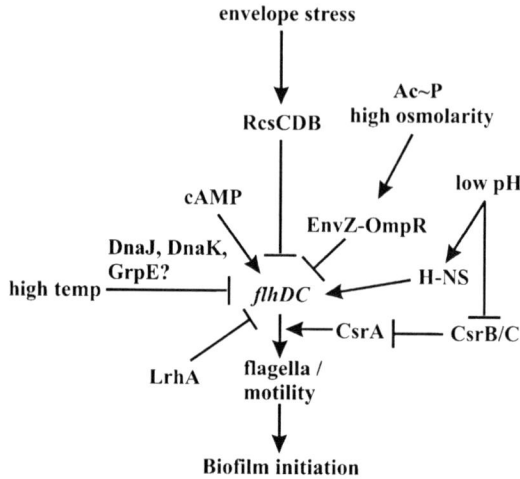

Fig. 2 Regulation of *E. coli* motility. The *flhDC* operon encodes a DNA binding protein (FlhD$_2$C$_2$) that serves as a central regulatory point to initiate the motility and chemotaxis cascade of gene expression, which is needed for optimal biofilm formation. Stressful conditions such as high concentrations of salts, sugars, or alcohols, high temperature, both low and high pH, or conditions of blocked DNA replication inhibit flagellum biosynthesis. The RcsCDB phosphorelay system, which somehow is activated by envelope stress, represses *flhDC*. Acetyl-phosphate and high osmolarity activate the EnvZ-OmpR two component signal transduction system, which represses *flhDC*. The heat shock chaperones DnaK, DnaJ, and GrpE are needed for flagellum gene expression, but may be limiting at high temperatures. In addition, *flhDC* transcription is under catabolite repression and is activated by cAMP-Crp. The RNA binding protein CsrA activates *flhDC* expression by binding to the untranslated leader and stabilizing this mRNA. However, the main effect of CsrA on biofilm formation is to repress expression of the adhesin PGA (see Fig. 5). LrhA, a LysR-type transcriptional regulator, represses motility as well as expression of type 1 fimbriae. In various species, c-di-GMP, which is synthesized by GGDEF domain-containing proteins and is degraded by EAL domain proteins, inhibits flagellum-based motility

is to repress expression of the polysaccharide adhesin PGA of *E. coli* K-12 (e.g., Wang et al. 2005) and overall it acts as a strong repressor of biofilm formation (Jackson et al. 2002a).

The temporal control of flagellum biogenesis also involves the Rcs phosphorelay and acetyl-phosphate (Fredericks et al. 2006). The Rcs (regulator of capsule synthesis) phosphorelay activates genes required for capsular biosynthesis and membrane proteins (Boulanger et al. 2005), while repressing genes required for flagellum biogenesis (Francez-Charlot et al. 2003). The Rcs regulon is thought to be activated by surface contact and envelope stress; however, the exact nature of the signal remains unknown (reviewed by Majdalani and Gottesman 2005).

Recent studies implicate the ubiquitous bacterial secondary messenger c-di-GMP (3′-5′-cyclic diguanylic acid) as a central regulator of motility and biofilm formation in diverse Gram-negative species. In general, this nucleotide, which is synthesized by GGDEF domain-containing proteins and is degraded by EAL or HD-GYP domain proteins, affects the transition from planktonic to sessile communities by promoting the production of adhesins and exopolysaccharides and inhibiting flagellum- and pilus-based motility (reviewed in Jenal and Malone 2006; Ryan et al. 2006). While c-di-GMP metabolizing proteins often contain sensory domains (e.g., PAS, GAF, CheY-like, and REC), only a few environmental cues are known or suspected to influence c-di-GMP metabolism, and with the exception of cellulose synthase, the mechanism of action of this nucleotide is unknown. As in the case of Csr regulation, c-di-GMP generally has opposite effects on biofilm formation and motility, consistent with the idea that while motility facilitates initiation of biofilm formation, it may be detrimental at later stages.

3.2 Surface Sensing?

Are bacteria able to sense contact with a surface and respond by expressing adhesins? The Cpx signaling system in *E. coli* has provided some circumstantial evidence for surface sensing. Cpx is a two-component system composed of CpxA, a sensor kinase/phosphatase, and CpxR, a DNA-binding response regulator (Raivio and Silhavy 1997). Studies by Otto and Silhavy (2002) showed that a *cpxR* mutant strain forms altered cell-surface interactions in comparison with the wild type strain and that Cpx-regulated gene expression is enhanced by surface attachment. The mechanism of surface sensing is unknown and may be indirect. Studies indicate that the Cpx system responds to misfolded proteins in the periplasm (Danese and Silhavy 1998). In a microtiter plate assay for biofilm formation, *cpxA* mutants that apparently have lost the phosphatase activity of the CpxA protein formed biofilm with less biomass than wild type strains (Dorel et al. 1999). This was due to decreased transcription of the curlin-encoding gene *csgA* (described Sect. 3.3.1). In uropathogenic *E. coli*, Cpx responds to misfolded pyelonephritis-associated P pilin subunits in the periplasm. In turn, DNA binding by CpxR, in conjunction with other transcription factors, induces transcription from the *papB* and *papI* promoters (Hung et al. 2001). Finally, transcriptome analysis in *E. coli* K-12 showed that *cpxP* is highly expressed in biofilms and affects biofilm structure (Beloin et al. 2004). Whether attachment to a surface leads to denaturation of certain envelope proteins and mediates the proposed surface-sensing by Cpx remains to be determined.

3.3 Environmental Effects on Surface Attachment Proteins

Bacteria make extensive use of proteinaceous extracellular fimbriae or pili, which permit them to establish surface contacts that promote biofilm formation. Fimbriae are generally under complex regulatory controls, often involving multiple physiological and/or environmental inputs. The following discussion presents some examples in which the environmental conditions and genetic regulation of fimbriae of *E. coli* and its relatives have been examined, illustrating the complexity of the regulatory networks involved in biofilm formation.

3.3.1 Curli

Proteinaceous extracellular fibers called curli were first observed in *E. coli* (Olsen et al. 1989) and have been shown to mediate adhesion, colonization, and biofilm formation in this and other species. In *Salmonella* spp., curli are also known as thin aggregative fimbriae (Romling et al. 1998). In *E. coli*, curli promote both initial adhesion and cell-cell interaction (Prigent-Combaret et al. 2000). A variety of environmental isolates of *E. coli* form biofilms according to their ability to express curli (Castonguay et al. 2006). Curli synthesis in *E. coli* is dependent on at least six genes located in the divergently transcribed *csgBA* and *csgDEFG* operons. CsgD activates transcription of the *csgBA* operon, which encodes CsgA, the structural subunit that is secreted outside of the cell, where CsgB nucleates it into a fiber (Barnhart and Chapman 2006).

Expression of curli is activated under conditions of low temperature; microaerophilic conditions; low nitrogen, phosphate, and iron; low osmolarity; and slow growth or starvation (Gerstel et al. 2003; Maurer et al. 1998; Olsen et al. 1993a, 1993b; reviewed in Barnhart and Chapman 2006) (Fig. 3). These features imply that curli are produced in the external environment, as opposed to in the mammalian host. However, in addition to abiotic surfaces, curli mediate bacterial binding to extracellular matrix proteins such as fibronectin and laminin (Barnhart and Chapman 2006), suggesting that they may be produced in anticipation or preparation for host attachment and colonization. Other studies have indicated that within a biofilm, curli fimbriae may be expressed at 37°C (Kikuchi et al. 2005). Of note, curli are not expressed in many laboratory strains of *E. coli*, due to silencing of the *csgD* promoter (Hammar et al. 1995).

Curli expression responds to environmental conditions through at least three different phosphorelay signaling systems. The EnvZ-OmpR two-component regulatory system activates *csgD* transcription and thereby promotes production of curli fimbriae and stable cell-surface interactions at low osmolarity (Vidal et al. 1998; Prigent-Combaret et al. 2001). However, in conditions of low osmolarity, there is a reduced level of the active response regulator, phosphorylated OmpR, due to the decreased kinase/phosphatase ratio of EnvZ (Cai and Inouye 2002). This would seem to suggest that an increase in osmolarity should result in higher *csgD* transcription and curli biosynthesis. However, high osmolarity has a negative effect on transcription of the curli genes (Prigent-Combaret et al. 2001). This apparent contradiction can be reconciled by the observation that the Cpx pathway, which represses transcription of curli, is induced by high osmolarity and masks OmpR activation (Prigent-Combaret et al. 2001). Whereas CpxR represses *csgD* in high salt concentrations, the nucleoid-associated protein H-NS mediates *csgD* repression in high sucrose, independently of CpxR (Jubelin et al. 2005). Activation of the Cpx pathway by curlin accumulation also results in the repression

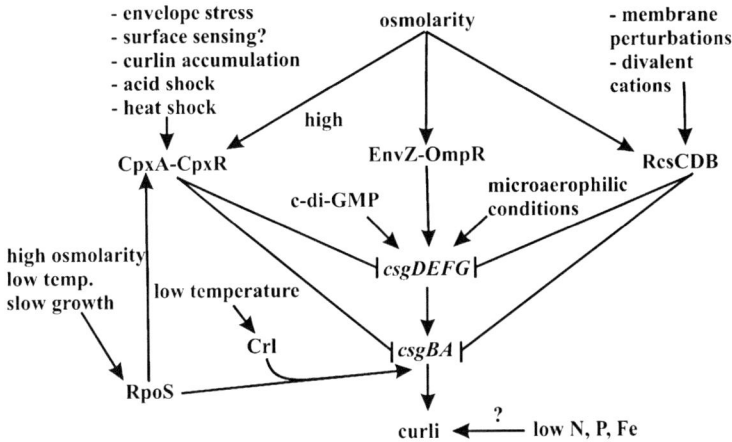

Fig. 3 Conditions affecting curli fimbriae in *E. coli*. Curli fimbriae aid in biofilm formation in certain *E. coli* strains and related species, and are produced through the expression of the divergent operons *csgDEFG* and *csgBA*. CsgD is a DNA-binding protein necessary for transcription of *csgBA*, which encodes the nucleation factor and pilin for curli fimbriae, respectively. Other Csg proteins are involved in pilus biogenesis. Both OmpR-P (activator) and CpxR-P (repressor) can simultaneously occupy the *csgDEFG* promoter. EnvZ-OmpR promotes *csgDEFG* transcription at low osmolarity, while CpxA-CpxR represses this operon under envelope stress and high osmolarity. H-NS has multiple effects on these pathways, one of which is to repress *csgD* in high sucrose, independently of CpxR. The RcsCDB phosphorelay system, which controls synthesis of capsule and flagella, also represses curli in response to membrane perturbations and high osmolarity. c-di-GMP activates production of both curli and cellulose in response to uncharacterized stimuli. Low temperature, nitrogen, phosphorus or iron limitation, slow growth, and microaerophilic conditions promote curli production. RpoS, in conjunction with Crl, activates transcription of the *csgBA* promoter in response to several of these conditions

of the *csgD* and *csgB* operons (Prigent-Combaret et al. 2001). In addition, the RcsCDB phosphorelay system, which controls the synthesis of capsule and flagella, also represses expression of curli (Vianney et al. 2005). A comprehensive model in which EnvZ-OmpR, Cpx, and Rcs regulate *csgD* transcription and curli gene expression in response to changes in osmolarity has been proposed (Jubelin et al. 2005).

Transcription from the *csgD* promoter is also regulated by other global transcription factors, including *rpoS*, *crl*, and *hns* (Romling et al. 1998). The stationary phase sigma factor RpoS (σ^s) directly activates transcription of the *csgBA* promoter in response to slow growth or other stresses (Hengge-Aronis 2002). The small protein Crl, which is preferentially expressed at low temperature and in stationary phase, interacts with the σ^s subunit and apparently promotes curli production by strengthening the association of σ^s with core RNA polymerase to enhance transcription initiation at *csgBA* (Bougdour et al. 2004). The protein H-NS has both direct and indirect effects on curli, depending on the environmental conditions (Jubelin et al. 2005). Apparently, integration host factor (IHF), H-NS, and OmpR form a nucleoprotein complex with the *csgD* promoter, resulting in elevated expression under microaerophilic growth conditions (Gerstel et al. 2003).

The regulatory nucleotide c-di-GMP (which is produced in response to complex regulatory cues) activates the production of both curli and cellulose in certain *E. coli* strains and in *Salmonella enterica* serovar Typhimurium (e.g., Kader et al. 2006; Weber et al. 2006). Curli and cellulose together produce a strong biofilm matrix facilitating attachment to hydrophilic and hydrophobic surfaces (Zogaj et al. 2003).

3.3.2 Type 1 Fimbriae

Type 1, or mannose-sensitive, fimbriae are rigid, 7-nm-wide and approximately 1-µm-long, rod-shaped surface structures found on the majority of *E. coli* strains and are widespread among the Enterobacteriaceae (Schembri et al. 2001). Type 1 fimbriae are important in the colonization of various host tissues by *E. coli* and in biofilm formation on abiotic surfaces (see the chapter by J.K. Hatt and P.N. Rather, this volume; Pratt and Kolter 1998). The FimH adhesive protein expressed on the tip of type 1 fimbriae binds to glycoproteins, including natural ligands such as uroplakins on urinary epithelial cells in urinary bladders and immunoglobulin A or mucin in intestines and lungs (e.g., Mulvey et al. 2000). A typical type 1 fimbriated bacterium has 200-500 peritrichously arranged fimbriae (Lowe et al. 1987).

Production of type 1 fimbriae requires a polycistronic operon comprising the seven structural genes (*fimAICDFGH*) and two monocistronic operons encoding the site-specific recombinases FimB and FimE. Transcription of type 1 fimbriae genes is phase variable due to FimB- and FimE-mediated inversion of a 314-bp DNA fragment that contains the promoter for the polycistronic *fim* operon (Klemm 1986). Within a cell population, type 1 fimbriae expression is activated at body temperature and is repressed by high osmolarity and low pH. These effects are mediated through altered switching frequency of the *fim* operon promoter (e.g., Gally et al. 1993; Schwan et al. 2002). Although the environmental signals remain to be shown for LrhA, this transcriptional regulator represses motility and chemotaxis genes and represses production of type 1 fimbriae by altering phase variation (Blumer et al. 2005). Furthermore, the alarmone ppGpp (guanosine 3′, 5′-bispyrophosphate), which is produced in response to amino acid or carbon starvation, activates expression of type 1 fimbriae and biofilm formation in uropathogenic *E. coli* through its role in expression of the FimB recombinase (Aberg et al. 2006). Acetyl-phosphate activates production of type 1 fimbriae, perhaps by serving as a phosphodonor for the FimZ response regulator (discussed in Wolfe et al. 2003). Acetyl-phosphate accumulates at the transition to stationary phase in the presence of excess carbon and/or the lack of oxygen (Wolfe 2005). Thus, the production of type 1 fimbriae on cells is complex; it is governed by nutritional status and repressed by stresses such as low pH, low temperature, and high osmolarity.

3.3.3 Antigen 43 and Related Proteins

Antigen 43, encoded by the *flu* locus, is an autoaggregation factor produced by ma. *E. coli* strains. It was originally discovered for its ability to cause bacterial aggregation (reviewed in Klemm et al. 2006). Antigen 43 is a member of the self-associating autotransporter (SAAT) group of proteins consisting of a signal peptide for transfer across the inner membrane, a translocator domain, and a secreted passenger domain. SAATs, including Ag43, TibA, and AIDA (adhesin involved in diffuse adherence), can interact with each other to cause formation of mixed bacterial aggregates. These proteins are anchored directly to the outer membrane and protrude only approximately 10 nm from the surface, resulting in closer cell-cell interactions than those seen with curli or other fimbriae. Expression of bulky surface structures that protrude beyond this distance in the bacterial envelope (e.g., type 1 pili or capsules) interferes sterically with Ag43-mediated aggregation (Klemm et al. 2006).

Ag43 expression undergoes phase variation controlled by OxyR and Dam (deoxyadenosine methylase). The cellular redox sensor OxyR represses Ag43 expression by binding to the *flu* promoter, while Dam activates Ag43 expression by methylating DNA that overlaps the OxyR binding (Wallecha et al. 2002; see the chapter by C. Beloin et al., this volume). OxyR plays an important role in sensing peroxides encountered during oxidative stress, although its oxidation state may not influence *flu* regulation (Wallecha et al. 2003). It activates protective measures, such as enzymes that detoxify reactive oxygen compounds or repair damage caused by them. Furthermore, Ag43-mediated cell aggregation confers protection from hydrogen peroxide killing (Schembri et al. 2003a). Ag43 and other SAAT proteins, including AIDA-I and TibA, also impair bacterial motility (Ulett et al. 2006). Several studies have indicated that Ag43 is induced specifically during biofilm growth, and its expression enhances *E. coli* biofilm formation (discussed in Klemm et al. 2004; Schembri et al. 2003b). In urinary track infections, Ag43 is expressed by *E. coli* cells that form biofilm-like structures within bladder cells (Anderson et al. 2003).

Finally, environmental pH affects antigen 43-mediated cellular aggregation, which occurs more rapidly as the pH decreases from 10 to 4 (Klemm et al. 2004). This strong effect of pH on cellular aggregation has been proposed to facilitate more rapid transit and thus improved survival in the stomach (Klemm et al. 2006).

4 Conversion from Temporary to Permanent Attachment: A Regulated Process?

During normal biofilm development, some species of bacteria bind to a surface reversibly or temporarily, followed by irreversible or permanent attachment. This phenomenon was first reported in the 1940s (reviewed in Stoodley et al. 2002a). Genes affecting this transition and biofilm development have been studied in *P. aeruginosa* and *Pseudomonas fluorescens* (Caiazza and O'Toole 2004; Hinsa et al. 2003) as well as in *E. coli* (Agladze et al. 2005). In these species, temporarily attached cells interact with a surface by a cell pole, whereas permanently attached cells are associated via the lateral cell surface. Mutants of *P. fluorescens* that failed to produce a large adhesive protein (LapA) and *E. coli* mutants that fail to produce a polysaccharide adhesin (PGA, described below) were similarly defective in the conversion from temporary to permanent attachment (Agladze et al. 2005; Hinsa et al. 2003). In *E. coli*, the kinetics of this transition process was monitored with an assay developed for this purpose (Agladze et al. 2005).

Conversion from temporary to permanent attachment has been proposed to be a regulated process, perhaps allowing the cell to sample its local environment before committing to a sessile lifestyle (Caiazza and O'Toole 2004). Furthermore, because cell attachment in both monolayers and more mature biofilms of *E. coli* exhibit distinct, nonrandom spatial organization, it has been suggested that proximity to neighboring cells might govern the conversion to permanent attachment (Agladze et al. 2003; 2005). *E. coli* mutants lacking the polysaccharide adhesin PGA exhibited aperiodic cell distribution and no apparent cell-cell adhesion. In theory, formation of such patterns could be guided by a reaction-diffusion or Turing process (e.g., Maini et al. 2006), based on the sensing of a bacterially synthesized inhibitor of attachment. Validation of such hypotheses will require an understanding of the putative signals in the local environment that are being recognized, the putative signal transduction pathways through which this information flows, and a better appreciation of the biochemistry of temporary and permanent attachment processes.

5 Environmental Effects on Matrix Polysaccharides

A hallmark of prototypical biofilms is that they are composed of cells embedded within a complex matrix (reviewed in Branda et al. 2005; Sutherland 2001 a, 2001b). While polypeptides, nucleic acids, lipids, and a host of small molecules are often present in biofilm matrices, polysaccharide, which may include multiple different polymers, is often the main component (e.g., Morikawa et al. 2006; Schooling and Beveridge 2006; Steinberger and Holden 2005; Whitchurch et al. 2002). Due to their roles in cellular interactions with surfaces and their direct exposure to cells of the immune system, matrix polysaccharides have become topics of considerable interest. However, an understanding of these polymers is limited, even for the best studied biofilms. Certain polysaccharides influence biofilm architecture, ion selectivity, resistance to desiccation, and other properties, but probably do not function as biofilm adhesins per se. Acidic polysaccharides, such as alginate of *P. aeruginosa*, colanic acid and K antigens of *E. coli*, and capsular polysaccharides of *Pantoea stewartii* and *Xanthomonas campestris* may be considered in this class; they are not essential for biofilm formation and may even be inhibitory under certain conditions (Crossman and Dow 2004; Hanna et al. 2003; Schembri et al. 2004; Stapper et al. 2004; von Bodman et al. 2003; Wozniak et al. 2003). In contrast, other polysaccharides serve as adhesins that assist cell-surface and/or cell-cell attachment. The conditions and regulatory factors that promote the synthesis of the latter polysaccharides drive biofilm formation. Polymers that fall into the latter category tend to be basic or neutral, and include β-1,6-*N*-acetyl-D-glucosamine polymers of staphylococci, *E. coli, Yersinia pestis, Bordetella* species, *Actinobacilli*, and *P. fluorescens* (Heilmann et al. 1996a, 1996b; Itoh et al. 2005; Litran et al. 2002; Maira-Wang et al. 2004; Parise et al. 2007; Kaplan et al. 2004), Psl and Pel of *P. aeruginosa* (Friedman et al. 2004; Jackson et al. 2004; Vasseur et al. 2005), cellulose, which is produced by many eubacteria (reviewed in Lasa 2006), and the extracellular D-glucans of *Streptococcus mutans* (Munro et al. 1995). Some examples that illustrate the complex regulation of poly-β-1,6-*N*-acetyl-D-glucosamine polymers in Gram-positive and -negative bacteria follow.

Poly-β-1,6-*N*-acetyl-D-glucosamine was discovered in *S. epidermidis* (Heilmann et al. 1996 a, 1996b; see the chapter by M. Otto, this volume) and later was found to serve as a biofilm adhesin in Gram-negative bacteria (Wang et al. 2004). This polymer is referred to as PIA (polysaccharide intercellular adhesin) or PNAG in *S. epidermidis* and *S. aureus*, respectively. Production of PIA/PNAG is dependent on the *ica* operon (*icaADBC*), which is regulated by a divergently transcribed gene (*icaR*) that encodes a transcriptional repressor, which responds to various environmental conditions (Conlon et al. 2002; Fig. 4).

Expression of the *icaADBC* operon is increased during growth in nutrient-rich or iron-limiting conditions and is induced by stressful stimuli such as heat, ethanol, and high concentrations of salt which increase *ica* expression and PIA production (Vuong et al. 2005 and references therein). The latter stressors are known to repress tricarboxylic acid (TCA) cycle activity, and the TCA cycle inhibitor fluorocitrate increases PIA production (Vuong et al. 2005). Furthermore, anaerobic conditions induce PIA production (Cramton et al. 2001). Subinhibitory concentrations of tetracycline and the semisynthetic streptogramin antibiotic quinupristin-dalfopristin enhance *icaADBC* expression nine- to 11-fold (Rachid et al. 2000). Ethanol induction of PIA synthesis is *icaR*-dependent (Conlon et al. 2002). Interestingly, glucose addition causes repression of *icaADBC*, but enhances PIA production, possibly via its precursor-product relationship with PIA (Dobinsky et al. 2003).

SarA is a global regulatory DNA-binding protein involved in expression of a variety of staphylococcal virulence genes. Transcription of *icaADBC*, which is essential for biofilm development in *S. aureus*, is activated by SarA binding (Tormo et al. 2005; Valle et al. 2003). In turn, *sarA* is activated by the stress response sigma factor, σ^B, which modulates responses to environmental stress and energy depletion. It is important to note that SarA, but not σ^B, is essential for biofilm development by *S. aureus* (Valle et al. 2003), suggesting that there are other means of activating *sarA* expression. σ^B also represses, possibly indirectly, *icaR* expression (Tormo et al. 2005), indicative of the complex interactions within this regulatory system.

The bacterial LuxS-dependent quorum-sensing systems are found in diverse species, and may permit bacteria to assess the overall microbial density of the environment (Schauder and Bassler 2001; Xavier and Bassler 2003). Biofilm formation in a *luxS* mutant strain of *S. epidermidis* was considerably enhanced, suggesting that the reaction product of the LuxS protein, autoinducer 2 (AI-2), represses *icaADBC* (Xu et al. 2006). Of note, quorum-sensing systems generally promote the

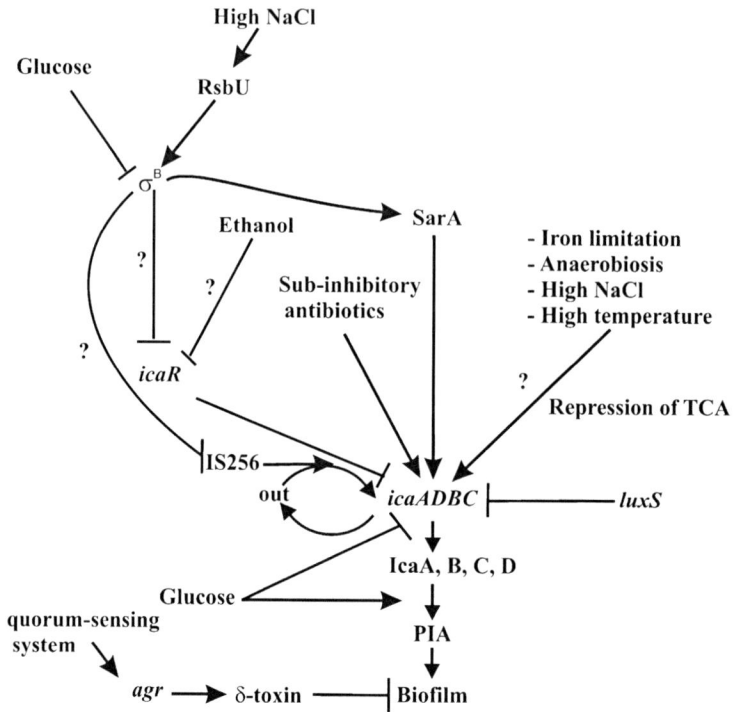

Fig. 4 Environmental influences on staphylococcal polysaccharide intercellular adhesin (PIA). The β-1,6-GlcNAc polymer PIA or PNAG, is required for cell-cell adhesion and biofilm formation in *S. epidermidis* and *S. aureus*. Production of PIA depends on *icaADBC* and is repressed by the divergently transcribed *icaR* gene. *icaABCD* expression is increased by growth in nutrient-replete, iron-limiting, anaerobic, and stress-inducing conditions. Several of these environmental conditions repress tricarboxylic acid (TCA) cycle activity. Subinhibitory concentrations of certain antibiotics also enhance *icaADBC* expression. IS256 causes phase variation by integrating into and excising from *icaADBC* or genes that affect its expression. The global regulator SarA activates transcription of *icaA* and is essential for biofilm development in *S. aureus*. In turn, *sarA* is activated by the general stress sigma factor σ^B, which also represses *icaR* and IS256 transposition. Glucose apparently represses *icaADBC* expression, but nevertheless enhances PIA production via a possible product-precursor relationship. The *agr* quorum sensing system negatively regulates biofilm development

expression of factors required for biofilm formation (Kirisits and Parsek 2006; Kong et al. 2006, Spoering and Gilmore 2006); although another example of quorum-sensing inhibition of biofilm formation is found in *Vibrio cholerae* (Hammer and Bassler 2003). A well-studied quorum-sensing system of *S. epidermidis* and *S. aureus, agr* (accessory gene regulator), also inhibits biofilm formation, but does not affect PIA levels (Vuong et al. 2003).

The IS256 insertion element is able to integrate into and inactivate or excise from *icaADBC* or genes that affect *ica* expression (e.g., *sarA*), thus constituting a phase-variable mode of regulation (Conlon et al. 2004; Ziebuhr et al. 1999). Transposition of IS256, but not transcription of the transposase, is repressed by σ^B (Valle et al. 2007). Valle and colleagues believe that environmental stress conditions activate σ^B and decrease the generation of biofilm-negative variants, in line with evidence indicating that NaCl and other stressors induce *ica*-dependent biofilm formation. The authors of this study also speculate that the IS256 element may modulate biofilm dispersal by affecting the proportion of biofilm-negative variants in a biofilm.

In *E. coli*, an understanding of biofilm regulation preceded the discovery of the *pgaABCD* structural genes, which in turn led to the identification of novel regulatory genes for biofilm formation (Fig. 5). An initial observation was that the global regulatory gene *csrA* of *E. coli* dramatically represses biofilm formation (Romeo et al. 1993), a phenotype that could not be explained by any previously known adhesin (Jackson et al. 2002a). A genetic screen for factors that cause hyper-biofilm formation in the *csrA* mutant led to discovery of the *pgaABCD* locus, which encodes gene products similar to the glycosyltransferase IcaA (PgaC) (Gerke et al. 1998)

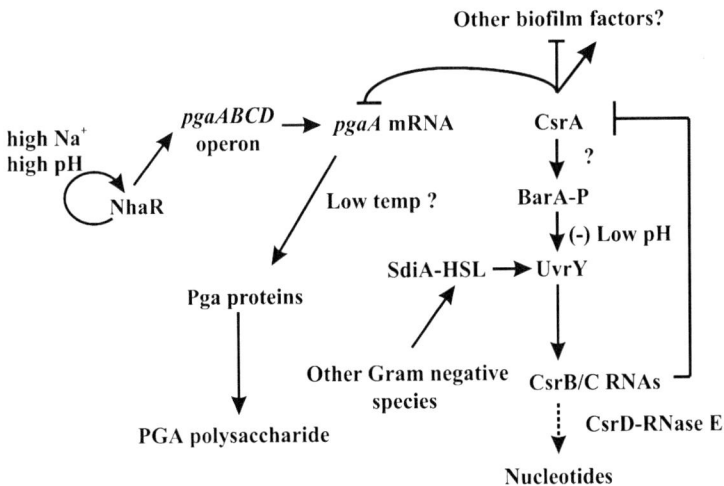

Fig. 5 Regulation of the biofilm adhesin PGA in *E. coli*. Poly-β-1,6-*N*-acetyl-glucosamine (*PGA*) synthesis is regulated at several levels. NhaR binds to the *pgaABCD* promoter and activates transcription in response to high pH or high sodium ion concentrations. CsrA protein binds to six sites in the leader of the *pgaABCD* transcript, which blocks ribosome binding and accelerates the turnover of this mRNA. Expression of *csrA* is activated as cultures approach the stationary phase, by unknown mechanism(s). In addition, CsrA is sequestered by the noncoding RNAs CsrB and CsrC. Transcription of these RNAs requires the BarA-UvrY two-component signal-transduction system, and CsrA itself. The signal for this system is not known, although BarA-UvrY signaling is blocked at low pH. SdiA activates *uvrY* transcription upon binding to *N*-acyl-homoserine lactones (*HSL*). *E. coli* does not produce HSL. Therefore, *csrB* and *csrC* transcription should be enhanced in the presence of Gram-negative species that produce such quorum-sensing compounds. CsrB and CsrC RNAs are degraded by a pathway involving a possible sensory protein, CsrD, and RNase E

and the *N*-deacetylase IcaB (PgaB) (Vuong et al. 2004). The mechanism of CsrA in this regulation is to bind to the *pgaABCD* mRNA leader at six sites, including sites that overlap the Shine-Dalgarno sequence and initiation codon, and thereby prevent ribosome binding (Wang et al. 2005). Translational repression likely results in the observed destabilization of this transcript by CsrA. Transcription of the *pgaABCD* operon is activated by the binding of the LysR family protein NhaR to the sole promoter of this operon in response to high pH or high Na^+ (Goller et al. 2006). The biosynthesis of PGA is also regulated by c-di-GMP (Suzuki et al. 2006; A. Pannuri, C. Goller, and T. Romeo; Y. Itoh and T. Romeo, unpublished studies) and is increased at low temperature (Wang et al. 2005; unpublished data). The latter findings are reminiscent of regulation in the homologous *hmsHFRS* system of *Y. pestis* (Bobrov et al. 2005; Perry et al. 2004; Simm et al. 2005).

The *hmsHFRS* operon and *hmsT* are required for *Y. pestis* biofilm formation in the gut of the flea vector and are important in the transmission of plague (see the chapter by Hinnebusch and Erickson). Hms-dependent biofilm formation is optimal at low temperature. The levels of HmsH, HmsR, and HmsT proteins are lower at 37°C than at 26°C, and temperature-dependent degradation of HmsH, HmsR, and HmsT proteins seems to be responsible for the Hms⁻ phenotype at 37°C (Perry et al. 2004). Additionally, biofilm formation is stimulated by HmsT, a protein that synthesizes c-di-GMP, and is inhibited by HmsP, which likely degrades c-di-GMP (Bobrov et al. 2005).

6   Conditions and Factors Mediating Biofilm Dispersal

No doubt, there are times when it is advantageous for cells to be able to escape from a biofilm. Entrapment within the biofilm environment limits bacterial growth (e.g., Rani et al. 2007). Furthermore, the transcriptome of mature biofilm, on average, has been suggested to be more similar to that of stationary phase cells than of exponentially growing cultures, although many changes in gene expression appear to be biofilm-specific (Beloin et al. 2004; Sauer et al. 2002; Waite et al. 2005). In addition, the biofilm matrix may prevent or at least deter cells from fleeing deleterious conditions. Dispersal processes are of interest because of their potential to promote spread of bacteria in the environment and because of the possibility to exploit these processes to combat detrimental biofilms. Release of cells or clumps of cells from biofilm can be accomplished by constitutive low level sloughing as well as active dispersion in which a substantial proportion of the population synchronously exits the biofilm. Several different cellular patterns of biofilm dispersal or escape have been documented under microscopic examination (reviewed by Hall-Stoodley and Stoodley 2005). In addition, the dissolution of cell attachments by surfactant production in *Bacillus subtilis*, *P. aeruginosa* and *S. epidermidis* may help to shape biofilm architecture (Boles et al. 2005; Branda et al. 2001; Davey et al. 2003; Vuong et al. 2003).

Environmental conditions that influence biofilm dispersal include nutrient availability, oxygen levels, pH, and specific compounds (Gjermansen et al. 2005;

Jackson et al. 2002a; Sauer et al. 2004; Thormann et al. 2005; Table 1). Changes in nutrient availability are a well-recognized determinant of dispersal. This is not surprising, given the importance of nutrient acquisition to bacterial survival. For example, early studies revealed that introduction of a rich medium to a tightly-aggregated *Acinetobacter* biofilm that had been grown under low nutrient conditions led to a more open, widely dispersed cell arrangement (James et al. 1995). Although the molecular genetics of biofilm dispersal has lagged behind that of formation, recent breakthroughs have paved the way for understanding the dispersal process, from the detection of environmental cues to signal transduction circuitry to the biochemical activities responsible for dispersal.

Biofilm dispersion in *P. aeruginosa* is perhaps the most studied and best understood process. Sauer and coworkers examined the proteome of this bacterium during active dispersion and found expression patterns that more closely resembled those of planktonic cells than biofilm cells (Sauer et al. 2002). Specific carbon nutrients, including succinate and glutamate, were found to trigger immediate large-scale release of cells (Sauer et al. 2004). Genes for motility, ribosomal proteins, and phage Pf1 were induced in the dispersed cells, while cells remaining attached contained elevated transcripts for pilus production and anaerobic nitrogen respiration. The latter activity indicates that insufficient oxygen was available for complete aerobic metabolism of the added carbon substrate in this biofilm.

Table 1 Molecular genetics of biofilm dispersal processes in Gram-negative bacteria

Organism	Environmental cue	Signal transduction	Output	Reference
P. aeruginosa	Carbon nutrients	BdlA, c–di–GMP	Adhesins?	Morgan et al. 2006
P. aeruginosa	Nitric oxide	?	Phage induction, other?	Barraud et al. 2006
P. aeruginosa	Quorum sensing (*las/rhl*)	?	Phage induction	Purevdorj– Gage et al. 2005
P. putida	Carbon starvation	c–di–GMP?	LapA protein? Polysaccharide?	Gjermansen et al. 2005, 2006
X. campestris	Quorum sensing	Rpf signal pathway	β–1,4– Mannanase	Dow et al. 2003
S. oneidensis	Anaerobic conditions	c–di–GMP?	Polysaccharide?	Thormann et al. 2005, 2006
E. coli	Quorum sensing, other?	Csr system	PGA, other?	Jackson Wang et al. 2002a; et al. 2004, 2005
A. actinomycete mcommitans	?	?	PGA hydrolase (dispersin B)	Itoh et al. 2005; Kaplan et al. 2003, 2004

The above observations support the recent discovery that trace amounts of nitric oxide (NO) or a metabolite thereof mediates dispersal (Barraud et al. 2006). This product of anaerobic respiration facilitated the seeding dispersal of cells from mature biofilm. In this phage Pf1-dependent process, mature biofilm structures appear to liquefy internally, involving both cell death and release of viable cells, and leaving behind hollow, shell-like structures. In addition, exposure to NO dispersed immature biofilms without causing cell death. The normal resistance of biofilm cells to certain antibacterial agents reverted back to the planktonic, sensitive phenotype during this dispersion process. Sauer and coworkers recently identified a gene encoding an apparent chemotaxis protein, BdlA, which is crucial for nutrient dispersal of *P. aeruginosa* (Morgan et al. 2006). A mutant lacking this protein also exhibited increased adherence and increased c-di-GMP levels. The latter observation is consistent with a rapidly expanding role of this nucleotide in stimulating bacterial exopolysaccharide synthesis and enhancing adherence properties of cells (reviewed in Jenal and Malone 2006; Romling and Amikam 2006), and the correlation of degradation of this nucleotide with biofilm dispersal (e.g., Gjermansen et al. 2006; Morgan et al. 2006 Thormann et al. 2006). It is tempting to suggest that BldA might regulate c-di-GMP levels in response to, perhaps even by binding to NO. This is consistent with the observations that (1) nutrient-induced dispersal leads to increased anaerobic respiration and (2) the BdlA protein structure includes a PAS domain, which is typically involved in signal detection. Another consideration is that lung infections of cystic fibrosis patients by *P. aeruginosa* become anaerobic. How these observations might apply to this host environment is still an open question (discussed in Romeo 2006).

There are parallels in other species that suggest dispersal processes that are related, though not identical, to those of *P. aeruginosa*. *Shewanella oneidensis* biofilm disperses rapidly under anoxic conditions and is likewise induced by an increase in c-di-GMP levels and possibly mediated via effects on exopolysaccharide production (Thormann et al. 2006). Is it possible that the cue for this process might not be the decrease of oxygen, but rather the production of NO or another product of anaerobic respiration? *Pseudomonas putida* responds to carbon starvation by inducing dispersal in a process that might involve c-di-GMP regulation, exopolysaccharide and a large proteinaceous adhesin that has also been studied in *P. fluorescens* (Gjermansen et al. 2005; 2006; Hinsa et al. 2003; 2006).

Studies in *E. coli* suggest that CsrA may facilitate dispersal (Jackson et al. 2002a). This RNA-binding protein, alternatively referred to as RsmA (repressor of stationary phase metabolites) in some species, posttranscriptionally represses production of the biofilm polysaccharide adhesin β-1,6-*N*-acetyl-D-glucosamine, or PGA, with dramatic effects on biofilm formation (Wang et al. 2005). The mechanism of CsrA in biofilm dispersal remains unknown, but could be based on inhibition of PGA synthesis if this polysaccharide is continuously removed by turnover or sloughing. The relatively slow rate of biofilm release (a few hours) that occurs in response to *csrA* induction suggests that the way in which CsrA affects dispersal may be different than in the preceding examples. CsrA activity is governed to a large extent by noncoding regulatory RNAs that sequester this protein, e.g., CsrB,

CsrC in *E. coli* (Gudapaty et al. 2001; Liu et al. 1997; Suzuki et al. 2002; Weilbacher et al. 2003). Thus, CsrA activity should increase as CsrB and CsrC synthesis decreases or their turnover increases. The environmental control of Csr RNAs is not well defined in any species. However, it typically involves transcriptional activation via BarA-UvrY or homologous two-component signal transduction systems, such as BarA-SirA, ExpS-ExpA, GacS-GacA, or VarS-VarA, and is connected to quorum-sensing pathways (Lenz et al. 2005; Suzuki et al. 2002; reviewed in Babitzke and Romeo 2007). Furthermore, while quorum-sensing systems often promote biofilm formation, they can activate biofilm dispersal in some species (e.g., Dow et al. 2003; Hammer and Bassler 2003; Yarwood et al. 2004).

7 Mixed Species Biofilms

The natural environments that most bacteria inhabit are typically complex and dynamic. Unfortunately, this complexity is not fully appreciated when growing organisms in monocultures under laboratory conditions. Biofilm communities associated with the plant rhizosphere (Ramey et al. 2004), intestinal mucosa (Eckburg et al. 2005), oral cavity and gingival crevices (Kroes et al. 1999; Kolenbrander 2000), and many other natural sites are inhabited by numerous different species in close proximity. Such environments are rich in biological stimuli to be processed by bacterial cells and used to direct biofilm development in response to changing conditions.

Studies using species-specific probes and microscopy have revealed complex spatial organization of species in natural biofilm communities (e.g., Bottari et al. 2006). Furthermore, co-culture experiments have demonstrated the importance of competition for nutrients and commensal metabolic networks in the dynamics of mixed-species biofilms (e.g., Christensen et al. 2002). Thus, spatial and metabolic interactions between species contribute to the organization of multispecies biofilms, and the production of a dynamic local environment (Battin et al. 2007; Tolker-Nielsen and Molin 2000). The distribution of cells and biomass in complex biofilms is influenced by the physiology of the organisms present, which in turn leads to the development of local nutrient gradients. Furthermore, mixed species biofilms can evolve rapidly and lead to stable interactions between species when driven by selective pressure for co-metabolism (Hansen et al. 2007). In the latter model system, a commensalistic relationship was established between *Acinetobacter* sp. strain C6 and *P. putida* KT2440 when the latter species evolved the ability to adhere and form biofilm close to *Acinetobacter* microcolonies, and thereby capture the metabolite benzoate.

Mutualistic relationships can also occur in mixed-species biofilms. For example, biofilm formation by *E. coli* PHL565 was synergistically enhanced by growth in mixed culture with *P. putida* MT2 (Castonguay et al. 2006). Particularly striking mutualistic effects on biofilm formation have been shown for species that inhabit dental plaque (e.g., Palmer et al. 2001). Conjugative plasmids have been demonstrated to induce bacterial biofilm development in co-culture experiments (e.g., Ghigo 2001), and biofilm formation increases the chance for lateral gene transfer and thus the risk for interspecies gene transfer and the consequent spread of virulence factors and antibiotic resistance (e.g., Weigel et al., 2007). In fact, many examples of synergistic induction of biofilm formation were observed when a large collection of nondomesticated *E. coli* strains were individually cocultivated with a laboratory strain or with each other. This was most often precipitated by conjugal transfer of natural plasmids carried by the isolates.

Quorum sensing can have somewhat unpredictable effects on biofilm formation (Merritt et al. 2003; Schauder and Bassler 2001; Waters and Bassler, 2005; Bassler and Losick 2006). While many quorum-sensing systems are relatively species-specific, the autoinducer-2 based (or LuxS-dependent) quorum-sensing system is widespread among eubacteria and may serve as a universal language for these organisms. Despite its profound implications, the impact of interspecies communication on biofilm development is presently not well understood.

8   Conclusions and Outlook

Recent advances in our understanding of the biofilm development cycle have indicated that in most cases, it is a dynamic process in which common environmental factors such as nutritional conditions, temperature, oxygen tension and osmolarity have strong influences. We have begun to understand the factors and pathways that respond to environmental cues and regulate the surface transformations that drive the biofilm development cycle. The distinctive conditions that govern biofilm development for a given species can provide important clues to its natural ecology and life cycle and vice-versa. Many surprises lay in store, and the rules for biofilm development seem to be made to be broken. For example, the minimalist bacterium *Mycoplasma pulmonis* lacks any two component signal transduction system or recognizable global regulator. Nevertheless, it is able to modulate biofilm formation through slipped-strand mispairing of the gene for an adhesive surface protein (Simmons et al. 2007).

At the present time, there are many unanswered or partially answered questions concerning the influence of the bacterial environment on biofilm development:

1. Which steps in development are most important for regulation and how are these steps regulated? The transition from reversible to irreversible attachment would seem to be an important site for regulation to occur, but this remains to be shown.
2. While a variety of environmental influences on biofilm are now known in a few model organisms, information on their relative importance and integration is lacking.
3. Systematic analysis of gene expression by array studies has provided much information concerning gene expression patterns during biofilm development.

How are these genes regulated and which of these genes are critical for the development process?

4. How does the presence of other microorganisms and growth in association with eukaryotic hosts influence biofilm formation by a given species? This is a complex biological question that likely differs for each species of interest. Nevertheless, it is critical for the development of new therapeutic strategies and other applications.

References

Aberg A, Shingler V, Balsalobre C (2006) (p)ppGpp regulates type 1 fimbriation of *Escherichia coli* by modulating the expression of the site-specific recombinase FimB. Mol Microbiol 60:1520-1533

Agladze K, Jackson D, Romeo T (2003) Periodicity of cell attachment patterns during *Escherichia coli* biofilm development. J Bacteriol 185:5632-5638

Agladze K, Wang X, Romeo T (2005) Spatial periodicity of *Escherichia coli* K-12 biofilm microstructure initiates during a reversible, polar attachment phase of development and requires the polysaccharide adhesin PGA. J Bacteriol 187:8237-8246

Anderson BN, Ding AM, Nilsson LM, Kusuma K, Tchesnokova V, Vogel V, Sokurenko EV, Thomas WE (2007) Weak rolling adhesion enhances bacterial surface colonization. J Bacteriol 189:1794-1802

Anderson GG, Palermo JJ, Schilling JD, Roth R, Heuser J, Hultgren SJ (2003) Intracellular bacterial biofilm-like pods in urinary tract infections. Science 301:105-107

Babitzke P, Romeo T (2007) CsrB ncRNA family: sequestration of RNA-binding regulatory proteins. Curr Opin Microbiol 10:156-163

Barnhart MM, Chapman MR (2006) Curli biogenesis and function. Annu Rev Microbiol 60:131-147

Barraud N, Hassett DJ, Hwang SH, Rice SA, Kjelleberg S, Webb JS (2006) Involvement of nitric oxide in biofilm dispersal of *Pseudomonas aeruginosa*. J Bacteriol 188:7344-7353

Bassler BL, Losick R (2006) Bacterially speaking. Cell 125: 237-246

Battin TJ, Sloan WT, Kjelleberg S, Daims H, Head IM, Curtis TP, Eberl L (2007) Microbial landscapes: new paths to biofilm research. Nat Rev Microbiol 5:76-81

Beloin C, Valle J, Latour-Lambert P, Faure P, Kzreminski M, Balestrino D, Haagensen JA, Molin S, Prensier G, Arbeille B, Ghigo JM (2004) Global impact of mature biofilm lifestyle on *Escherichia coli* K-12 gene expression. Mol Microbiol 51(3):659-674

Blumer C, Kleefeld A, Lehnen D, Heintz M, Dobrindt U, Nagy G, Michaelis K, Emody L, Polen T, Rachel R, Wendisch VF, Unden G (2005) Regulation of type 1 fimbriae synthesis and biofilm formation by the transcriptional regulator LrhA of *Escherichia coli*. Microbiology 151: 3287-3298

Bobrov AG, Kirillina O, Perry RD (2005) The phosphodiesterase activity of the HmsP EAL domain is required for negative regulation of biofilm formation in *Yersinia pestis*. FEMS Microbiol Lett 247:123-130

Boles BR, Thoendel M, Singh PK (2005) Rhamnolipids mediate detachment of *Pseudomonas aeruginosa* from biofilms. Mol Microbiol 57:1210-1223

Bottari B, Ercolini D, Gatti M, Neviani E (2006) Application of FISH technology for microbiological analysis: current state and prospects. Appl Microbiol Biotechnol 73(3):485-494

Bougdour A, Lelong C, Geiselmann J (2004) Crl, a low temperature-induced protein in *Escherichia coli* that binds directly to the stationary phase σ subunit of RNA polymerase. J Biol Chem 279: 19540-19550

Boulanger A, Francez-Charlot A, Conter A, Castanie-Cornet MP, Cam K, Gutierrez C (2005) Multistress regulation in *Escherichia coli*: expression of *osmB* involves two independent promoters responding either to σ^s or to the RcsCDB His-Asp phosphorelay. J Bacteriol 187:3282-3286

Branda SS, Gonzalez-Pastor JE, Ben-Yehuda S, Losick R, Kolter R (2001) Fruiting body formation by *Bacillus subtilis*. Proc Natl Acad Sci U S A 98:11621-11626

Branda SS, Vik A, Friedman L, Kolter R (2005) Biofilms: the matrix revisited. Trends Microbiol 13:20-26

Cai SJ, Inouye M (2002) EnvZ-OmpR interaction and osmoregulation in *Escherichia coli*. J Biol Chem 277: 24155-24161

Caiazza NC, O'Toole GA (2004) SadB is required for the transition from reversible to irreversible attachment during biofilm formation by *Pseudomonas aeruginosa* PA14. J Bacteriol 186:4476-4485

Castonguay MH, van der Schaaf S, Koester W, Krooneman J, van der Meer W, Harmsen H, Landini P (2006) Biofilm formation by *Escherichia coli* is stimulated by synergistic interactions and co-adhesion mechanisms with adherence-proficient bacteria. Res Microbiol 157:471-478

Cerca N, Pier GB, Vilanova M, Oliveira R, Azeredo J (2005) Quantitative analysis of adhesion and biofilm formation on hydrophilic and hydrophobic surfaces of clinical isolates of *Staphylococcus epidermidis*. Res Microbiol 156:506-514

Chavant P, Martinie B, Meylheuc T, Bellon-Fontaine MN, Hebraud M (2002) *Listeria monocytogenes* LO28: surface physicochemical properties and ability to form biofilms at different temperatures and growth phases. Appl Environ Microbiol 68:728-737

Christensen BB, Haagensen JAJ, Heydorn A, Molin S (2002) Metabolic commensalism and competition in a two-species microbial consortium. Appl Environ Microbiol 68:2495-2502

Conlon KM, Humphreys H, O'Gara JP (2002) *icaR* encodes a transcriptional repressor involved in environmental regulation of *ica* operon expression and biofilm formation in *Staphylococcus epidermidis*. J Bacteriol 184:4400-4408

Conlon KM, Humphreys H, O'Gara JP (2004) Inactivations of *rsbU* and *sarA* by IS256 represent novel mechanisms of biofilm phenotypic variation in *Staphylococcus epidermidis*. J Bacteriol 186:6208-6219

Cramton SE, Ulrich M, Gotz F, Doring G (2001) Anaerobic conditions induce expression of polysaccharide intercellular adhesin in *Staphylococcus aureus* and *Staphylococcus epidermidis*. Infect Immun 69:4079-4085

Crossman L, Dow JM (2004) Biofilm formation and dispersal in *Xanthomonas campestris*. Microbes Infect 6:623-629

Danese PN, Silhavy TJ (1998) CpxP, a stress-combative member of the Cpx regulon. J Bacteriol 180(4):831-839

Davey ME, Caiazza NC, O'Toole GA (2003) Rhamnolipid surfactant production affects biofilm architecture in *Pseudomonas aeruginosa* PAO1. J Bacteriol 185:1027-1036

Dobinsky S, Kiel K, Rohde H, Bartscht K, Knobloch JKM, Horstkotte MA, Mack D (2003) Glucose-related dissociation between *icaADBC* transcription and biofilm expression by *Staphylococcus epidermidis*: evidence for an additional factor required for polysaccharide intercellular adhesin synthesis. J Bacteriol 185: 2879-2886

Donlan RM (2002) Biofilms: microbial life on surfaces. Emerg Infect Dis 8:881-890

Donlan RM, Costerton JW (2002) Biofilms: survival mechanisms of clinically relevant microorganisms. Clin Microbiol Rev 15:167-193

Dorel C, Vidal O, Prigent-Combaret C, Vallet I, Lejeune P (1999) Involvement of the Cpx signal transduction pathway of *E. coli* in biofilm formation. FEMS Microbiol Lett 178:169-175

Dow JM, Crossman L, Findlay K, He YQ, Feng JX, Tang JL (2003) Biofilm dispersal in *Xanthomonas campestris* is controlled by cell-cell signaling and is required for full virulence to plants. Proc Natl Acad Sci U S A 100:10995-11000

Dunne WM Jr (2002) Bacterial adhesion: seen any good biofilms lately? Clin Microbiol Rev 15:155-166

Eckburg PB, Bik EM, Bernstein CN, Purdom E, Dethlefsen L, Sargent M, Gill SR, Nelson KE, Relman DA (2005) Diversity of the human intestinal microbial flora. Science 308:1635-1638

Francez-Charlot A, Laugel B, Van Gemert A, Dubarry N, Wiorowski F, Castanie-Cornet MP, Gutierrez C, Cam K (2003) RcsCDB His-Asp phosphorelay system negatively regulates the *flhDC* operon in *Escherichia coli*. Mol Microbiol 49:823-832

Fredericks CE, Shibata S, Aizawa SI, Reimann SA, Wolfe AJ (2006) Acetyl phosphate-sensitive regulation of flagellar biogenesis and capsular biosynthesis depends on the Rcs phosphorelay. Mol Microbiol 61:734-747

Friedman L, Kolter R (2004) Two genetic loci produce distinct carbohydrate-rich structural components of the *Pseudomonas aeruginosa* biofilm matrix. J Bacteriol 186:4457-4465

Gally DL, Bogan JA, Eisenstein BI, Blomfield IC (1993) Environmental regulation of the *fim* switch controlling type 1 fimbrial phase variation in *Escherichia coli* K-12: effects of temperature and media. J Bacteriol 175:6186-6193

Geesey GG (2001) Bacterial behavior at surfaces. Curr Opin Microbiol 4:296-300

Gerke C, Kraft A, Sussmuth R, Schweitzer O, Gotz F (1998) Characterization of the *N*-acetylglucosaminyltransferase activity involved in the biosynthesis of the *Staphylococcus epidermidis* polysaccharide intercellular adhesin. J Biol Chem 273(29):18586-18593

Gerstel U, Park C, Romling U (2003) Complex regulation of *csgD* promoter activity by global regulatory proteins. Mol Microbiol 49:639-654

Ghigo JM (2001) Natural conjugative plasmids induce bacterial biofilm development. Nature 412:442-445

Gjermansen M, Ragas P, Sternberg C, Molin S, Tolker-Nielsen T (2005) Characterization of starvation-induced dispersion in *Pseudomonas putida* biofilms. Environ Microbiol 7:894-906

Gjermansen M, Ragas P, Tolker-Nielsen T (2006) Proteins with GGDEF and EAL domains regulate *Pseudomonas putida* biofilm formation and dispersal. FEMS Microbiol Lett 265(2):215-224

Goller C, Wang X, Itoh Y, Romeo T (2006) The cation-responsive protein NhaR of *Escherichia coli* activates *pgaABCD* transcription, required for production of the biofilm adhesin poly-β-1,6-*N*-acetyl-D-glucosamine. J Bacteriol 188:8022-8032

Gudapaty S, Suzuki K, Wang X, Babitzke P, Romeo T (2001) Regulatory interactions of Csr components: the RNA binding protein CsrA activates *csrB* transcription in *Escherichia coli*. J Bacteriol 183:6017-6027

Hall-Stoodley L, Stoodley P (2005) Biofilm formation and dispersal and the transmission of human pathogens. Trends Microbiol 13:7-10

Hammar M, Arnqvist A, Bian, Z, Olsen A, Normark S (1995) Expression of two *csg* operons is required for production of fibronectin- and congo red-binding curli polymers in *Escherichia coli* K-12. Mol Microbiol 18:661-670

Hammer BK, Bassler BL (2003) Quorum sensing controls biofilm formation in *Vibrio cholerae*. Mol Microbiol 50:101-104

Hanna A, Berg M, Stout V, Razatos A (2003) Role of capsular colanic acid in adhesion of uropathogenic *Escherichia coli*. Appl Environ Microbiol 69(8):4474-4481

Hansen SK, Rainey PB, Haagensen JA, Molin S (2007) Evolution of species interactions in a biofilm community. Nature 445:533-536

Heilmann C, Gerke C, Perdreau-Remington F, Götz F (1996a) Characterization of Tn*917* insertion mutants of *Staphylococcus epidermidis* affected in biofilm formation. Infect Immun 64:277-282

Heilmann C, Schweitzer O, Gerke C, Vanittanakom N, Mack D, Gotz F (1996b) Molecular basis of intercellular adhesion in the biofilm-forming *Staphylococcus epidermidis*. Mol Microbiol 20(5):1083-1091

Hengge-Aronis R (2002) Signal transduction and regulatory mechanisms involved in control of the σˢ (RpoS) subunit of RNA polymerase. Microbiol Mol Biol Rev 66:373-395

Hinsa SM, O'Toole GA (2006) Biofilm formation by *Pseudomonas fluorescens* WCS365: a role for LapD. Microbiology 152:1375-1383

Hinsa SM, Espinosa-Urgel M, Ramos JL, O'Toole GA (2003) Transition from reversible to irreversible attachment during biofilm formation by *Pseudomonas fluorescens* WCS365 requires an ABC transporter and a large secreted protein. Mol Microbiol 49:905-918

Hung DL, Raivio TL, Jones CH, Silhavy TJ, Hultgren SJ (2001) Cpx signaling pathway monitors biogenesis and affects assembly and expression of *P. pili*. EMBO J 20:1508-1518

Isberg RR, Barnes P (2002) Dancing with the host: flow-dependent bacterial adhesion. Cell 110:1-4

Itoh Y, Wang X, Hinnebusch BJ, Preston JF 3rd, Romeo T (2005) Depolymerization of beta-1,6-*N*-acetyl-D-glucosamine disrupts the integrity of diverse bacterial biofilms. J Bacteriol 187:382-387

Jackson DW, Suzuki K, Oakford L, Simecka JW, Hart ME, Romeo T (2002a) Biofilm formation and dispersal under the influence of the global regulator CsrA of *Escherichia coli*. J Bacteriol 184:290-301

Jackson DW, Simecka JW, Romeo T (2002b) Catabolite repression of *Escherichia coli* biofilm formation. J Bacteriol 184:3406-3410

Jackson KD, Starkey M, Kremer S, Parsek MR, Wozniak DJ (2004) Identification of *psl*, a locus encoding a potential exopolysaccharide that is essential for *Pseudomonas aeruginosa* PAO1 biofilm formation. J Bacteriol 186:4466-4475

James GA, Korber DR, Caldwell DE, Costerton JW (1995) Digital image analysis of growth and starvation responses of a surface-colonizing *Acinetobacter* sp. J Bacteriol 177:907-915

Jenal U, Malone J (2006) Mechanisms of cyclic-di-GMP signaling in bacteria. Annu Rev Genet 40:385-407

Jubelin G, Vianney A, Beloin C, Ghigo JM, Lazzaroni JC, Lejeune P, Dorel C (2005) CpxR/OmpR interplay regulates curli gene expression in response to osmolarity in *Escherichia coli*. J Bacteriol 187:2038-2049

Kader A, Simm R, Gerstel U, Morr M, Romling U (2006) Hierarchical involvement of various GGDEF domain proteins in rdar morphotype development of *Salmonella enterica* serovar Typhimurium. Mol Microbiol 60:602-616

Kaplan JB, Ragunath C, Ramasubbu N, Fine DH (2003) Detachment of *Actinobacillus actinomycetemcomitans* biofilm cells by an endogenous beta-hexosaminidase activity. J Bacteriol 185:4693-4698

Kaplan JB, Velliyagounder K, Ragunath C, Rohde H, Mack D, Knobloch JK, Ramasubbu N (2004) Genes involved in the synthesis and degradation of matrix polysaccharide in *Actinobacillus actinomycetemcomitans* and *Actinobacillus pleuropneumoniae* biofilms. J Bacteriol 186:8213-8220

Kikuchi T, Mizunoe Y, Takade A, Naito S, Yoshida S (2005) Curli fibers are required for development of biofilm architecture in *Escherichia coli* K-12 and enhance bacterial adherence to human uroepithelial cells. Microbiol Immunol 49:875-884

Kirisits MJ, Parsek MR (2006) Does *Pseudomonas aeruginosa* use intercellular signalling to build biofilm communities? Cell Microbiol 8:1841-1849

Klemm P (1986) Two regulatory *fim* genes, *fimB* and *fimE*, control the phase variation of type 1 fimbriae in *Escherichia coli*. EMBO J 5:1389-1393

Klemm P, Hjerrild L, Gjermansen M, Schembri MA (2004) Structure-function analysis of the self-recognizing Antigen 43 autotransporter protein from *Escherichia coli*. Mol Microbiol 51:283-296

Klemm P, Vejborg RM, Sherlock O (2006) Self-associating autotransporters, SAATs: functional and structural similarities. Int J Med Microbiol 296:187-195

Kolenbrander PE (2000) Oral microbial communities: biofilms, interactions, and genetic systems. Annu Rev Microbiol 54:413-437

Kong KF, Vuong C, Otto M (2006) Staphylococcus quorum sensing in biofilm formation and infection. Int J Med Microbiol 296:133-139

Kroes I, Lepp PW, Relman DA (1999) Bacterial diversity within the human subgingival crevice. Proc Natl Acad Sci U S A 96:14547-14552

Lasa I (2006) Towards the identification of the common features of bacterial biofilm development. Int Microbiol 9:21-28

Lenz DH, Miller MB, Zhu J, Kulkarni RV, Bassler BL (2005) CsrA and three redundant small RNAs regulate quorum sensing in *Vibrio cholerae*. Mol Microbiol 58:1186-1202

Liu MY, Gui G, Wei B, Preston JF 3rd, Oakford L, Yuksel U, Giedroc DP, Romeo T (1997) The RNA molecule CsrB binds to the global regulatory protein CsrA and antagonizes its activity in *Escherichia coli*. J Biol Chem 272:17502-17510

Lowe MA, Holt SC, Eisenstein BI (1987) Immunoelectron microscopic analysis of elongation of type 1 fimbriae in *Escherichia coli*. J Bacteriol 169:157-163

Macnab RM (2003) How bacteria assemble flagella. Annu Rev Microbiol 57:77-100

Mack D, Davies AP, Harris LG, Rohde H, Horstkotte MA, Knobloch JK (2007) Microbial interactions in *Staphylococcus epidermidis* biofilms. Anal Bioanal Chem 387:399-408

Maini PK, Baker RE, Chuong CM (2006) Developmental biology. The Turing model comes of molecular age. Science 314:1397-1398

Maira-Litran T, Kropec A, Abeygunawardana C, Joyce J, Mark Iii G, Goldmann DA, Pier GB (2002) Immunochemical properties of the Staphylococcal poly-N-acetylglucosamine surface polysaccharide. Infect Immun 70:4433-4440

Majdalani N, Gottesman S (2005) The Rcs phosphorelay: a complex signal transduction system. Annu Rev Microbiol 59:379-405

Maurer JJ, Brown TP, Steffens WL, Thayer SG (1998) The occurrence of ambient temperature-regulated adhesins, curli, and the temperature-sensitive hemagglutinin *tsh* among avian *Escherichia coli*. Avian Dis 42:106-118

Maurer LM, Yohannes E, Bondurant SS, Radmacher M, Slonczewski JL (2005) pH regulates genes for flagellar motility, catabolism, and oxidative stress in *Escherichia coli* K-12. J Bacteriol 187:304-319

Merritt J, Qi F, Goodman SD, Anderson MH, Shi W (2003) Mutation of *luxS* affects biofilm formation in *Streptococcus mutans*. Infect Immun 71:1972-1979

McClaine JW, Ford RM (2002) Reversal of flagellar rotation is important in initial attachment of *Escherichia coli* to glass in a dynamic system with high- and low-ionic-strength buffers. Appl Environ Microbiol 68:1280-1289

Mondragon V, Franco B, Jonas K, Suzuki K, Romeo T, Melefors O, Georgellis D (2006) pH-dependent activation of the BarA-UvrY two-component system in *Escherichia coli*. J Bacteriol 188: 8303-8306

Morgan R, Kohn S, Hwang SH, Hassett DJ, Sauer K (2006) BdlA, a chemotaxis regulator essential for biofilm dispersion in *Pseudomonas aeruginosa*. J Bacteriol 188:7335-7343

Morikawa M, Kagihiro S, Haruki M, Takano K, Branda S, Kolter R, Kanaya S (2006) Biofilm formation by a *Bacillus subtilis* strain that produces {gamma}-polyglutamate. Microbiology 152:2801-2807

Mulvey MA, Schilling JD, Martinez JJ, Hultgren SJ (2000) From the cover: bad bugs and beleaguered bladders: interplay between uropathogenic *Escherichia coli* and innate host defenses. Proc Natl Acad Sci U S A 97:8829-8835

Munro CL, Michalek SM, Macrina FL (1995) Sucrose-derived exopolymers have site-dependent roles in *Streptococcus mutans*-promoted dental decay. FEMS Microbiol Lett 128:327-332

Murga R, Miller JM, Donlan RM (2001) Biofilm formation by Gram-negative bacteria on central venous catheter connectors: effect of conditioning films in a laboratory model. J Clin Microbiol 39:2294-2297

Olsen A, Jonsson A, Normark S (1989) Fibronectin binding mediated by a novel class of surface organelles on *Escherichia coli*. Nature 338(6217):652-655

Olsen A, Arnqvist A, Hammar M, Normark S (1993a) Environmental regulation of curli production in *Escherichia coli*. Infect Agents Dis 2:272-274

Olsen A, Arnqvist A, Hammar M, Sukupolvi S, Normark S (1993b) The RpoS sigma factor relieves H-NS-mediated transcriptional repression of *csgA*, the subunit gene of fibronectin-binding curli in *Escherichia coli*. Mol Microbiol 7:523-536

O'Toole GA, Kolter R (1998) Flagellar and twitching motility are necessary for *Pseudomonas aeruginosa* biofilm development. Mol Microbiol 30: 295-304

Otto K, Silhavy TJ (2002) Surface sensing and adhesion of *Escherichia coli* controlled by the Cpx-signaling pathway. Proc Natl Acad Sci U S A 99:2287-2292

Palmer RJ, Kazmerzak K Jr, Hansen MC, Kolenbrander PE (2001) Mutualism versus independence: strategies of mixed-species oral biofilms in vitro using saliva as the sole nutrient source. Infect Immun 69:5794-5804

Parise G, Mishra M, Itoh Y, Romeo T, Deora R (2007) Role of a putative polysaccharide locus in Bordetella biofilm development. J Bacteriol 189:750-760

Park S, Wolanin PM, Yuzbashyan EA, Silberzan P, Stock JB, Austin RH (2003) Motion to form a quorum. Science 301:188

Patti JM, Allen BL, McGavin MJ, Hook M (1994) MSCRAMM-mediated adherence of microorganisms to host tissues. Annu Rev Microbiol 48:585-617

Perry RD, Bobrov AG, Kirillina O, Jones HA, Pedersen LL, Abney J, Fetherston JD (2004) Temperature regulation of the hemin storage (Hms+) phenotype of *Yersinia pestis* is posttranscriptional. J Bacteriol 186:1638-1647

Pratt LA, Kolter R (1998) Genetic analysis of *Escherichia coli* biofilm formation: roles of flagella, motility, chemotaxis and type I pili. Mol Microbiol 30:285-293

Prigent-Combaret C, Vidal O, Dorel C, Lejeune P (1999) Abiotic surface sensing and biofilm-dependent regulation of gene expression in *Escherichia coli*. J Bacteriol 181:5993-6002

Prigent-Combaret C, Prensier G, Le Thi TT, Vidal O, Lejeune P, Dorel C (2000) Developmental pathway for biofilm formation in curli-producing *Escherichia coli* strains: role of flagella, curli and colanic acid. Environ Microbiol 2:450-464

Prigent-Combaret C, Brombacher E, Vidal O, Ambert A, Lejeune P, Landini P, Dorel C (2001) Complex regulatory network controls initial adhesion and biofilm formation in *Escherichia coli* via regulation of the *csgD* gene. J Bacteriol 183:7213-7223

Purevdorj-Gage B, Costerton WJ, Stoodley P (2005) Phenotypic differentiation and seeding dispersal in non-mucoid and mucoid *Pseudomonas aeruginosa* biofilms. Microbiology 151:1569-1576

Rachid S, Ohlsen K, Witte W, Hacker J, Ziebuhr W (2000) Effect of subinhibitory antibiotic concentrations on polysaccharide intercellular adhesin expression in biofilm-forming *Staphylococcus epidermidis*. Antimicrob Agents Chemother 44:3357-3363

Raivio TL, Silhavy TJ (1997) Transduction of envelope stress in *Escherichia coli* by the Cpx two-component system. J Bacteriol 179:7724-7733

Ramey BE, Koutsoudis M, von Bodman SB, Fuqua C (2004) Biofilm formation in plant-microbe associations. Curr Opin Microbiol 7:602-609

Rani SA, Pitts B, Beyenal H, Veluchamy RA, Lewandowski Z, Davison WM, Buckingham-Meyer K, Stewart PS (2007) Spatial patterns of DNA replication, protein synthesis and oxygen concentration within bacterial biofilms reveal diverse physiological states. J Bacteriol 189:4223-4233

Reisner A, Holler BM, Molin S, Zechner EL (2006a) Synergistic effects in mixed *Escherichia coli* biofilms: conjugative plasmid transfer drives biofilm expansion. J Bacteriol 188: 3582-3588

Reisner A, Krogfelt KA, Klein BM, Zechner EL, Molin S (2006b) In vitro biofilm formation of commensal and pathogenic *Escherichia coli* strains: impact of environmental and genetic factors. J Bacteriol 188:3572-3581

Rickard AH, Gilbert P, High NJ, Kolenbrander PE, Handley PS (2003) Bacterial coaggregation: an integral process in the development of multi-species biofilms. Trends Microbiol 11:94-100

Romeo T (1998) Global regulation by the small RNA-binding protein CsrA and the noncoding-RNA CsrB. Mol Microbiol 29:1321-1330

Romeo T (2006) When the party is over: a signal for dispersal of *Pseudomonas aeruginosa* biofilms. J Bacteriol 188:7325-7327

Romeo T, Gong M, Liu MY, Brun-Zinkernagel AM (1993) Identification and molecular characterization of *csrA*, a pleiotropic gene from *Escherichia coli* that affects glycogen biosynthesis, gluconeogenesis, cell size, and surface properties. J Bacteriol 175:4744-4755

Romling U, Amikam D (2006) Cyclic di-GMP as a second messenger. Curr Opin Microbiol 9:218-228

Romling U, Bian Z, Hammar M, Sierralta WD, Normark S (1998) Curli fibers are highly conserved between *Salmonella typhimurium* and *Escherichia coli* with respect to operon structure and regulation. J Bacteriol 180:722-731

Rupp CJ, Fux CA, Stoodley P (2005) Viscoelasticity of *Staphylococcus aureus* biofilms in response to fluid shear allows resistance to detachment and facilitates rolling migration. Appl Environ Microbiol 71:2175-2178

Ryan RP, Fouhy Y, Lucey JF, Dow JM (2006) Cyclic di-GMP signaling in bacteria: recent advances and new puzzles. J Bacteriol 188: 8327-8334

Sauer K (2003) The genomics and proteomics of biofilm formation. Genome Biol 4:219

Sauer K, Camper AK, Ehrlich GD, Costerton JW, Davies DG (2002) *Pseudomonas aeruginosa* displays multiple phenotypes during development as a biofilm. J Bacteriol 184:1140-1154

Sauer K, Cullen MC, Rickard AH, Zeef LA, Davies DG, Gilbert P (2004) Characterization of nutrient-induced dispersion in *Pseudomonas aeruginosa* PA01 bioflm. J Bacteriol 186:7312-7326

Schauder S, Bassler BL (2001) The languages of bacteria. Genes Dev 15:1468-1480

Schembri MA, Christiansen G, Klemm P (2001) FimH-mediated autoaggregation of *Escherichia coli*. Mol Microbiol 41:1419-1430

Schembri MA, Hjerrild L, Gjermansen M, Klemm P (2003a) Differential expression of the *Escherichia coli* autoaggregation factor antigen 43. J Bacteriol 185:2236-2242

Schembri MA, Kjaergaard K, Klemm P (2003b) Global gene expression in *Escherichia coli* biofilms. Mol Microbiol 48:253-267

Schembri MA, Dalsgaard D, Klemm P (2004) Capsule shields the function of short bacterial adhesins. J Bacteriol 186:1249-1257

Schooling SR, Beveridge TJ (2006) Membrane vesicles: an overlooked component of the matrices of biofilms. J Bacteriol 188:5945-5957

Schwan WR, Lee JL, Lenard FA, Matthews BT, Beck MT (2002) Osmolarity and pH growth conditions regulate *fim* gene transcription and type 1 pilus expression in uropathogenic *Escherichia coli*. Infect Immun 70:1391-1402

Shi W, Zhou Y, Wild J, Adler J, Gross CA (1992) DnaK, DnaJ, GrpE are required for flagellum synthesis in *Escherichia coli*. J Bacteriol 174:6256-6263

Shin S, Park C (1995) Modulation of flagellar expression in *Escherichia coli* by acetyl phosphate and the osmoregulator OmpR. J Bacteriol 177:4696-4702

Shrout JD, Chopp DL, Just CL, Hentzer M, Givskov M, Parsek MR (2006) The impact of quorum sensing and swarming motility on *Pseudomonas aeruginosa* biofilm formation is nutritionally conditional. Mol Microbiol 62:1264-1277

Silverman M, Simon M (1974) Characterization of *Escherichia coli* flagellar mutants that are insensitive to catabolite repression. J Bacteriol 120:1196-1203

Simm R, Fetherston JD, Kader A, Romling U, Perry RD (2005) Phenotypic convergence mediated by GGDEF-domain-containing proteins. J Bacteriol 187:6816-6823

Singh PK, Parsek MR, Greenberg EP, Welsh MJ (2002) A component of innate immunity prevents bacterial biofilm development. Nature 417: 552-555

Soutourina OA, Bertin PN (2003) Regulation cascade of flagellar expression in Gram-negative bacteria. FEMS Microbiol Rev 27:505-523

Soutourina O, Kolb A, Krin E, Laurent-Winter C, Rimsky S, Danchin A, Bertin P (1999) Multiple control of flagellum biosynthesis in *Escherichia coli*: role of H-NS protein and the cyclic AMP-catabolite activator protein complex in transcription of the *flhDC* master operon. J Bacteriol 181:7500-7508

Soutourina OA, Krin E, Laurent-Winter C, Hommais F, Danchin A, Bertin PN (2002) Regulation of bacterial motility in response to low pH in *Escherichia coli*: the role of H-NS protein. Microbiology 148:1543-1551

Spoering AL, Gilmore MS (2006) Quorum sensing and DNA release in bacterial biofilms. Curr Opin Microbiol 9:133-137

Stapper AP, Narasimhan G, Ohman DE, Barakat J, Hentzer M, Molin S, Kharazmi A, Hoiby N, Mathee K (2004) Alginate production affects *Pseudomonas aeruginosa* biofilm development and architecture, but is not essential for biofilm formation. J Med Microbiol 53:679-690

Stanley NR, Lazazzera BA (2004) Environmental signals and regulatory pathways that influence biofilm formation. Mol Microbiol 52(4):917-924

Steinberger RE, Holden PA (2005) Extracellular DNA in single- and multiple-species unsaturated biofilms. Appl Environ Microbiol 71:5404-5410

Stoodley P, Lewandowski Z, Boyle JD, Lappin-Scott HM (1999) The formation of migratory ripples in a mixed species bacterial biofilm growing in turbulent flow. Environ Microbiol 1:447-455

Stoodley P, Sauer K, Davies, DG, Costerton JW (2002a) Biofilms as complex differentiated communities. Annu Rev Microbiol 56:187-209

Stoodley P, Cargo R, Rupp CJ, Wilson S, Klapper I (2002b) Biofilm material properties as related to shear-induced deformation and detachment phenomena. J Ind Microbiol Biotechnol 29:361-367

Sutherland IW (2001a) Biofilm exopolysaccharides: a strong and sticky framework. Microbiology 147:3-9

Sutherland IW (2001b) The biofilm matrix - an immobilized but dynamic microbial environment. Trends Microbiol 9:222-227

Suzuki K, Wang X, Weilbacher T, Pernestig AK, Melefors O, Georgellis D, Babitzke P, Romeo T (2002) Regulatory circuitry of the CsrA/CsrB and BarA/UvrY systems of *Escherichia coli*. J Bacteriol 184:5130-5140

Suzuki K, Babitzke P, Kushner SR, Romeo T (2006) Identification of a novel regulatory protein (CsrD) that targets the global regulatory RNAs CsrB and CsrC for degradation by RNase E. Genes Dev 20: 2605-2617

Thomas WE, Nilsson LM, Forero M, Sokurenko EV, Vogel V (2004) Shear-dependent 'stick-and-roll' adhesion of type 1 fimbriated *Escherichia coli*. Mol Microbiol 53:1545-1557

Thormann KM, Saville RM, Shukla S, Spormann AM (2005) Induction of rapid detachment in *Shewanella oneidensis* MR-1 biofilms. J Bacteriol 187:1014-1021

Thormann KM, Duttler S, Saville RM, Hyodo M, Shukla S, Hayakawa Y, Spormann AM (2006) Control of formation and cellular detachment from *Shewanella oneidensis* MR-1 biofilms by cyclic di-GMP. J Bacteriol 188:2681-2691

Tieszer C, Reid G, Denstedt J (1998) Conditioning film deposition on ureteral stents after implantation. J Urol 160: 876-881

Tolker-Nielsen T, Molin S (2000) Spatial organization of microbial biofilm communities. Microb Ecol 40:75-84

Tormo MA, Marti M, Valle J, Manna AC, Cheung AL, Lasa, I, Penades JR (2005) SarA is an essential positive regulator of *Staphylococcus epidermidis* biofilm development. J Bacteriol 187:2348-2356

Ulett GC, Webb RI, Schembri MA (2006) Antigen-43-mediated autoaggregation impairs motility in *Escherichia coli*. Microbiology 152:2101-2110

Valle J, Toledo-Arana A, Berasain C, Ghigo JM, Amorena B, Penades JR, Lasa I (2003) SarA and not σ^B is essential for biofilm development by *Staphylococcus aureus*. Mol Microbiol 48:1075-1087

Valle J, Vergara M, Merino N, Penadés JR, Lasa I (2007) σ^B regulates IS256-mediated *Staphylococcus aureus* biofilm phenotypic variation. J Bacteriol Epub in advance of print

Van Loosdrecht MC, Heijnen JJ, Eberl H, Kreft J, Picioreanu C (2002) Mathematical modelling of biofilm structures. Antonie Van Leeuwenhoek 81:245-256

Vasseur P, Vallet-Gely I, Soscia C, Genin S, Filloux A (2005) The *pel* genes of the *Pseudomonas aeruginosa* PAK strain are involved at early and late stages of biofilm formation. Microbiology 151:985-997

Vianney A, Jubelin G, Renault S, Dorel C, Lejeune P, Lazzaroni JC (2005) *Escherichia coli tol* and *rcs* genes participate in the complex network affecting curli synthesis. Microbiology 151:2487-2497

Vidal O, Longin R, Prigent-Combaret C, Dorel C, Hooreman M, Lejeune P (1998) Isolation of an *Escherichia coli* K-12 mutant strain able to form biofilms on inert surfaces: involvement of a new *ompR* allele that increases curli expression. J Bacteriol 180:2442-2449

Von Bodman SB, Ball JK, Faini MA, Herrera CM, Minogue TD, Urbanowski ML, Stevens AM (2003) The quorum sensing negative regulators EsaR and ExpREcc, homologues within the LuxR family, retain the ability to function as activators of transcription. J Bacteriol 185:7001-7007

Vuong C, Gerke C, Somerville GA, Fischer ER, Otto M (2003) Quorum-sensing control of biofilm factors in *Staphylococcus epidermidis*. J Infect Dis 188:706-718

Vuong C, Kocianova S, Voyich JM, Yao Y, Fischer ER, DeLeo FR, Otto M (2004) A crucial role for exopolysaccharide modification in bacterial biofilm formation, immune evasion, and virulence. J Biol Chem 279:54881-54886

Vuong C, Kidder JB, Jacobson ER, Otto M, Proctor RA, Somerville GA (2005) *Staphylococcus epidermidis* polysaccharide intercellular adhesin production significantly increases during tricarboxylic acid cycle stress. J Bacteriol 187:2967-2973

Waite RD, Papakonstantinopoulou A, Littler E, Curtis MA (2005) Transcriptome analysis of Pseudomonas aeruginosa growth: comparison of gene expression in planktonic cultures and developing and mature biofilms. J Bacteriol 187:6571-6576

Wallecha A, Munster V, Correnti J, Chan T, van der Woude M (2002) Dam- and OxyR-dependent phase variation of *agn43*: essential elements and evidence for a new role of DNA methylation. J Bacteriol 184:3338-3347

Wallecha A, Correnti J, Munster V, van der Woude M (2003) Phase variation of Ag43 is independent of the oxidation state of OxyR. J Bacteriol 185:2203-2209

Wang X, Preston JF 3rd, Romeo T (2004) The *pgaABCD* locus of *Escherichia coli* promotes the synthesis of a polysaccharide adhesin required for biofilm formation. J Bacteriol 186:2724-2734

Wang X, Dubey AK, Suzuki K, Baker CS, Babitzke P, Romeo T (2005) CsrA post-transcriptionally represses *pgaABCD*, responsible for synthesis of a biofilm polysaccharide adhesin of *Escherichia coli*. Mol Microbiol 56:1648-1663

Waters CM, Bassler BL (2005) Quorum sensing: cell-to-cell communication in bacteria. Annu Rev Cell Dev Biol 21:319-346

Weber H, Pesavento C, Possling A, Tischendorf G, Hengge R (2006) Cyclic-di-GMP-mediated signaling within the σ^s network of *Escherichia coli*. Mol Microbiol 62:1014-1034

Wei BL, Brun-Zinkernagel AM, Simecka JW, Pruss BM, Babitzke P, Romeo T (2001) Positive regulation of motility and *flhDC* expression by the RNA-binding protein CsrA of *Escherichia coli*. Mol Microbiol 40:245-256

Weigel LM, Donlan RM, Shin DH, Jensen B, Clark NC, McDougal LK, Zhu W, Musser KA, Thompson J, Kohlerschmidt D, Dumas N, Limberger RJ, Patel JB (2007) High-level vancomycin-resistant *Staphylococcus aureus* isolates associated with a polymicrobial biofilm. Antimicrob Agents Chemother 51:231-238

Weilbacher T, Suzuki K, Dubey AK, Wang X, Gudapaty S, Morozov I, Baker CS, Georgellis D, Babitzke P, Romeo T (2003) A novel sRNA component of the carbon storage regulatory system of *Escherichia coli*. Mol Microbiol 48:657-670

Whitchurch CB, Tolker-Nielsen T, Ragas PC, Mattick JS (2002) Extracellular DNA required for bacterial biofilm formation. Science 295:1487

Wolfe AJ (2005) The acetate switch. Microbiol Mol Biol Rev 69:12-50

Wolfe AJ, Chang DE, Walker JD, Seitz-Partridge JE, Vidaurri MD, Lange CF, Pruss BM, Henk MC, Larkin JC, Conway T (2003) Evidence that acetyl phosphate functions as a global signal during biofilm development. Mol Microbiol 48:977-988

Wozniak DJ, Wyckoff TJ, Starkey M, Keyser R, Azadi P, O'Toole GA, Parsek MR (2003) Alginate is not a significant component of the extracellular polysaccharide matrix of PA14 and PAO1 *Pseudomonas aeruginosa* biofilms. Proc Natl Acad Sci U S A 100:7907-7912

Xavier KB, Bassler BL (2003) LuxS quorum sensing: more than just a numbers game. Curr Opin Microbiol 6:191-197

Xu L, Li H, Vuong C, Vadyvaloo V, Wang J, Yao Y, Otto M, Gao Q (2006) Role of the *luxS* quorum-sensing system in biofilm formation and virulence of *Staphylococcus epidermidis*. Infect Immun 74:488-496

Yarwood JM, Bartels DJ, Volper EM, Greenberg EP (2004) Quorum sensing in *Staphylococcus aureus* biofilms. J Bacteriol 186:1838-1350

Ziebuhr W, Krimmer V, Rachid S, Lossner I, Gotz F, Hacker J (1999) A novel mechanism of phase variation of virulence in *Staphylococcus epidermidis*: evidence for control of the polysaccharide intercellular adhesin synthesis by alternating insertion and excision of the insertion sequence element IS256. Mol Microbiol 32:345-356

ZoBell CE (1943) The effect of solid surfaces upon bacterial activity. J Bacteriol 46:39-56

Zogaj X, Bokranz W, Nimtz M, Romling U (2003) Production of cellulose and curli fimbriae by members of the family Enterobacteriaceae isolated from the human gastrointestinal tract. Infect Immun 71:4151-4158

Quorum Sensing and Microbial Biofilms

Y. Irie and M. R. Parsek(✉)

Abstract Some bacterial species engage in two well-documented social behaviors: the formation of surface-associated communities known as biofilms, and intercellular signaling, or quorum sensing. Recent studies have begun to reveal how these two social behaviors are related in different species. This chapter will review the role quorum sensing plays in biofilm formation for different species. In addition, different aspects of quorum sensing in the context of multispecies biofilms will be discussed.

1 Introduction

Microbiology has traditionally followed a guideline of isolating and studying pure cultures of bacterial species. This has accelerated our understanding of bacterial physiology and molecular biology, particularly in the context of pathogenesis as

M. R. Parsek
Department of Microbiology, University of Washington, HSB Room K-343B,
1959 NE Pacific St., Box 357242, Seattle, WA 98195-7242, USA
parsem@u.washington.edu

T. Romeo (ed.), *Bacterial Biofilms.*
Current Topics in Microbiology and Immunology 322.
© Springer-Verlag Berlin Heidelberg 2008

dictated by Koch's postulates (Kaufmann and Schaible 2005). However, conventional pure culture microbiology fails to inform us about aspects of bacterial interactions and group behavior important in the environment and in disease.

Many bacteria and simple eukaryotes are often found growing as surface-associated aggregates, commonly referred to as biofilms. Biofilms have been recognized as a common form of microbial growth on aquatic surfaces in natural, clinical, and industrial environments.

Although natural and laboratory biofilms formed by different species have been shown to exhibit a wide variety of structural characteristics, most appear to be encased in a self-produced extracellular matrix. The contents of the matrix can vary from organism to organism, but are usually abundant in polysaccharides, nucleic acids, and proteins (Sutherland 2001). Biofilm formation has been suggested to result from a developmental programme of gene expression. Indeed, biofilm development and maintenance have been shown to require a wide range of genetic determinants and to involve bacterial subpopulations carrying out different functions. One of the regulatory mechanisms suggested to play a significant role coordinating biofilm formation for many species is intercellular signaling, or quorum sensing (QS) (Parsek and Greenberg 2005).

Microorganisms secrete a wide variety of small molecules that can be self-recognized in a concentration-dependent manner and subsequently induce or repress expression of QS-controlled genes. These QS signals are often referred to as auto-inducers (AIs) and can be classified based upon their structures (Camilli and Bassler 2006). In this review, we will discuss some of the primary classes of QS systems and how (or if) they contribute to biofilm development in different bacterial systems. Additionally, the potential role of QS in multispecies biofilms will be discussed.

2 Different Systems for Sensing a Quorum

The term quorum sensing was coined to describe microbial signaling mediated by self-recognized, secreted molecules, also known as auto-inducers (Fuqua et al. 1994). There are several molecular mechanisms for intercellular signaling in microbes. We will briefly describe a few different, commonly studied quorum sensing systems.

2.1 Acyl Homoserine Lactones

The first described acyl homoserine lactone (AHL) QS system was in *Vibrio fischeri* (Nealson 1977). *V. fischeri* is a marine species that can bioluminesce in the light organs of various marine animals, such as the Hawaiian bobtail squid

Euprymna scolopes (Ruby 1996). *V. fischeri* was found to bioluminesce at high cell densities in liquid batch culture (Nealson 1977). Due to an accumulation of the AHL signal, 3-oxo-hexanoyl homoserine lactone (Eberhard et al. 1981; Kaplan and Greenberg 1985).

AHL signaling has subsequently been described in a wide number of Gram-negative α-, β-, and γ-proteobacteria species (McDougald et al. 2006). AHL signals consist of a homoserine lactone moiety that is linked by an amide bond to an acyl side chain. As shown in Fig. 1, the acyl side-chain lengths and degree of substitution can vary from one QS system to another. AHLs can exhibit a wide range of diffusion characteristics. Short side-chain AHLs diffuse freely across cell membranes, as long side-chain AHLs may partition to the membrane, requiring active efflux for signal export (Pearson et al. 1999). AHL synthesis is primarily catalyzed by a single enzyme belonging to the LuxI family. While signals are perceived by cytoplasmic DNA-binding regulatory proteins belonging to the LuxR family (Parsek and Greenberg 2000).

2.2 Peptide Auto-inducers

While AHL signaling has been found exclusively in Gram-negative bacterial species, many Gram-positive species have been shown to utilize peptides for QS (Fig. 1). Competence signal peptides (CSP) are examples of QS molecules frequently used by streptococcal species. An accumulation of CSP induces autolysis, releasing chromosomal DNA into the environment (Steinmoen et al. 2002). Subsequent uptake of DNA by neighboring cells is thought to promote horizontal gene transfer (Thomas and Nielsen 2005). Natural competence is therefore thought to be advantageous as a form of group behavior, but CSPs and other peptide-based QS molecules are also implicated to regulate other group-associated behaviors in different Gram-positive species, such as biofilm formation (Li et al. 2001) and bacteriocin production (van der Ploeg 2005).

There is a basic molecular scheme for peptide-based signaling (Kleerebezem et al. 1997). A pro-peptide signal is translated in the cytoplasm. This signal can be processed both before and during transport across the cell membrane. Signals usually vary in size from 5 to 87 amino acids depending on the system. Additionally, signals often have unusual, modified amino acids and in certain cases are cyclized via lactone and thiolactone linkages. These modifications/lactonizations are thought to promote the stability of the signal in the extracellular environment (Horswill et al. 2007). There can be several differences and variations upon this scheme. For instance, *Streptococcus pneumoniae* CSP has been shown to be unmodified (Håvarstein et al. 1995). Signal molecule export is mainly carried out by an ABC transporter. Unlike AHLs, signal detection usually occurs at the surface of the cell by a membrane-associated sensor kinase, which then transduces a signal to a DNA-binding response regulator.

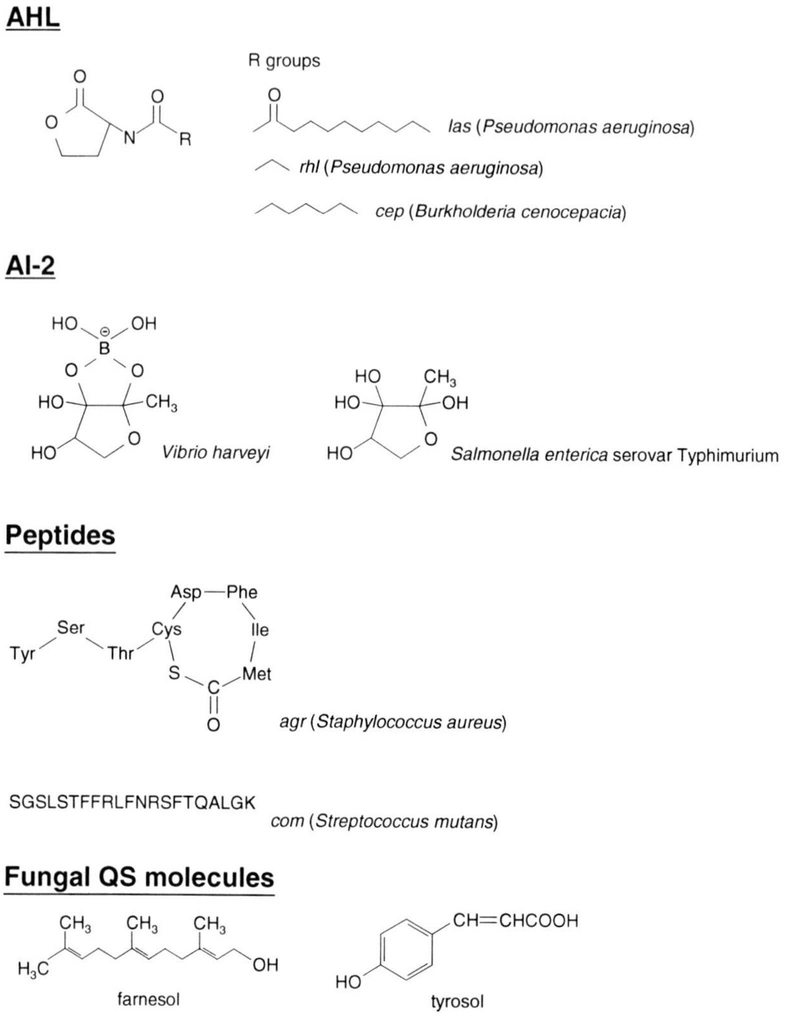
AHL
O
O
O
N
R
R groups
O
las (Pseudomonas aeruginosa)
rhl (Pseudomonas aeruginosa)
cep (Burkholderia cenocepacia)
AI-2
HO
OH
B
O
O
HO
CH3
HO
O
Vibrio harveyi
HO
CH3
HO
OH
HO
O
Salmonella enterica serovar Typhimurium
Peptides
Asp
Phe
Ser
Cys
Ile
Tyr
Thr
S
C
Met
O
agr (Staphylococcus aureus)
SGSLSTFFRLFNRSFTQALGK
com (Streptococcus mutans)
Fungal QS molecules
CH3
CH3
CH3
H3C
OH
farnesol
CH=CHCOOH
HO
tyrosol

Fig. 1 Representative chemical structures of QS molecules

2.3 *Auto-inducer 2*

Auto-inducer 2(AI-2) is a QS signal produced by some Gram-positive and Gram-negative bacterial species. To date, only two AI-2 structures have been solved: those of *Vibrio harveyi* (Chen et al. 2002) and *Salmonella enterica* serovar Typhimurium (Miller et al. 2004) (Fig. 1). While the two structures are distinct, possibly allowing for species specificities, cross-species signaling appears to be prevalent (Camilli and Bassler 2006).

A key step in AI-2 synthesis is catalyzed by a highly conserved enzyme LuxS (Pei and Zhu 2004). The *luxS* gene is found in a vast number of bacterial species, implicating its importance for basic biological function (Xavier and Bassler 2003; Vendeville et al. 2005). AI-2 is derived from the precursor compound, 4,5-dihydroxy-2,3-pentane-dione (DPD). LuxS converts *S*-ribosyl homocysteine to homocysteine and DPD. Once outside the cell, DPD can undergo a number of spontaneous chemical rearrangements to form different furans, including the two known AI-2 structures. AI-2 is perceived by the cell in different manners depending on the system. In the *V. harveyi* system, AI-2 is recognized at the cell surface by cell membrane-associated receptor LuxP (Bassler et al. 1994), while in the *S. enterica* sv. Typhimurium system AI-2 is first transported inside the cell prior to initiating a signaling cascade (Taga et al. 2001).

Various species appear to use the AI-2 QS system to regulate different functions, and furthermore, LuxS catalysis has an alternate role in cell metabolism as an integral component of the activated methyl cycle (Vendeville et al. 2005).

2.4   Fungal QS Systems

Eukaryotic microbial QS systems have recently been discovered and characterized in the dimorphic fungal species *Candida albicans* and *Saccharomyces cerevisiae*. Several QS molecules have been found, including farnesol (Hornby et al. 2001), tyrosol (Chen et al. 2004), phenylethanol, and tryptophol (Chen and Fink 2006). Farnesol is the best studied and appears to block yeast-to-mycelium conversion of *C. albicans* (Hornby et al. 2001). The timing of response to farnesol also regulates dimorphism by determining the commitment to morphotype conversion, but the molecular mechanisms are largely unknown (Nickerson et al. 2006). Interestingly, QS response kinetics of *C. albicans* to farnesol and tyrosol were different (Alem et al. 2006) and may represent a check-point for regulation during development.

3   Rationale for Using Quorum Sensing to Control Biofilm Formation

One can envisage different ways in which QS might influence biofilm formation. For example, QS-regulated functions might serve to initiate biofilm formation. Inducing concentrations of QS signals might precede starvation and other types of stress associated with crowded planktonic bacterial populations. To protect themselves from such types of stress, bacteria may form biofilms, a lifestyle that is characteristically more stress-resistant. Most biofilm systems have demonstrated enhanced resistance to external insults such as antibiotics, shear force, and the host immune system (Lewis 2001; Davies 2003; Jesaitis et al. 2003).

QS may also function to control the population size in a biofilm. Biofilm formation has been described as a developmental cycle (Fig. 2), and QS may serve as

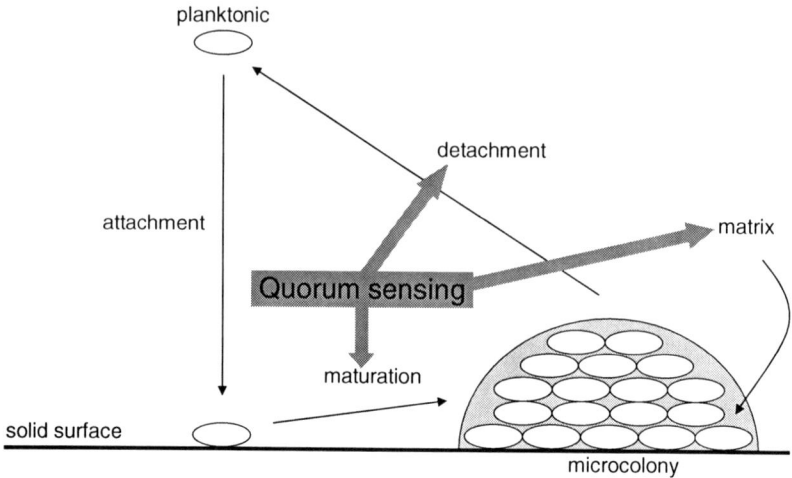

Fig. 2 Developmental biology of biofilms. Unattached free-swimming bacteria (planktonic) initially attach to a solid surface, eventually maturing into structured aggregates called microcolonies. Biofilms are composed of these microcolonies, often encased in extracellular polymeric substances known as the matrix. Some biofilm bacteria detach from microcolonies and become planktonic, presumably to colonize a new surface. QS is believed to be involved in regulating different steps of biofilm development, depending on the organisms and growth conditions

the checkpoint for reinitiating the cycle by promoting dispersion or dissolution of a subpopulation of cells. In this case, dispersing cells might escape the nutritional stress that accompanies or follows inducing concentrations of QS signal. For non-motile species, QS might serve to regulate population density in a biofilm using a different mechanism. Some Gram-positive bacteria initiate autolysis in response to reaching a quorum (Steinmoen et al. 2002).

Finally, QS may induce behaviors in biofilm cells (as they transition from a QS-uninduced state to a QS-induced state) that alter the course of biofilm development, such as the production of secreted factors, such as exopolysaccharides or other adhesins. Alternatively, QS might induce or repress group activities such as surface motility, which in turn could have a profound impact on biofilm structure.

4 Examples of QS-Regulated Biofilm Phenotypes

The above points all represent plausible scenarios of how and when QS may impact biofilms. Thus, perhaps it is not surprising that a survey of the current literature reveals that QS can influence biofilm development in a myriad of ways for different species (Table 1). These studies are not necessarily definitive, they were generally conducted for biofilms grown under a single culturing condition. Even identical strains of the same organism can form biofilms with profoundly different structural characteristics when grown under different nutritional conditions, and the effect of QS on biofilms can be nutritionally conditional (Yarwood et al. 2004; Shrout et al. 2006).

Another potential complication in interpreting the results in Table 1 is that many species have multiple, integrated quorum-sensing systems. Identifying the primary quorum-sensing system responsible for biofilm-related phenotypes can be difficult. For instance, in *Pseudomonas aeruginosa*, there are at least three primary quorum-sensing signals: two AHLs (butyryl homserine lactone and 3-oxo dodecanoyl homserine lactone) and a quinoline-like signal, PQS. The *las* QS system controls expression of the *rhl* QS system and both AHL systems can regulate expression of the PQS system (Latifi et al. 1996; Pesci et al. 1997).

As shown in Table 1, QS has been shown to be involved at multiple stages of biofilm formation in different species. Some QS systems appear to promote biofilm

Table 1 Effects of disabled QS system on biofilm formation in representative organisms

Organism	Disabled QS system	Effect on biofilm	References
Aeromonas hydrophila	*ahy* (AHL)	Defective maturation of biofilm	Lynch et al. 2002
Burkholderia cenocepacia	*cep* (AHL)	More susceptible to ciprofloxacin (double mutant)	Huber et al. 2001
	cci (AHL)	More susceptible to SDS (single mutants) Impaired maturation (*cep*)	Tomlin et al. 2005
Candida albicans	farnesol	Deficient in biofilm dispersal	Kruppa et al. 2004
Klebsiella pneumoniae	AI-2	Delayed biofilm development	Balestrino et al. 2005
Listeria monocytogenes	AI-2	*luxS* Mutant 58% more biofilm than wild type	Belval et al. 2006
Pseudomonas aeruginosa	*las* (AHL)	Flat, unstructured biofilm, more sensitive to SDS (*las*)	Davies et al. 1998
	rhl (AHL)	More susceptible to tobramycin and H_2O_2 (double mutant) Varies with growth conditions	Bjarnsholt et al. 2005a
Pseudomonas putida	*ppu* (AHL)	Formation of more structured biofilm with distinct microcolonies and water channels	Steidle et al. 2002
Serratia liquefaciens	*swr* (AHL)	Thinner biofilm, lacking cell aggregates and cell chains	Labbate et al. 2004
Serratia marcescens	*swr* (AHL)	No biofilm dispersal	Rice et al. 2005
Staphylococcus aureus	*agr* (peptide)	Varies with growth conditions	Yarwood et al. 2004

.nation, while others appear to be involved in biofilm dispersion. In *Vibrio cholerae*, QS negatively regulates production of the *vps* exopolysaccharide (Hammer and Bassler 2003; Zhu and Mekalanos 2003). Expression of *vps* has been shown to promote biofilm formation (Yildiz and Schoolnik 1999). Thus, at high cell densities, QS induction would dissuade biofilm formation. QS mutations result in upregulated matrix production, and therefore enhanced biofilm production (Hammer and Bassler 2003; Zhu and Mekalanos 2003). A completely opposite phenomenon has been recently observed in *P. aeruginosa*, where QS positively regulates *pel*, a major biofilm-related exopolysaccharide operon (Sakuragi and Kolter 2007). In *Serratia*, the conserved AHL system (*swr*) works differently between two species of the same genus. *Serratia liquefaciens swr* appears to promote biofilm formation (Labbate et al. 2004), while in *Serratia marcescens swr* promotes biofilm dispersal (Rice et al. 2005).

QS has also been shown to regulate biofilm formation in simple eukaryotes such as *C. albicans*. Alem et al. reported that early in biofilm development, the QS signal tyrosol is responsible for promoting hyphal formation, while during later stages, another QS signal, farnesol, promotes dispersal of yeast cells from the biofilm (Alem et al. 2006). This suggests that the syntheses of tyrosol and farnesol are differential throughout biofilm development, and cells respond according to which QS signal accumulates at particular developmental stages.

Although QS has been shown to be important during biofilm formation for a wide variety of species, in many cases the QS-regulated functions responsible for observed biofilm phenotypes have not been identified. This is an emerging challenge for researchers, particularly since this question may be complicated. In the case of *P. aeruginosa*, several QS-regulated functions have been shown to influence biofilm development (Kirisits and Parsek 2006).

5 Relationships Between QS, Biofilms, and Infection

It has been suggested that a majority of chronic bacterial infections are characterized by biofilm formation (Parsek and Singh 2003; Hall-Stoodley et al. 2004). One of the paradigm biofilm diseases are the chronic airway infections that afflict people who suffer from cystic fibrosis (CF). *P. aeruginosa* is one of the major CF pathogens and exhibits many characteristics of biofilms in the airways, including the formation of large cellular aggregates and distinct AHL production patterns (Singh et al. 2000).

Virulence factors required for acute infection are often repressed during chronic infections for species capable of causing both types of infection. Acute virulence factors can stimulate the host immune system and cause damage to host tissues, while establishing chronic infection necessitates avoiding the host immune response and maintaining a stalemate with the host, where invasive tissue damage is minimized. Biofilm formation may be an important mechanism for host immune modulation and/or virulence factor downregulation (Irie et al. 2004; Kuchma et al. 2005).

V. cholerae is an interesting example of integrating QS and biofilm formation during its pathogenic life cycle. *V. cholerae* does not cause a chronic infection, but causes acute gastroenteritis, which leads to severe diarrheal symptoms. *V. cholerae* has an aquatic environmental niche, and infection is thought to be only transient in its lifestyle, promoting its proliferation and dissemination. *V. cholerae* most likely resides in biofilm communities while in the environment (Watnick and Kolter 1999; Watnick et al. 2001). Upon ingestion, bacteria first encounter stomach acid, and it has been shown that biofilm cells are more resistant to acid than planktonic cells (Zhu and Mekalanos 2003). Subsequently, QS downregulates extracellular matrix production (Hammer and Bassler 2003; Zhu and Mekalanos 2003) and dislodges the bacteria from biofilms once inside the intestinal lumen, leading to more widespread colonization. Bacterial proliferation in the intestines is thought to induce QS once again, which downregulates adhesins required for biofilm formation (Miller et al. 2002; Zhu et al. 2002) as well as the matrix production, allowing the bacteria to detach from the intestinal epithelia and exit the host via diarrheal flow.

Studying biofilm infections in the laboratory has been problematic. Very few biofilm models of animal infections have been established, and even fewer have tested virulence of QS mutants in such models. Deletion of *luxS* in *Staphylococcus epidermidis* resulted in enhanced biofilm formation in vitro, and enhanced virulence in the catheter-associated infection model (Xu et al. 2006). Similar observations were made with *agr* QS mutants of *S. epidermidis* (Vuong et al. 2004). On the other hand, decreased chronic virulence has been observed for QS mutants in other species, including *S. marcescens* (Coulthurst et al. 2004), *Vibrio vulnificus* (Kim et al. 2003), and *Neisseria meningitidis* (Winzer et al. 2002),

Although it is unknown whether QS-regulated biofilm is important for enteropathogenic *Escherichia coli* (EPEC) in vivo during infection, there is evidence that QS mutants are deficient in biofilm formation in vitro (Moreira et al. 2006). The primary QS system in EPEC that appears to control important adhesins for in vitro biofilm formation is AI-3, an alternative class of QS molecule distinct from previously reported AHLs and AI-2 (Sperandio et al. 2003). AI-3 also controls other virulence factors such as the LEE (locus of enterocyte effacement) pathogenicity island, which encodes for type III secretion system (T3SS) and toxins secreted by T3SS. Furthermore, EPEC AI-3 receptor QseC also binds to epinephrine and norepinephrine (Clarke et al. 2006), which are secreted into the intestinal lumen by the host, providing evidence for interkingdom signaling pathway. It is intriguing to speculate whether the AI-3/epinephrine/norepinephrine QS signaling pathway is inducing in vivo biofilm formation in the intestinal lumen during EPEC infection of the host.

6   QS in Multispecies Biofilm Communities

Biofilms found in many clinical, industrial, and natural environments are frequently mixed species. In these communities, the high cell densities might result in high QS signal concentrations. This is valuable information, both for the species producing the signals and other species occupying the same habitat. There is growing interest in the importance of interspecies QS in shaping multispecies communities. In this section we will discuss some important considerations for QS in mixed-species systems.

6.1 Synergistic Interactions

Studies of oral bacterial communities have been the most progressive in terms of mixed-species biofilms. There are as many as 500 species found in dental biofilms, and colonization by different species appear to be spatiotemporally dynamic (Kolenbrander et al. 2002). AI-2 QS has been suggested to be important for promoting mixed-species interactions in the oral cavity. For example, the AI-2 QS system is required for the oral commensal species, *Streptococcus gordonii,* to form a mixed biofilm with the oral pathogen *Porphyromonas gingivalis* (McNab et al. 2003). Interestingly, the bacteria are capable of forming dual-species biofilm even if the other species is unable to produce AI-2, suggesting a cooperative usage of the AI-2 signal. Only when both AI-2 signal are inactivated are co-culture biofilms disrupted.

There are examples of interspecies QS communication among AHL-producing organisms important in CF lung infections. Interspecies AHL QS has been demonstrated in *P. aeruginosa–Burkholderia cenocepacia* dual-species biofilms (Riedel et al. 2001). In this system, *B. cenocepacia* could perceive *P. aeruginosa* signal, but *P. aeruginosa* did not respond to *B. cenocepacia* AHL. This study suggests that AHLs produced by *P. aeruginosa* may promote AHL-regulated virulence factor production by *B. cenocepacia.*

There are other mixed-species biofilm systems that have enhanced phenotypes only when multiple species are present, such as biofilms formed by bacteria isolated from marine algae (Burmølle et al. 2006). These data are suggestive of cooperative or synergistic QS interactions, but the specific molecular mechanisms of interspecies communications are still undetermined.

6.2 Antagonistic Interactions

Microorganisms occupying the same niche are constantly competing for common resources. Competition within mixed-species biofilms may be fierce, with high cell numbers of competing species spatially fixed within close proximity of one another. Bacteriocin production (Tait and Sutherland 2002) and lowering of pH (Burne and Marquis 2000) are two of the mechanisms within biofilm communities that may help some species to compete. Bacteriocin production can be regulated by QS (Kleerebezem 2004) and may represent one of the mechanisms through which QS is important for competition. *P. aeruginosa* appears to utilize QS to help achieve population dominance over *Agrobacterium tumefaciens* in biofilm growth

(An et al. 2006). Yeast species appear to take a direct approach at dealing with competitors, using their QS molecule farnesol. Farnesol-exposed *Staphylococcus aureus* displays enhanced antibiotic susceptibility, as well as decreased biofilm formation (Jabra-Rizk et al. 2006).

QS might also provide established bacterial biofilms a means to prevent predation by simple eukaryotes. In addition to promoting biofilm dispersion in *S. marcescens*, *swr* QS system mediates resistance against protozoan grazing (Queck et al. 2006). It is unclear, however, whether QS promotes active resistance against grazing. It also remains to be determined whether a similar observation can be made for biofilms under attack by the predatory bacterial species *Bdellovibrio bacteriovorus* (Kadouri and O'Toole 2005).

Another case of microbial warfare in the context of biofilms has been studied for the soil microorganisms *P. aeruginosa* and *C. albicans* interactions. *P. aeruginosa* can degrade farnesol, the QS signal required for *C. albicans* filamentation and mature biofilm formation (Cantwell et al. 1978). *P. aeruginosa* can also use QS-regulated virulence factors to kill *C. albicans*. *C. albicans* exposed to the *las* QS AHL signal does not filament, remaining in the yeast morphotype (Hogan et al. 2004). Yeast-only *C. albicans* has been demonstrated to be impaired for biofilm formation (Lewis et al. 2002; Ramage et al. 2002). The yeast form of *C. albicans* is, however, more resistant to killing by *P. aeruginosa* (Hogan and Kolter 2002), enhancing their survival in this environment.

6.3 *Listening but Not Talking*

Some microbes have been shown to specifically respond to QS signals that they do not produce. This may constitute an effective means of gearing one's physiology to neighboring competitors. For example, despite the inability to produce AI-2, *P. aeruginosa* upregulates virulence factors such as phenazine, elastase, and rhamnolipid in its presence. This may be significant in mixed-species environments such as CF airway infections (Duan et al. 2003). AI-2 molecules have indeed been found in CF sputum where *P. aeruginosa* is thought to produce biofilms. *P. aeruginosa* may produce virulence factors to eliminate competing microbial species colonizing the respiratory tract. Similarly, *S. enterica*, despite their inability to produce AHL QS signal, encodes a probable AHL-responsive transcriptional regulator (SdiA). This allows *S. enterica* to detect and respond to AHL molecules produced by other bacteria (Ahmer 2004; Henke and Bassler 2004).

Bacteria are not the only organisms that are eavesdropping. In the ocean, zoospores of the green seaweed *Ulva* are attracted to AHL molecules produced by bacteria (Tait et al. 2005). Biofouling is often initiated by bacterial biofilms colonizing a surface, followed by eukaryotes such as algae (Beech et al. 2005). It is unclear why algae are specifically attracted to bacterial biofilms, but it has been reported that several green algae have developmental defects in the absence of periphytic bacteria (Provasoli and Pintner 1980; Stratmann et al. 1996).

6.4   Interfering with QS Signaling

Recent work has demonstrated an abundance of organisms capable of degrading QS signals. In spatially structured microbial communities, QS signal degradation may prevent signal propagation from one region of a biofilm to another. This may result in signaling being spatially confined to different regions of the community. One example of QS signal degradation is the ability of *P. aeruginosa* to degrade *C. albicans* signal, farnesol (Cantwell et al. 1978).

Two of the most common quorum quenchers of AHL systems are AHL-lactonases and AHL-acylases (Dong and Zhang 2005). AHL-lactonases can hydrolyze the lactone ring while AHL-acylases hydrolyze the amide linkages. An AHL-lactonase was initially isolated from *Bacillus* (Dong et al. 2000) and has subsequently been found in several other bacterial species. AHL-acylases were initially discovered in *Variovorax paradoxus*, which can utilize AHLs as a carbon source (Leadbetter and Greenberg 2000). *P. aeruginosa* (Huang et al. 2003) and *Ralstonia* species (Lin et al. 2003) have also been discovered to produce AHL-acylases.

Interference of QS signaling can also be achieved through the use of QS signal mimics (McDougald et al. 2006). For example, the *P. aeruginosa las* QS molecule can inhibit the *Chromobacterium violaceum* AHL QS system (McClean et al. 1997). Another example is *Xanthomonas campestris* pathovar campestris, which produces cis-11-methyl-2-dodecenoic acid (DSF) (Barber et al. 1997), which is a structural analog of farnesol (Wang et al. 2004). DSF can inhibit germ tube formation of *C. albicans*, but farnesol does not interfere with *X. campestris* (Oh et al. 2001). Finally, another classic example is the truncated peptides of some *S. aureus agr* QS systems, which can inhibit QS in other *S. aureus* strains (Lyon et al. 2000).

Eukaryotes also possess the ability to interfere with bacterial QS. Human airway epithelia can degrade *P. aeruginosa las* QS molecules through the activity of secreted paraoxygenases (Chun et al. 2004). Interestingly, the *rhl* QS molecule was not inactivated by the epithelia. Since the *P. aeruginosa las* QS system controls the *rhl* QS system (Latifi et al. 1996), inactivation of *las* may be more effective for the host to reduce the virulence of the pathogen.

Another example of eukaryotic inactivation of bacterial QS was discovered in the Australian marine macroalga, *Delisea pulchra*, which uses QS signal mimics called furanones to interfere with AHL signaling in Gram-negative bacterial species, such as *P. aeruginosa* (Manefield et al. 1999; Hentzer et al. 2002). This is thought to inhibit microbial biofilm formation on the surface of its leaves.

7   Summary

Studies of social behavior in bacteria have begun to clearly shape how microbiologists perceive microbial communities. The relationship between QS and biofilm formation has the potential to shape these communities. Closely related species seemingly employ QS for very different purposes during biofilm development.

Many environments are colonized by biofilms consisting of multiple species. In this context, the likelihood of interspecies interactions in the form of QS signaling may be high. Cooperative QS signaling has been demonstrated in several multispecies biofilm systems (McNab et al. 2003; Burmølle et al. 2006). In some cases, QS appears to be utilized by microbes to compete with one another. Biofilms probably represent a common relevant context for these types of interactions.

Microbial biofilms have strong relevance to chronic bacterial infections (Parsek and Singh 2003; Hall-Stoodley et al. 2004). Since biofilm formation for many organisms are QS-mediated, therapeutic strategies targeting QS systems are attracting attention in the drug development fields. Microbes have long utilized various anti-QS strategies for competition against other species (Zhang and Dong 2004). Understanding these molecular mechanisms may be fruitful in developing therapeutic strategies against pathogenic species. Already, some quorum-quenching chemical compounds have demonstrated success in inhibiting microbial biofilms (Dong and Zhang 2005). Unlike modern antibiotics, QS-directed therapies are not designed to cause bactericidal or bacteriostatic effects, and thus, emergence of resistance may be less problematic. Anti-QS measures have been demonstrated to be effective for plant infections. Heterologously expressed AHL-lactonase rendered tobacco and potato plants significantly more resistant to *Erwinia carotovora* (Dong et al. 2001).

As the studies of microbial communities continues, understanding how different species percieve and respond to one another will be crucial to understanding community composition and function. The most relevant context for such studies may be microbial biofilm.

References

Ahmer BM (2004) Cell-to-cell signalling in *Escherichia coli* and *Salmonella enterica*. Mol Microbiol 52:933–945

Alem MA, Oteef MD, Flowers TH, Douglas LJ (2006) Production of tyrosol by *Candida albicans* biofilms and its role in quorum sensing and biofilm development. Eukaryot. Cell 5:1770–1779

An D, Danhorn T, Fuqua C, Parsek MR (2006) Quorum sensing and motility mediate interactions between *Pseudomonas aeruginosa* and *Agrobacterium tumefaciens* in biofilm cocultures. Proc Natl Acad Sci U S A 103:3828–3833

Balaban N, Cirioni O, Giacometti A et al (2007) Treatment of *Staphylococcus aureus* Biofilm Infection by the Quorum-Sensing Inhibitor RIP. Antimicrob Agents Chemother 51:2226–2229

Balaban N, Stoodley P, Fux CA, Wilson S, Costerton JW, Dell'Acqua G (2005) Prevention of staphylococcal biofilm-associated infections by the quorum sensing inhibitor RIP. Clin Orthop Relat Res (437):48–54

Balestrino D, Haagensen JA, Rich C, Forestier C (2005) Characterization of type 2 quorum sensing in *Klebsiella pneumoniae* and relationship with biofilm formation. J Bacteriol 187:2870–2880

Barber CE, Tang JL, Feng JX et al (1997) A novel regulatory system required for pathogenicity of *Xanthomonas campestris* is mediated by a small diffusible signal molecule. Mol Microbiol 24:555–566

Bassler BL, Wright M, Silverman MR (1994) Multiple signalling systems controlling expression of luminescence in *Vibrio harveyi*: sequence and function of genes encoding a second sensory pathway. Mol Microbiol 13:273–286

Beech IB, Sunner JA, Hiraoka K (2005) Microbe-surface interactions in biofouling and biocorrosion processes. Int Microbiol 8:157–168

Belval CS, Gal L, Margiewes S, Garmyn D, Piveteau P, Guzzo J (2006) Assessment of the roles of LuxS, *S*-ribosyl homocysteine, and autoinducer 2 in cell attachment during biofilm formation by *Listeria monocytogenes* EGD-e. Appl Environ Microbiol 72:2644–2650

Bjarnsholt T, Jensen PO, Bermolle M et al (2005a) *Pseudomonas aeruginosa* tolerance to tobramycin, hydrogen peroxide and polymorphonuclear leukocytes is quorum-sensing dependent. Microbiology 151:373–383

Bjarnsholt T, Jensen PO, Rasmussen TB et al (2005b) Garlic blocks quorum sensing and promotes rapid clearing of pulmonary *Pseudomonas aeruginosa* infections. Microbiology 151:3873–3880

Burmølle M, Webb JS, Rao D, Hansen LH, Sørensen SJ, Kjelleberg S (2006) Enhanced biofilm formation and increased resistance to antimicrobial agents and bacterial invasion are caused by synergistic interactions in multispecies biofilms. Appl Environ Microbiol 72:3916–3923

Burne RA, Marquis RE (2000) Alkali production by oral bacteria and protection against dental caries. FEMS Microbiol Lett 193:1–6

Camilli A, Bassler BL (2006) Bacterial small-molecule signaling pathways. Science 311:1113–1116

Cantwell SG, Lau EP, Watt DS, Fall RR (1978) Biodegradation of acyclic isoprenoids by *Pseudomonas* species. J Bacteriol 135:324–333

Chen H, Fink GR (2006) Feedback control of morphogenesis in fungi by aromatic alcohols. Genes Dev 20:1150–1161

Chen H, Fujita M, Feng Q, Clardy J, Fink GR (2004) Tyrosol is a quorum-sensing molecule in *Candida albicans*. Proc Natl Acad Sci U S A 101:5048–5052

Chen X, Schauder S, Potier N et al (2002) Structural identification of a bacterial quorum-sensing signal containing boron. Nature 415:545–549

Chun CK, Ozer EA, Welsh MJ, Zabner J, Greenberg EP (2004) Inactivation of a *Pseudomonas aeruginosa* quorum-sensing signal by human airway epithelia. Proc Natl Acad Sci U S A 101:3587–3590

Clarke MB, Hughes DT, Zhu C, Boedeker EC, Sperandio V (2006) The QseC sensor kinase: a bacterial adrenergic receptor. Proc Natl Acad Sci U S A 103:10420–10425

Coulthurst SJ, Kurz CL, Salmond GP (2004) *luxS* mutants of *Serratia* defective in autoinducer-2-dependent 'quorum sensing' show strain-dependent impacts on virulence and production of carbapenem and prodigiosin. Microbiology 150:1901–1910

Davies D (2003) Understanding biofilm resistance to antibacterial agents. Nat Rev Drug Discov 2:114–122

Davies DG, Parsek MR, Pearson JP, Iglewski BH, Costerton JW, Greenberg EP (1998) The involvement of cell-to-cell signals in the development of a bacterial biofilm. Science 280:295–298

Dong Y-H, Zhang L-H (2005) Quorum sensing and quorum-quenching enzymes. J Microbiol 43:S101–S109

Dong YH, Wang LH, Xu JL, Zhang HB, Zhang XF, Zhang LH (2001) Quenching quorum-sensing-dependent bacterial infection by an *N*-acyl homoserine lactonase. Nature 411:813–817

Dong YH, Xu JL, Li XZ, Zhang LH (2000) AiiA, an enzyme that inactivates the acylhomoserine lactone quorum-sensing signal and attenuates the virulence of *Erwinia carotovora*. Proc Natl Acad Sci U S A 97:3526–3531

Duan K, Dammel C, Stein J, Rabin H, Surette MG (2003) Modulation of *Pseudomonas aeruginosa* gene expression by host microflora through interspecies communication. Mol Microbiol 50:1477–1491

Eberhard A, Burlingame AL, Eberhard C, Kenyon GL, Nealson KH, Oppenheimer NJ (1981) Structural identification of autoinducer of *Photobacterium fischeri* luciferase. Biochemistry 20:2444–2449

Fuqua WC, Winans SC, Greenberg EP (1994) Quorum sensing in bacteria: the LuxR-LuxI family of cell density-responsive transcriptional regulators. J Bacteriol 176:269–275

Håvarstein LS, Coomaraswamy G, Morrison DA (1995) An unmodified heptadecapeptide pheromone induces competence for genetic transformation in *Streptococcus pneumoniae*. Proc Natl Acad Sci U S A 92:11140–11144

Hall-Stoodley L, Costerton JW, Stoodley P (2004) Bacterial biofilms: from the natural environment to infectious diseases. Nat Rev Microbiol 2:95–108

Hammer BK, Bassler BL (2003) Quorum sensing controls biofilm formation in *Vibrio cholerae*. Mol Microbiol 50:101–104

Henke JM, Bassler BL (2004) Bacterial social engagements. Trends Cell Biol. 14:648–656

Hentzer M, Riedel K, Rasmussen TB et al (2002) Inhibition of quorum sensing in *Pseudomonas aeruginosa* biofilm bacteria by a halogenated furanone compound. Microbiology 148:87–102

Hogan DA, Kolter R (2002) *Pseudomonas-Candida* interactions: an ecological role for virulence factors. Science 296:2229–2232

Hogan DA, Vik A, Kolter R (2004) A *Pseudomonas aeruginosa* quorum-sensing molecule influences *Candida albicans* morphology. Mol Microbiol 54:1212–1223

Hornby JM, Jensen EC, Lisec AD et al (2001) Quorum sensing in the dimorphic fungus *Candida albicans* is mediated by farnesol. Appl Environ Microbiol 67:2982–2992

Horswill AR, Stoodley P, Stewart PS, Parsek MR (2007) The effect of the chemical, biological, and physical environment on quorum sensing in structured microbial communities. Anal Bioanal Chem 387:371–380

Huang JJ, Han JI, Zhang LH, Leadbetter JR (2003) Utilization of acyl-homoserine lactone quorum signals for growth by a soil pseudomonad and *Pseudomonas aeruginosa* PAO1. Appl Environ Microbiol 69:5941–5949

Huber B, Riedel K, Hentzer M et al (2001) The *cep* quorum-sensing system of *Burkholderia cepacia* H111 controls biofilm formation and swarming motility. Microbiology 147:2517–2528

Irie Y, Mattoo S, Yuk MH (2004) The Bvg virulence control system regulates biofilm formation in *Bordetella bronchiseptica*. J Bacteriol 186:5692–5698

Jabra-Rizk MA, Meiller TF, James CE, Shirtliff ME (2006) Effect of farnesol on *Staphylococcus aureus* biofilm formation and antimicrobial susceptibility. Antimicrob Agents Chemother 50:1463–1469

Jesaitis AJ, Franklin MJ, Berglund D et al (2003) Compromised host defense on *Pseudomonas aeruginosa* biofilms: characterization of neutrophil and biofilm interactions. J Immunol 171:4329–4339

Kadouri D, O'Toole GA (2005) Susceptibility of biofilms to *Bdellovibrio bacteriovorus* attack. Appl Environ Microbiol 71:4044–4051

Kaplan HB, Greenberg EP (1985) Diffusion of autoinducer is involved in regulation of the *Vibrio fischeri* luminescence system. J Bacteriol 163:1210–1214

Kaufmann SH, Schaible UE (2005) 100th anniversary of Robert Koch's Nobel Prize for the discovery of the tubercle bacillus. Trends Microbiol 13:469–475

Kim SY, Lee SE, Kim YR et al (2003) Regulation of *Vibrio vulnificus* virulence by the LuxS quorum-sensing system. Mol Microbiol 48:1647–1664

Kirisits MJ, Parsek MR (2006) Does *Pseudomonas aeruginosa* use intercellular signalling to build biofilm communities? Cell Microbiol 8:1841–1849

Kleerebezem M (2004) Quorum sensing control of lantibiotic production; nisin and subtilin autoregulate their own biosynthesis. Peptides 25:1405–1414

Kleerebezem M, Quadri LE, Kuipers OP, de Vos WM (1997) Quorum sensing by peptide pheromones and two-component signal-transduction systems in Gram-positive bacteria. Mol Microbiol 24:895–904

Kolenbrander PE, Andersen RN, Blehert DS, Egland PG, Foster JS, Palmer RJ, Jr. (2002) Communication among oral bacteria. Microbiol Mol Biol Rev 66:486–505

Kruppa M, Krom BP, Chauhan N, Bambach AV, Cihlar RL, Calderone RA (2004) The two-component signal transduction protein Chk1p regulates quorum sensing in *Candida albicans*. Eukaryot Cell 3:1062–1065

Kuchma SL, Connolly JP, O'Toole GA (2005) A three-component regulatory system regulates biofilm maturation and type III secretion in *Pseudomonas aeruginosa*. J Bacteriol 187:1441–1454

Labbate M, Queck SY, Koh KS, Rice SA, Givskov M, Kjelleberg S (2004) Quorum sensing-controlled biofilm development in *Serratia liquefaciens* MG1. J Bacteriol 186:692–698

Latifi A, Foglino M, Tanaka K, Williams P, Lazdunski A (1996) A hierarchical quorum-sensing cascade in *Pseudomonas aeruginosa* links the transcriptional activators LasR and RhlR (VsmR) to expression of the stationary-phase sigma factor RpoS. Mol Microbiol 21:1137–1146

Leadbetter JR, Greenberg EP (2000) Metabolism of acyl-homoserine lactone quorum-sensing signals by *Variovorax paradoxus*. J Bacteriol 182:6921–6926

Lewis K (2001) Riddle of biofilm resistance. Antimicrob Agents Chemother 45:999–1007

Lewis RE, Lo HJ, Raad II, Kontoyiannis DP (2002) Lack of catheter infection by the *efg1/efg1 cph1/cph1* double-null mutant, a *Candida albicans* strain that is defective in filamentous growth. Antimicrob Agents Chemother 46:1153–1155

Li Y-H, Hanna MN, Svensäter G, Ellen RP, Cvitkovitch DG (2001) Cell density modulates acid adaptation in *Streptococcus mutans*: implications for survival in biofilms. J Bacteriol 183:6875–6884

Lin YH, Xu JL, Hu J et al (2003) Acyl-homoserine lactone acylase from *Ralstonia* strain XJ12B represents a novel and potent class of quorum-quenching enzymes. Mol Microbiol 47:849–860

Lynch MJ, Swift S, Kirke DF, Keevil CW, Dodd CE, Williams P (2002) The regulation of biofilm development by quorum sensing in *Aeromonas hydrophila*. Environ Microbiol 4:18–28

Lyon GJ, Mayville P, Muir TW, Novick RP (2000) Rational design of a global inhibitor of the virulence response in *Staphylococcus aureus*, based in part on localization of the site of inhibition to the receptor-histidine kinase, AgrC. Proc Natl Acad Sci U S A 97:13330–13335

Manefield M, de Nys R, Kumar M et al (1999) Evidence that halogenated furanones from *Delisea pulchra* inhibit acylated homoserine lactone (AHL)-mediated gene expression by displacing the AHL signal from its receptor protein. Microbiology 145:283–291

McClean KH, Winson MK, Fish L et al (1997) Quorum sensing and *Chromobacterium violaceum*: exploitation of violacein production and inhibition for the detection of *N*-acylhomoserine lactones. Microbiology 143:3703–3711

McDougald D, Rice SA, Kjelleberg S (2006) Bacterial quorum sensing and interference by naturally occurring biomimics. Anal Bioanal Chem 387:445–453

McNab R, Ford SK, El-Sabaeny A, Barbieri B, Cook GS, Lamont RJ (2003) LuxS-based signaling in *Streptococcus gordonii*: autoinducer 2 controls carbohydrate metabolism and biofilm formation with *Porphyromonas gingivalis*. J Bacteriol 185:274–284

Miller MB, Skorupski K, Lenz DH, Taylor RK, Bassler BL (2002) Parallel quorum sensing systems converge to regulate virulence in *Vibrio cholerae*. Cell 110:303–314

Miller ST, Xavier KB, Campagna SR et al (2004) *Salmonella typhimurium* recognizes a chemically distinct form of the bacterial quorum-sensing signal AI-2. Mol Cell 15:677–687

Moreira CG, Palmer K, Whiteley M et al (2006) Bundle-forming pili and EspA are involved in biofilm formation by enteropathogenic *Escherichia coli*. J Bacteriol 188:3952–3961

Nealson KH (1977) Autoinduction of bacterial luciferase. Occurrence, mechanism and significance. Arch Microbiol 112:73–79

Nickerson KW, Atkin AL, Hornby JM (2006) Quorum sensing in dimorphic fungi: farnesol and beyond. Appl Environ Microbiol 72:3805–3813

Oh KB, Miyazawa H, Naito T, Matsuoka H (2001) Purification and characterization of an autoregulatory substance capable of regulating the morphological transition in *Candida albicans*. Proc Natl Acad Sci U S A 98:4664–4668

Parsek MR, Greenberg EP (2000) Acyl-homoserine lactone quorum sensing in Gram-negative bacteria: a signaling mechanism involved in associations with higher organisms. Proc Natl Acad Sci U S A 97:8789–8793

Parsek MR, Greenberg EP (2005) Sociomicrobiology: the connections between quorum sensing and biofilms. Trends Microbiol 13:27–33

Parsek MR, Singh PK (2003) Bacterial biofilms: an emerging link to disease pathogenesis. Annu Rev Microbiol 57:677–701

Pearson JP, Van Delden C, Iglewski BH (1999) Active efflux and diffusion are involved in transport of *Pseudomonas aeruginosa* cell-to-cell signals. J Bacteriol 181:1203–1210

Pei D, Zhu J (2004) Mechanism of action of *S*-ribosylhomocysteinase (LuxS). Curr Opin Chem Biol 8:492–497

Pesci EC, Pearson JP, Seed PC, Iglewski BH (1997) Regulation of *las* and *rhl* quorum sensing in *Pseudomonas aeruginosa*. J Bacteriol 179:3127–3132

Provasoli L, Pintner IJ (1980) Bacteria induced polymorphism in an axenic laboratory strain of *Ulva lactuca* (Chlorophyceae). J Phycol 16:196–201

Queck S-Y, Weitere M, Moreno AM, Rice SA, Kjelleberg S (2006) The role of quorum sensing mediated developmental traits in the resistance of *Serratia marcescens* biofilms against protozoan grazing. Environ Microbiol 8:1017–1025

Ramage G, VandeWalle K, López-Ribot JL, Wickes BL (2002) The filamentation pathway controlled by the Efg1 regulator protein is required for normal biofilm formation and development in *Candida albicans*. FEMS Microbiol Lett 214:95–100

Rasmussen TB, Bjarnsholt T, Skindersoe ME et al (2005) Screening for quorum-sensing inhibitors (QSI) by use of a novel genetic system, the QSI selector. J Bacteriol 187:1799–1814

Rice SA, Koh KS, Queck SY, Labbate M, Lam KW, Kjelleberg S (2005) Biofilm formation and sloughing in *Serratia marcescens* are controlled by quorum sensing and nutrient cues. J Bacteriol 187:3477–3485

Riedel K, Hentzer M, Geisenberger O et al (2001) *N*-acylhomoserine-lactone-mediated communication between *Pseudomonas aeruginosa* and *Burkholderia cepacia* in mixed biofilms. Microbiology 147:3249–3262

Ruby EG (1996) Lessons from a cooperative, bacterial-animal association: the *Vibrio fischeri-Euprymna scolopes* light organ symbiosis. Annu Rev Microbiol 50:591–624

Sakuragi Y, Kolter R (2007) Quorum sensing regulation of the biofilm matrix genes (*pel*) of *Pseudomonas aeruginosa*. J Bacteriol 189:5383–5386

Shrout JD, Chopp DL, Just CL, Hentzer M, Givskov M, Parsek MR (2006) The impact of quorum sensing and swarming motility on *Pseudomonas aeruginosa* biofilm formation is nutritionally conditional. Mol Microbiol 62:1264–1277

Singh PK, Schaefer AL, Parsek MR, Moninger TO, Welsh MJ, Greenberg EP (2000) Quorum-sensing signals indicate that cystic fibrosis lungs are infected with bacterial biofilms. Nature 407:762–764

Smith EE, Buckley DGD, Wu Z et al (2006) Genetic adaptation by *Pseudomonas aeruginosa* to the airways of cystic fibrosis patients. Proc Natl Acad Sci U S A 103:8487–8492

Sperandio V, Torres AG, Jarvis B, Nataro JP, Kaper JB (2003) Bacteria-host communication: the language of hormones. Proc Natl Acad Sci U S A 100:8951–8956

Steidle A, Allesen-Holme M, Riedel K et al (2002) Identification and characterization of an *N*-acylhomoserine lactone-dependent quorum-sensing system in *Pseudomonas putida* strain IsoF. Appl Environ Microbiol 68:6371–6382

Steinmoen H, Knutsen E, Håvarstein LS (2002) Induction of natural competence in *Streptococcus pneumoniae* triggers lysis and DNA release from a subfraction of the cell population. Proc Natl Acad Sci U S A 99:7681–7686

Stratmann J, Paputsoglu G, Oertel W (1996) Differentiation of *Ulva mutabilis* (Chlorophyta) gametangia and gamete release are controlled by extracellular inhibitors. J Phycol 32:1009–1021

Sutherland IW (2001) The biofilm matrix – an immobilized but dynamic microbial environment. Trends Microbiol 9:222–227

Taga ME, Semmelhack JL, Bassler BL (2001) The LuxS-dependent autoinducer AI-2 controls the expression of an ABC transporter that functions in AI-2 uptake in *Salmonella typhimurium*. Mol Microbiol 42:777–793

Tait K, Joint I, Daykin M, Milton DL, Williams P, Cámara M (2005) Disruption of quorum sensing in seawater abolishes attraction of zoospores of the green alga *Ulva* to bacterial biofilms. Environ Microbiol 7:229–240

Tait K, Sutherland IW (2002) Antagonistic interactions amongst bacteriocin-producing enteric bacteria in dual species biofilms. J Appl Microbiol 93:345–352

Telford G, Wheeler D, Williams P et al (1998) The *Pseudomonas aeruginosa* quorum-sensing signal molecule *N*-(3-oxododecanoyl)-L-homoserine lactone has immunomodulatory activity. Infect Immun 66:36–42

Thomas CM, Nielsen KM (2005) Mechanisms of, and barriers to, horizontal gene transfer between bacteria. Nat Rev Microbiol 3:711–721

Tomlin KL, Malott RJ, Ramage G, Storey DG, Sokol PA, Ceri H (2005) Quorum-sensing mutations affect attachment and stability of *Burkholderia cenocepacia* biofilms. Appl Environ Microbiol 71:5208–5218

van der Ploeg JR (2005) Regulation of bacteriocin production in *Streptococcus mutans* by the quorum-sensing system required for development of genetic competence. J Bacteriol 187:3980–3989

Vendeville A, Winzer K, Heurlier K, Tang CM, Hardie KR (2005) Making 'sense' of metabolism: autoinducer-2, LuxS and pathogenic bacteria. Nat Rev Microbiol 3:383–396

Vuong C, Kocianova S, Yao Y, Carmody AB, Otto M (2004) Increased colonization of indwelling medical devices by quorum-sensing mutants of *Staphylococcus epidermidis in vivo*. J Infect Dis 190:1498–1505

Wang LH et al (2004) A bacterial cell-cell communication signal with cross-kingdom structural analogues. Mol Microbiol 51:903–912

Watnick PI, Kolter R (1999) Steps in the development of a *Vibrio cholerae* El Tor biofilm. Mol Microbiol 34:586–595

Watnick PI, Lauriano CM, Klose KE, Croal L, Kolter R (2001) The absence of a flagellum leads to altered colony morphology, biofilm development and virulence in *Vibrio cholerae* O139. Mol Microbiol 39:223–235

Winzer K, Sun YH, Green A et al (2002) Role of *Neisseria meningitidis luxS* in cell-to-cell signaling and bacteremic infection. Infect Immun 70:2245–2248

Xavier KB, Bassler BL (2003) LuxS quorum sensing: more than just a numbers game. Curr Opin Microbiol 6:191–197

Xu L, Li H, Vuong C et al (2006) Role of the *luxS* quorum-sensing system in biofilm formation and virulence of *Staphylococcus epidermidis*. Infect. Immun. 74:488–496

Yarwood JM, Bartels DJ, Volper EM, Greenberg EP (2004) Quorum sensing in *Staphylococcus aureus* biofilms. J Bacteriol 186:1838–1850

Yildiz FH, Schoolnik GK (1999) *Vibrio cholerae* O1 El Tor: identification of a gene cluster required for the rugose colony type, exopolysaccharide production, chlorine resistance, and biofilm formation. Proc Natl Acad Sci U S A 96:4028–4033

Zhang LH, Dong YH (2004) Quorum sensing and signal interference: diverse implications. Mol Microbiol 53:1563–1571

Zhu J, Mekalanos JJ (2003) Quorum sensing-dependent biofilms enhance colonization in *Vibrio cholerae*. Dev Cell 5:647–656

Zhu J, Miller MB, Vance RE, Dziejman M, Bassler BL, Mekalanos JJ (2002) Quorum-sensing regulators control virulence gene expression in *Vibrio cholerae*. Proc Natl Acad Sci U S A 99:3129–3134

Innate and Induced Resistance Mechanisms of Bacterial Biofilms

G. G. Anderson and G. A. O'Toole(✉)

Contents

Abstract Bacterial biofilms are highly recalcitrant to antibiotic treatment, which holds serious consequences for therapy of infections that involve biofilms. The genetic mechanisms of this biofilm antibiotic resistance appear to fall into two general classes: innate resistance factors and induced resistance factors. Innate mechanisms are activated as part of the biofilm developmental pathway, the factors being integral parts of biofilm structure and physiology. Innate pathways include decreased diffusion of antibiotics through the biofilm matrix, decreased oxygen and nutrient availability accompanied by altered metabolic activity, formation of persisters, and other specific molecules not fitting into the above groups. Induced resistance factors include those resulting from induction by the antimicrobial agent itself. Biofilm antibiotic resistance is likely manifested as an intricate mixture of innate and induced mechanisms. Many researchers are currently trying to overcome

G. A. O'Toole

Dartmouth Medical School, Department of Microbiology and Immunology,
Hanover, NH, 03755 USA
George.O'Toole@Dartmouth.edu

T. Romeo (ed.), *Bacterial Biofilms.*
Current Topics in Microbiology and Immunology 322.
© Springer-Verlag Berlin Heidelberg 2008

this extreme biofilm antibiotic resistance by developing novel therapies aimed at disrupting biofilms and killing the constituent bacteria. These studies have led to the identification of several molecules that effectively disturb biofilm physiology, often by interrupting bacterial quorum sensing. In this manner, manipulation of innate and induced resistance pathways holds much promise for treatment of biofilm infections.

1 Introduction

One of the most confounding aspects of bacterial biofilm formation is the increased resistance of the constituent microbes to antibiotics and other stressors. A biofilm lifestyle affords bacteria a 10- to 1,000-fold increase in antibiotic resistance compared to their planktonic counterparts (Mah and O'Toole 2001). Particularly in the clinic, increased resistance holds serious consequences for infection control, treatment regimes, and disease progression. Biofilms can form on medical implants (Donlan and Costerton 2002), leading to increased morbidity and mortality of affected individuals. Often, removal of the contaminated implant is the only effective treatment. Biofilms that form during specific disease states, such as in the lungs of cystic fibrosis patients, can also be extremely difficult to eliminate (Chernish and Aaron 2003; Gibson et al. 2003; Hoiby et al. 2005).

Despite decades of research, very little is known about the molecular mechanisms of antibiotic resistance in biofilms. Traditional antibiotic resistance (of planktonic bacteria) usually involves inactivation of the antibiotic, modification of targets, and exclusion of the antibiotic (Patel 2005). The actions typically require the acquisition of specific genetic factors, such as genes for β-lactamase or efflux pumps. However, research to date does not support a large role for these mechanisms in biofilm resistance. In this chapter, we define biofilm antibiotic resistance as the ability of biofilm bacteria to survive antibiotic treatment by using its existing complement of genes. This regulation can occur as an innate result of growing in a biofilm or be induced by the antimicrobial agent itself. Indeed, several innate biofilm phenomena and antibiotic-induced factors have been revealed that provide explanations for the ability of bacterial biofilms to survive under antibiotic or other chemical pressures (Costerton et al. 1999; Donlan and Costerton 2002; Dunne 2002; Mah and O'Toole 2001; Patel 2005; Stewart and Costerton 2001). Biofilm antimicrobial resistance is most likely the result of a complex mixture of these innate and induced factors.

In this chapter, we will discuss these innate and induced factors, with particular emphasis on how these pathways influence biofilm antibiotic resistance. First, we will describe innate antibiotic resistance mechanisms of bacterial biofilms: that is, growth in a biofilm resulting in altered genetic regulatory patterns that are an integral part of the biofilm lifestyle. Some of these regulated factors also protect biofilm bacteria from antibiotic killing. Accordingly, we will describe limited antibiotic diffusion through the biofilm, decreased growth and altered metabolism, and formation of specialized "persisters" as important innate biofilm phenomena that impact antibiotic resistance. Next, we will discuss the evidence for induced resistance factors, or in other words, antibiotic-induced expression of resistance factors. Following this discussion of innate and induced antibiotic resistance factors, we will describe novel therapeutic mechanisms that are being developed to more effectively target biofilm bacteria by disrupting these innate and induced pathways. Finally, we will briefly mention certain industrial applications for which increased biofilm resistance actually benefits the outcome of the application.

2   Innate Mechanisms: Why Wait?

Bacterial biofilm formation, in general, is accompanied by global genetic regulatory changes that occur as planktonic bacteria enter a community lifestyle. Many of these changes render the constituent bacteria resistant to antibiotics. In other words, biofilm antimicrobial resistance, in large part, can be thought of as an innate attribute resulting from conversion to a biofilm lifestyle.

Research has identified the influence of several different innate biofilm factors affecting antibiotic resistance (Costerton et al. 1999; Donlan and Costerton 2002; Dunne 2002; Mah and O'Toole 2001; Patel 2005; Stewart and Costerton 2001). First, the biofilm matrix may act as a diffusion barrier, preventing antibiotics from reaching their targets. Second, establishment of microenvironments within biofilms, such as reduced oxygen zones, leads to slow growth of the bacteria. Third, a small subpopulation of bacteria within the biofilm seems to differentiate into persisters, with greatly reduced susceptibility to antibiotics. Finally, several resistance genes have been identified that are specifically regulated within biofilms. Studies have only recently begun to elucidate the genetic regulation of these innate biofilm antibiotic resistance mechanisms. These molecular details are vital to our understanding of the ability of biofilms to thwart treatment.

2.1   Diffusion Confusion

As antimicrobial agents contact a biofilm, the first obstacle they encounter is the biofilm matrix. Antibiotics must traverse through this thick mixture of exopolysaccharide (EPS), DNA, and protein in order to reach their targets, and it is thought that the matrix acts as a diffusion barrier, limiting access to the biofilm bacteria (Fig. 1) (Donlan and Costerton 2002). A decrease in the levels of antibiotics reaching the bacteria would result in an apparent increase in resistance. Indeed, recent mathematical modeling predicted that while limited antibiotic diffusion may lead to death of the outer layer of bacteria, it provides a chance for a subpopulation of

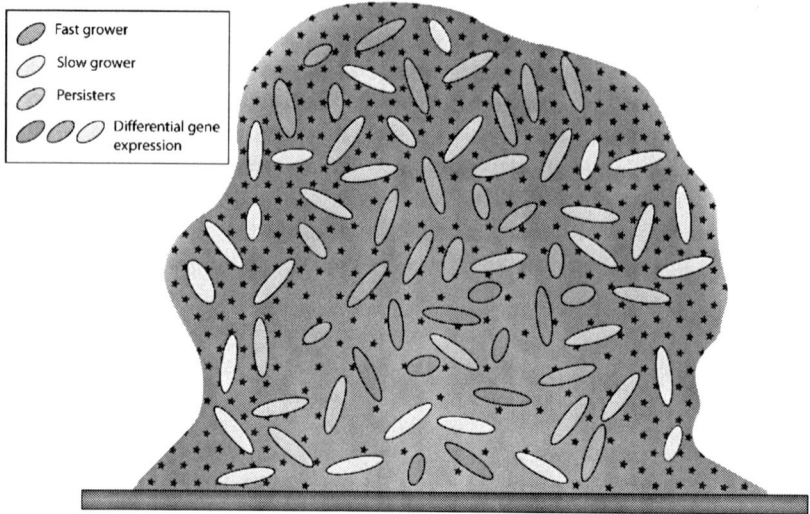

Fig. 1 Innate biofilm antibiotic resistance mechanisms. The single biofilm macrocolony shown here is made up of bacteria (*ovals*) surrounded by an extracellular matrix (*multicolored background*). *Small dark dots* represent the antibiotic molecules to which the biofilm has been exposed. Limited antibiotic diffusion through the matrix (depicted as a decreasing dot density toward the core of the microcolony) might protect bacteria buried deep within the biofilm from antibiotic action. Oxygen and nutrient concentrations also decrease as the deeper parts of the biofilms are approached, symbolized by a color gradient from *red* (aerobic and high nutrient concentrations) to *green* (anaerobic and low nutrient concentrations). These gradients slow the growth of the innermost bacteria (*tan*), and thus facilitate survival in the presence of antibiotic that typically kill only fast growing microorganisms (*magenta*). The *red* to *green* gradient also represents other possible variations within the heterogeneous biofilm, such as pH. Persister cells, also considered nongrowing or slow-growing, are represented by *blue ovals* scattered throughout the biofilm. Finally, the *green ovals* denote biofilm bacteria expressing specific biofilm activated resistance genes, such as *ndvB*. Differential expression of these genes (different shades of *green*) in response to environmental gradients in the community might influence the antibiotic resistance state of individual bacteria within the biofilm

bacteria buried deeper within the biofilm to enact adaptive changes to counter the insult (Szomolay et al. 2005). However, limited antibiotic diffusion does not appear to be a universal trait shared by all biofilms, and, as detailed below, the data conflict over whether the biofilm matrix is a major contributing factor influencing biofilm resistance (Patel 2005).

2.1.1 Antibiotic Trapping

A few studies have found that the biofilm matrix can limit penetration of antimicrobials. Alginate, an EPS produced by *Pseudomonas aeruginosa*, has been intensely studied for its ability to trap antimicrobial agents. This ability appears to be related to the anionic nature of the exopolymer. Cationic molecules can thus be retained within the matrix and prevented from acting upon the biofilm bacteria. In one study, alginate solutions inhibited disruption of membrane vesicles by cationic antimicrobial peptides, which can spontaneously insert into membranes (Chan et al. 2005). Additionally, incubation of these peptides with alginate led to dissolution of secondary structure and aggregation of the peptides. Alginate has also been shown to afford protection from cationic quaternary ammonium compounds, acting as a hydrophobic shield that decreases activity of these biocides (Campanac et al. 2002). Further, alginate can bind positively charged antibiotics, such as aminoglycosides, and inhibit their activity. Tobramycin, for instance, binds quite well to alginate (Nichols et al. 1988). A recent study found that diffusion of tobramycin through colony biofilms of *P. aeruginosa* was severely delayed; however, antibiotic eventually penetrated through to the distal regions of the biofilm in sufficient concentrations to kill resident microorganisms (Walters et al. 2003).

Anionic extracellular polymeric substances can also bind and sequester toxic cationic heavy metals. Metal chelation has been demonstrated for secreted polymeric molecules from a number of different microorganisms, including *Bacillus licheniformis*, *Xanthomonas campestris*, and freshwater-sediment bacteria (Kaplan et al. 1987; McLean et al. 1990; Mittelman and Geesey 1985; Teitzel and Parsek 2003). Thus, adsorption of positively charged antimicrobial agents by anionic matrix components appears to be an effective survival tool employed by certain bacteria.

2.1.2 Stimulation of EPS by Antibiotics

Intriguingly, a few antibiotics can actually stimulate EPS production. For instance, subinhibitory concentrations of tetracycline, quinupristin-dalfopristin, and erythromycin activated expression of genes encoding for polysaccharide intercellular adhesin in *Staphylococcus epidermidis*, as determined by β-galactosidase transcriptional fusions to the *ica* operon promoter (Rachid et al. 2000). Polysaccharide intercellular adhesin is vital for *S. epidermidis* biofilm formation, and this antibiotic effect corresponded with an increase in biofilm formation on polystyrene microtiter plates. Similarly, subinhibitory concentrations of various β-lactam antibiotics stimulated β-galactosidase expression from a *cps-lacZ* transcriptional fusion in *Escherichia coli* (Sailer et al. 2003). The *cps* genes in *E. coli* encode for enzymes in the production pathway for colanic acid, an EPS important for biofilm formation. In *P. aeruginosa* biofilms, alginate expression was highly upregulated by subinhibitory imipenem treatment (Bagge et al. 2004b). Hoffman et al. have also found that subinhibitory aminoglycoside concentrations enhance biofilm formation in *P. aeruginosa* and *E. coli*, although this effect appears to act by increasing bacterial biomass rather than stimulating matrix formation (Hoffman et al. 2005). These researchers further identified a gene in *P. aeruginosa* (which they named *arr* for aminoglycoside response regulator) that appears to mediate this effect by modulating cyclic-di-GMP levels within the cell. Thus, subinhibitory antibiotic concentrations seem to enhance biofilm formation in certain cases. It is intriguing to speculate that limited antibiotic diffusion through the biofilm matrix, coupled with a corresponding decrease in antimicrobial concentration, might actually stimulate biofilm formation in some instances by creating a positive feedback loop.

2.1.3 Free Diffusion

However, a large number of studies have shown that many antibiotics can freely diffuse through biofilms. In the case of *Klebsiella pneumoniae* colony biofilms, penetration of ampicillin was severely abrogated (Anderl et al. 2000). However, ampicillin could freely diffuse through β-lactamase-deficient *K. pneumoniae* colony biofilms, demonstrating that the matrix per se does not inhibit ampicillin diffusion, but that β-lactamase secreted by the bacteria inactivated the antibiotic (Anderl et al. 2000). Ciprofloxacin also exhibited unrestrained diffusion through these biofilms, and both ciprofloxacin and ampicillin could reach distal surfaces of biofilms and kill bacteria in these locations (Anderl et al. 2000, 2003; Zahller and Stewart 2002). Likewise, ciprofloxacin diffused relatively uninhibited through *P. aeruginosa* colony biofilms (Walters et al. 2003), rifampin easily penetrated *S. epidermidis* colony biofilms (Zheng and Stewart 2002), and tetracycline reached every bacterial cell within flow-cell grown *E. coli* biofilms (Stone et al. 2002). In most of these cases, the edges of the biofilms experienced small reductions in bacterial numbers, but the presence of antibiotic throughout the community did not drastically impact viability. Thus, while decreased penetration and diffusion of antimicrobials through the biofilm matrix might influence biofilm survival in some cases, this mechanism appears to be far from universal. Additional mechanisms must exist to account for increased biofilm antibiotic resistance.

2.2 Limited Growth Potential

While disagreement remains about the efficacy of the biofilm matrix as a diffusion barrier to antibiotics, altered microenvironments within the biofilm clearly play a role in antibiotic protection. Oxygen limitation in particular has been extensively investigated, and numerous studies have revealed the presence of hypoxic zones deep within biofilms. A recent microarray study of *E. coli* biofilms found an upregulation of the *cydAB* and b2997-*hybABC* gene clusters, which are known to be transcribed in oxygen-limiting conditions (Schembri et al. 2003). Similarly, nutrient diffusion through biofilms is restricted. Oxygen and nutrient deprivation consequently result in a decrease in bacterial metabolic activity and cessation of bacterial growth (Donlan and Costerton 2002; Dunne 2002). Indeed, experimental measurements have revealed a severe reduction in bacterial growth rates within biofilms compared to planktonic cultures (Anderl et al. 2003; Borriello et al. 2004). Even in planktonic cultures of *P. aeruginosa* and *K. pneumoniae*, deprivation of oxygen or nutrients, respectively, has resulted in slow growth and antibiotic resistance

(Anderl et al. 2003; Field et al. 2005). Because antibiotics typic
rapidly growing bacteria, slow or nongrowing microorganisms would
from killing (Fig. 1) (Brown et al. 1988).

2.2.1 Oxygen Limitation, Metabolism, and Antibiotic Killing

Several studies have shown a correlation between oxygen limitation, metabolic activity, and protection from antibiotic killing in biofilms. Alkaline phosphatase activity and expression of green fluorescent protein (GFP), as measures of general bacterial protein production, have been used to show restriction of bacterial metabolism to the medium-exposed edge of *P. aeruginosa* biofilms (Borriello et al. 2004; Walters et al. 2003; Xu et al. 1998). In these same studies, oxygen microelectrodes were utilized to analyze the dissolved oxygen at various depths within the biofilm. Intriguingly, oxygen penetration was also restricted to the medium-exposed edge, suggesting that decreased oxygen tension throughout the rest of the biofilm inhibited metabolic activity and, consequently, increased antibiotic resistance (Walters et al. 2003; Xu et al. 1998). Similarly, diffusion of glucose and oxygen was inhibited through intact *K. pneumoniae* biofilms, which corresponded to a decrease in bacterial growth and resistance to ampicillin (Anderl et al. 2003). In both of these cases, antibiotics completely permeated the biofilm, yet the drugs only affected the biofilm edge (Anderl et al. 2003; Walters et al. 2003). Thus, limited metabolic activity within these biofilms, created by oxygen and nutrient gradients, protects the constituent bacteria from antibiotic killing.

2.2.2 Anaerobic Metabolic Pathways

Discussion of the metabolic pathways used during anaerobic growth can shed some light on the genetic mechanisms governing the reduced killing of slow-growing bacteria. *P. aeruginosa*, for instance, can utilize NO_3^- and NO_2^- for anaerobic respiration (Hassett et al. 2002). These processes are carried out by the sequential actions of the *nar*, *nir*, *nor*, and *nos* genetic loci, which reduce the nitrogenous substances to N_2. *P. aeruginosa* tightly regulates these genes in order to prevent buildup of toxic intermediates in the pathway (such as the production of nitric oxide). In fact, altered regulation of these loci in mutants of the quorum sensing gene *rhlR* under anaerobic conditions leads to rapid cell death (Hassett et al. 2002). Consequently, drugs targeting quorum sensing or nitrogen utilization pathways may be efficacious in destroying tenacious biofilms. Intriguingly, treating mature *P. aeruginosa* biofilms under anaerobic conditions with a combination of NO_3^- and either ciprofloxacin or tobramycin significantly enhanced killing of the microorganisms compared to antibiotic treatment alone (Borriello et al. 2006). However, these effects were not apparent in younger biofilms (Borriello et al. 2004). Obviously, the age and metabolic state of the biofilm plays a major role in determining its susceptibility to antibiotic treatment.

2.2.3 Stationary Phase and Stress Response Similarities

The slow growth and altered metabolic activity apparent in biofilms have led some researchers to suggest that the biofilm bacteria are in a stationary-phase state (Anderl et al. 2003). One of the hallmark features of stationary-phase bacteria is the activity of *rpoS*, the stationary-phase sigma factor instrumental in regulating expression of stress response factors. Microarray analysis of *E. coli* biofilms revealed the upregulation of nearly 50% of all *rpoS*-regulated genes (Schembri et al. 2003). In the same study, an *rpoS* mutant failed to form a biofilm. On the other hand, a *P. aeruginosa rpoS* mutant formed a much larger and more antibiotic resistant biofilm than wild type (Whiteley et al. 2001).

Additional studies have further implicated stress response factors as integral components of bacterial biofilms. For instance, microarray analysis of tobramycin-treated wild type *P. aeruginosa* biofilms showed upregulation of the stress response chaperones *groES* and *dnaK* (Whiteley et al. 2001). Studies with *K. pneumoniae* demonstrated expression of catalase in stationary-phase planktonic cells and in biofilms, but not in exponentially growing planktonic bacteria (Anderl et al. 2003). Catalase breaks down hydrogen peroxide and consequently protects expressing microorganisms from destruction. In *P. aeruginosa*, the constitutive catalase gene *katA* and the hydrogen peroxide inducible catalase gene *katB* were found to be important in resistance and adaptation to hydrogen peroxide stress in biofilms (Elkins et al. 1999). Thus, stress responses activated within bacterial biofilms may impact bacterial resistance to biocides and potentially to other antimicrobial agents.

In summary, it is clear that altered metabolism within biofilms promotes the creation of a bacterial subpopulation with altered sensitivity to antibiotics (Fig. 1). By decreasing the growth rate and activating vigorous stress responses, biofilms increase their chances of surviving antimicrobial treatment. In this sense, these metabolic changes represent a vital innate biofilm antibiotic resistance mechanism.

2.3 A Persisting Problem

The phenomenon of persistence was recognized in the mid-1940s in experiments in which cultures of penicillin-sensitive bacteria survived treatment with penicillin. The subpopulation of surviving bacteria has been referred to as persisters. Persister cells have been proposed as an additional innate mechanism for biofilm antibiotic resistance (Lewis 2005). In the persister theory, a small subpopulation of bacteria, whether in biofilms or planktonic culture, differentiates into dormant, spore-like cells that survive after extreme antibiotic treatment (Fig. 1). Differentiation into this dormant state has been hypothesized to be the result of phenotypic variation rather than a stable genetic change (Keren et al. 2004a).

2.3.1 Genetic Factors Influencing Persister Formation

Interestingly, the results of recent studies suggest that, while persisters m phenotypic variants, specific genetic elements are required to form the persister state. Studies by Spoering, Vulíc, and Lewis implicated altered genetic activation of the glycerol-3-phosphate regulated genes *glpD*, *glpABC*, and *plsB* in *E. coli* as a mechanism of persister development (Spoering et al. 2006). The *glpD* gene was initially found to be important for this developmental pathway because plasmid-driven expression of the gene could increase the formation of ampicillin-resistant persisters in the exponential phase by approximately tenfold. Mutating the *glpD* gene or other genes involved in glycerol-3-phosphate metabolism, including *glpABC* or *plsB*, decreased tolerance to ampicillin by greater than 100-fold, indicating a role for glycerol-3-phosphate metabolism in persister formation. However, it was not reported whether these mutations altered the growth rate of the cell or the minimum inhibitory concentration for ampicillin. Further, given glycerol-3-phosphate's central metabolic role, these mutations did not provide any direct mechanistic insight into how persisters might be generated.

One mechanism proposed to explain the ability of persisters to resist the action of antibiotics is similar to a mechanism long hypothesized for biofilm resistance, namely a slowed growth rate. Indeed, persisters exhibit slow or no growth, as observed by microscopy of *E. coli* in a microfluidic device (Balaban et al. 2004). This decreased growth rate may inhibit antimicrobial action, as discussed above in Sect. 2.2. However, persisters can survive even after treatment with ofloxacin, which exerts bactericidal activity against nongrowing microorganisms (Kaldalu et al. 2004; Spoering and Lewis 2001), suggesting that limited growth rate alone cannot account for the increased antibiotic resistance of persisters. Alternatively, global transcriptional profiling by microarray analysis of persister cells revealed activation of numerous stress response pathways (Kaldalu et al. 2004; Keren et al. 2004b), potentially implicating these cells as hardy, stress-resistant microorganisms.

Another major factor influencing formation of persisters appears to be chromosomal toxin/antitoxin (TA) systems (Lewis 2005), which have previously been associated with programmed cell death in bacteria. Several TA modules were upregulated by microarray analysis of persisters in *E. coli*, including *dinJ/yafQ*, *relBE*, and *mazEF* (Keren et al. 2004b; Shah et al. 2006). Overexpression of the *relE* toxin gene, in particular, led to tolerance of high levels of such disparate antibiotics as ofloxacin, cefotaxime, and tobramycin (Keren et al. 2004b). The *hipBA* TA locus has also been found to be important for formation and maintenance of persisters, and mutation of the *hipA* toxin gene can enrich for persisters within in an *E. coli* population (Harrison et al. 2005; Keren et al. 2004a; Keren et al. 2004b; Moyed and Bertrand 1983). It has been suggested that these TA modules actually induce stasis of the bacterial cell by inhibiting the activity of a particular cellular machine, such as the ribosome (Keren et al. 2004b). It was proposed that this inert state then prevents the deleterious functions induced by antibiotics. For instance, an aminoglycoside cannot induce the formation of misfolded proteins if its target ribosome has been rendered static. In this sense, persister bacteria are considered antibiotic-tolerant rather than antibiotic-resistant (Keren et al. 2004b; Lewis 2005). Evidence for this induced stasis comes from studies demonstrating that, while overexpression of the *relE* or *chpAK* toxin genes in *E. coli* rapidly reduced colony-forming units, subsequent transcription of the *relB* or *chpAI* antitoxins, respectively, led to a restoration of colony formation on agar plates (Pedersen et al. 2002). In other words, the toxin-expressing bacteria were nongrowing, yet non-dead, and addition of antitoxin resuscitated these cells. Thus, random fluctuations of toxin and antitoxin levels may modulate the formation and awakening of dormant persisters.

2.3.2 Persister Controversies

Intriguingly, persister research has led to several claims about biofilm antibiotic resistance in opposition to generally accepted biofilm tenets. Specifically disputed is the widely held, and well-supported, hypothesis that biofilms are more resistant to antimicrobial killing than planktonic bacteria. This argument has led some researchers to solely examine planktonic cultures for phenotypic and genotypic analysis of persisters. For instance, Spoering and Lewis concluded from their studies that stationary-phase *P. aeruginosa* was equally or more resistant than biofilm cultures to several antibiotics (Spoering and Lewis 2001). This effect was quantified as greater bacterial CFU after 6 h of antibiotic challenge and was hypothesized to be the result of equal or greater persister formation in the planktonic stationary-phase bacteria compared to the biofilm population. Similarly, Harrison et al. discovered that planktonic and biofilm populations of *E. coli* required similar levels of amikacin and ceftrioxone to effect complete eradication of the population in 2 h. However, in this latter study, *E. coli* biofilms were more resistant to tobramycin than planktonic phase cells at 2 h. Further, increasing the incubation time to 24 h revealed a much greater antibiotic resistance of biofilms to all three antibiotics compared to planktonic cells. In other words, planktonic bacteria were more sensitive to lower antibiotic concentrations when treated for longer periods of time. This result leads one to wonder whether increased antibiotic incubation periods could have produced a similar effect in the work by Spoering and Lewis and similar studies (Brooun et al. 2000; Spoering and Lewis 2001).

An additional concern in these studies is the variance in bacterial numbers between planktonic and biofilm populations at the start of antibiotic treatment. Thus, while Spoering and Lewis found a greater number of surviving stationary planktonic-phase bacteria compared to biofilm bacteria after antibiotic treatment, they also started with a significantly greater number of stationary planktonic phase bacteria then biofilm bacteria. In effect, in the stationary phase cultures, the units of antibiotic per bacterial cell were markedly decreased relative to biofilm bacteria, and this difference might have led to an apparent increase in antibiotic resistance. In a later study of *E. coli* resistance to metal oxyanions, Harrison et al. equalized planktonic and biofilm bacterial numbers before antibiotic challenge and found that this action did not significantly alter the MIC. However, in these conditions for the planktonic bacteria, they reported that "the proportion of surviving cells was smaller than the fraction of survivors recovered from biofilms" (Harrison et al. 2005). As with the increased incubation time mentioned above, it would be intriguing to determine the effect of starting with similar bacterial numbers using the system as described by Spoering and Lewis.

Based on the results of these studies, it may be misleading to consider biofilm antibiotic resistance as a stationary-phase persister phenomenon. Alternatively, perhaps planktonic persisters have differentiated into a more biofilm-like phenotype, although there is no data to support this theory at this time. Recent microarray studies of *E. coli* suggested that the persister transcriptional profile represents a unique physiological state, distinct from exponential phase or stationary-phase bacteria (Shah et al. 2006). While no comparison was made to biofilms, it is intriguing to speculate that the persister phenotype is similar to the biofilm state. Indeed, the most highly expressed gene in persisters compared to nonpersisters in this microarray analysis was *ygiU*, which is also induced in biofilms and acts as a global regulator influencing biofilm formation (Shah et al. 2006). Further, mathematical modeling has predicted a steady accumulation of persisters as a biofilm matures and ages (Roberts and Stewart 2005). Thus, despite inconsistencies in persister literature, persister formation remains an intriguing concept as a supporting mechanism of biofilm antibiotic resistance.

In summary, innate formation of persisters might represent a common mechanism used by a wide range of bacteria during biofilm formation. Creation of this tenacious population within the biofilm may drastically inhibit the complete eradication of the biofilm during even prolonged, high-level antibiotic treatment (Fig. 1). However, at this stage, it is unclear what relationship, if any, can be drawn between planktonic persisters and biofilm resistance, and furthermore, the mechanisms(s) by which persisters form and/or confer increased antibiotic tolerance.

2.4   The Specifics

Many biofilm antibiotic resistance factors cannot appropriately be categorized into any previously described overarching phenomena, such as persister formation or decreased diffusion. Instead, certain biofilm-specific gene products may exert smaller, unique functions that enhance the overall antibiotic resistance of the biofilm. These factors have been the most difficult to uncover.

Perhaps the best example of a biofilm-specific factor is the *ndvB* gene of *P. aeruginosa*, identified in a screen for genes important for tobramycin resistance (Mah et al. 2003). The *ndvB* gene appears to encode for an enzyme involved in the synthesis of cyclic glucans. These glucans bind to tobramycin and prevent bacterial cell death, most likely by sequestering the antibiotic. An isogenic mutant in *ndvB* was much more sensitive to tobramycin than wild type (Mah et al. 2003). Interestingly, this effect was seen only in a biofilm and not when the bacteria were grown planktonically. Further, reverse transcriptase PCR on RNA isolated from type cultures demonstrated that *ndvB* was expressed in a biofilm but not expressed when bacteria were grown planktonically (Mah et al. 2003). Thus, *ndvB* is a factor involved in antibiotic resistance specifically within a biofilm (Fig. 1). Similar studies may reveal biofilm-specific factors in biofilms constructed by other microorganisms.

2.5 A Quick Recap

As discussed above, the process of biofilm formation apparently leads to innate mechanisms of antibiotic resistant bacteria. That is, some mechanisms of resistance appear to be part and parcel of growing in a biofilm. Inhibited diffusion through the matrix, reduced metabolism by nutrient limitation, and formation of dormant persisters all appear to impact the development of a protective environment within the biofilm. Working in combination, these pathways might confer a multilayered network of security for the constituent bacteria. Further exposition of the genetic pathways that lead to these innate phenomena may very well result in improved treatment regimes for disruption and elimination of bacterial biofilms.

3   On Cue: Induced Mechanisms

As with any environmental change, antibiotic treatment can alter regulatory patterns within bacteria. Antibiotic treatment can be a harsh stress, even for bacteria within a biofilm. Consequently, one would predict that there must be some antibiotic-regulated genes that influence antibiotic resistance or sensitivity of biofilm bacteria. As previously mentioned, antibiotics can activate regulatory pathways, leading to a profound effect upon the biofilm matrix and achieved biomass (Bagge et al. 2004b; Hoffman et al. 2005; Rachid et al. 2000; Sailer et al. 2003), and it is likely that numerous genetic loci are activated upon treatment with antibiotics (Fig. 2). These induced factors may work synergistically with innate factors to enhance survival in the face of strong antimicrobial stresses.

Very little work has been done to identify antibiotic-induced factors in biofilms. However, recent microarray analyses have yielded some intriguing clues. Imipenem-treated *P. aeruginosa* biofilms strongly expressed alginate genes and the chromosomal β-galactosidase gene *ampC* (Bagge et al. 2004b). Expression of *ampC* was restricted to the outer edges of microcolonies, as determined by epifluorescence and confocal scanning laser microscopy of an *ampC-GFP* transcriptional fusion (Bagge et al. 2004a). In a separate study, tobramycin treatment of *P. aeruginosa* biofilms resulted in upregulation of *PA1541* and *PA3920*, two possible antibiotic efflux systems (Whiteley et al. 2001). Although no functional data have been generated, both of these studies identified a number of hypothetical genes that were upregulated or repressed upon antibiotic treatment (Bagge et al. 2004b; Whiteley et al. 2001), suggesting that many more factors are potentially involved in biofilm resistance than have currently been identified.

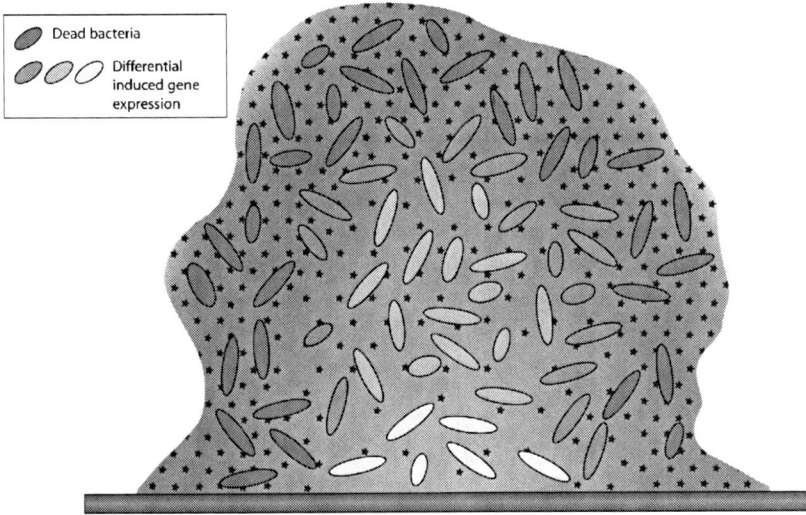

Fig. 2 Induced biofilm antibiotic resistance. Treatment with antibiotics might induce expression of bacterial resistance factors. As in Fig. 1, antibiotic molecules are shown as *dark dots*, and the biofilm bacteria are shown as *multicolored ovals*. Outer bacteria, with limited time to adapt to high antibiotic concentrations, may be rapidly killed (*gray*). Slowed antibiotic diffusion through the microcolony, due to matrix inhibition or other factors, might lead to the establishment of an antimicrobial gradient (*red* to *green* background and decreasing density of stars). This gradient may produce differential gene expression of antibiotic induced factors throughout the biofilm (*orange* to *yellow* bacteria). It is important to note that these induced factors may complement innate biofilm resistance factors, such that biofilm antibiotic resistance is the result of an intricate mixture of innate and induced factors

Efflux pumps and β-lactamases are some of the key mechanisms used by planktonic bacteria to overcome antibiotic challenge. Previous research has generally disregarded these factors and other planktonically associated systems as not important for biofilm antibiotic resistance, and much experimental evidence has supported this view (Patel 2005; Stewart and Costerton 2001). Nevertheless, as mentioned above, β-lactamases and possibly efflux pumps might exert some influence during a biofilm lifestyle (Anderl et al. 2000; Bagge et al. 2004a, 2004b; Whiteley et al. 2001). In reconciling these conflicting observations, it is intriguing to reflect on the numerous hypothetical genes that are differentially regulated during antibiotic exposure. It seems likely that novel orthologs will be discovered that might play a large role in antibiotic resistance specifically within biofilms. For example, *E. coli* alone contains genes for 37 proposed efflux pumps (Pages et al. 2005). It is possible that some of these genes are important for survival in a planktonic state, while others may be biofilm-specific. The putative efflux genes mentioned above, which were discovered by microarray analysis of tobramycin-treated *P. aeruginosa* biofilms (*PA1541* and *PA3920*), may be examples of such biofilm-specific orthologs (Whiteley et al. 2001). It is also possible that redundant function of similar proteins within the same bacterium may have obscured the activity of previously tested gene products. For instance, in another study, a *P. aeruginosa* mutant with deletions in both the *mexAB-oprM* and the *mexCD-oprJ* efflux pumps could not establish biofilms in the presence of azithromycin, while the single mutation constructs of each behaved as wild type (Gillis et al. 2005).

It is clear that much more research is needed to expose additional and/or novel antibiotic-induced factors in biofilms. The multifactorial nature of biofilm antibiotic resistance has hindered identification of these pathways, and much remains to be elucidated about induced factors in biofilm resistance. Discovery of these unknown factors will lead to new and better treatments for biofilm related infections.

4 Disruptive Behavior: Novel Therapeutics

Considering the extremely robust defense mechanisms of biofilms, designing novel therapeutics may seem like a daunting task. However, some have accepted this challenge and in the process have devised some clever and creative solutions.

4.1 Quorum-Sensing Inhibitors

One area of intense interest is the development of inhibitors of bacterial quorum sensing (Rasmussen and Givskov 2006). Quorum-sensing systems are a vital component in community behavior and biofilm formation for a wide range of diverse bacteria, and treatment with quorum sensing inhibitors could lead to a severe abrogation of biofilm formation. Many large screening projects are currently underway to identity such inhibitors. These endeavors have led to the discovery of three types of molecules: those that block production of the quorum-sensing signal, enzymes or other factors that degrade the signal, and signal analogs that disrupt quorum sensing by blocking binding of the true signal, thus preventing activation of the receptor (Rasmussen and Givskov 2006).

Identification of signal analogs has been a particularly productive endeavor. Many eukaryotes, as a microbial defense mechanism, produce secondary metabolites and other compounds that interfere with quorum sensing and other bacterial processes (Steinberg et al. 1997). The marine alga *Delisea pulchra*, for instance, secretes a class of molecules called furanones (Steinberg et al. 1997). Furanones are structurally quite similar to the acylhomoserine lactone class of quorum-sensing signals, and thus disrupt community behavior of bacteria that utilize this class of autoinducers (Rasmussen and Givskov 2006). The effects of furanones on bacteria and biofilms are many and varied. Treatment of *Serratia liquifaciens* cultures with furanone abrogated swarming motility by inhibiting expression of the quorum-sensing regulated gene *swrA*, involved in production of the swarming surfactant serrawettin W2 (Rasmussen et al. 2000). Furanone also inhibited quorum-sensing regulated virulence of *Vibrio harvey* and *P. aeruginosa* (Hentzer et al. 2002; Manefield et al. 2000). Furanone compounds penetrated *P. aeruginosa* microcolonies, affected biofilm architecture, and enhanced bacterial detachment from established biofilms. A furanone derivative could even inhibit the growth, swarming, and biofilm formation of the Gram-positive microorganism *Bacillus subtilis* (Ren et al. 2002; Ren et al. 2004). Thus, by interfering with cell–cell communication, furanones can perturb a number of functions of a wide range of different bacteria. The different effects on these several bacterial species most likely relates to differences in regulatory circuitry activated by quorum sensing in these microorganisms. Still, it is clear that furanone compounds inhibit community behaviors.

Several other inhibitors of bacterial quorum sensing have also been discovered. Screens of *Penicillium* extracts revealed two molecules, patulin and penicillic acid, that inhibited quorum-sensing regulation in *P. aeruginosa* (Rasmussen et al. 2005). Patulin also exhibited efficacy as a treatment for *P. aeruginosa* pulmonary infection in a mouse model. Intriguingly, this study found a synergistic effect on in vitro biofilm clearance when patulin and tobramycin were used in combination (Rasmussen et al. 2005). Synergy has also been observed between RNAIII-inhibiting peptide (RIP) and a number of different antibiotics during clearance of device-related *S. epidermidis* infections in vivo (Balaban et al. 2003a). RIP, a modified version of a heptapeptide isolated from cultures of *Staphylococcus xylosus*, prevented phosphorylation of target of RNAIII activating protein (TRAP), which under normal circumstances would activate the *agr* regulatory system of *Staphylococcus* species (Balaban et al. 2003a, 2003b). This hindrance resulted in decreased adherence and biofilm formation of both *S. aureus* and *S. epidermidis* on a variety of abiotic materials as well as mammalian cells in culture. Taken together, these studies point to a profound effect of natural compounds on bacterial quorum sensing. Especially considering antibiotic synergy, quorum-sensing inhibitor molecules have shown great potential for treatment of bacterial biofilms.

4.2 Non-Quorum-Sensing Inhibitors

Additional antibiofilm molecules have been discovered that appear to affect bacterial mechanisms other than quorum sensing. Another molecule that interferes with *S. aureus* biofilm formation is farnesol, produced by *Candida albicans* (Jabra-Rizk et al. 2006). Farnesol compromised membrane integrity of *Staphylococcus aureus* biofilm bacteria and acted synergistically in reducing the minimum inhibitory concentration of gentamicin for both methicillin-sensitive and methicillin-resistant *S. aureus*. In a separate study, Ren et al. screened thousands of natural plant extracts and discovered that ursolic acid disrupts biofilms formed by *E. coli*, *P. aeruginosa*, and *V. harvey* (Ren et al. 2005). It was demonstrated that quorum sensing was not involved in this effect. While the exact mechanism of inhibition remained elusive, microarray profiling implicated motility, heat shock, cysteine synthesis, and sulfur metabolism as affected by ursolic acid treatment. Finally, subinhibitory concentrations of the macrolide antibiotic clarithromycin inhibited twitching motility of *P. aeruginosa* (Wozniak and Keyser 2004). While macrolides have generally not exhibited activity against *Pseudomonas*, clarithromycin treatment altered *P. aeruginosa* biofilm architecture, raising the possibility of utilizing macrolides in combination with other antibiotics for biofilm eradication.

4.3   Mechanical Means

On the basis of combating biofilm antibiotic resistance by enhanced or more efficient delivery of antimicrobial agents, much research has been focused on engineering better materials and methods for treatment of biofilms (Smith 2005). For instance, electrical, ultrasound, and photodynamic stimulation can disrupt biofilms and enhance the efficiency of certain antimicrobial agents. Further, coating surfaces with antimicrobial agents has shown efficacy in preventing biofilm formation. Similarly, implantable, biodegradable matrices, scaffolds, microparticles, and gels can release high concentrations of drugs in a controlled fashion in vivo. Liposomes have also been used to enhance the concentration and targeting of antimicrobials to biofilms. This strategy has been useful for delivery of drugs to intracellular pathogens as well. Aerosolization of antibiotics has been shown to be quite effective for direct application of these drugs to the respiratory system. In particular, aerosolized tobramycin, and more recently nebulized hypertonic saline, have achieved clinical efficacy in treating *P. aeruginosa* lung infection in patients with cystic fibrosis (Donaldson et al. 2006; Elkins et al. 2006; Gibson et al. 2003). In this manner, higher concentrations of drug can be delivered directly to the site of infection.

Treatment strategies for biofilms are constantly evolving. The synergy between natural compounds and traditional antibiotics seems quite promising for future clinical applications. Coupled with improved delivery mechanisms, these molecules may prove to be a boon to the medical field. Indeed, much progress has already been achieved, as seen with aerosolized delivery of tobramycin. While much research is still needed, novel treatments and biofilm inhibitory molecules are constantly being identified. These potential therapies offer much hope for the future of combating biofilm infections.

5   Beneficial Biofilms

In some instances, formation of highly resistant biofilms has proven to be an advantage. Particularly in industrial settings, chemically resistant bacterial biofilms provide a hardy platform for a number of applications involving high concentrations of toxic metals or other chemicals (Morikawa 2006). Studies have revealed the utility of biofilms in the synthesis of ethanol, poly-3-hydroxybutyrate, benzaldehyde, and other chemicals (Kunduru and Pometto 1996; Li et al. 2006; Zhang et al. 2004). Biofilms have also assisted wastewater treatment, phenol bioremediation, biodegradation of 2,4- and 2,6-dinitrophenol, and bioremediation of toxic metal contamination of environmental sites (Lendenmann et al. 1998; Luke and Burton 2001; Nicolella et al. 2000; Singh and Cameotra 2004). These applications highlight the usefulness of extremely resistant biofilms in chemical synthesis and breakdown. In these cases, biofilms can be used as a tool for beneficial purposes.

6 Conclusion

The enigma of extreme biofilm resistance has puzzled researchers since the beginnings of biofilm research. Conventional antibiotic resistance mechanisms do not seem to influence biofilm survival, and dispersion of the biofilm bacteria leads to reversion to an antibiotic-sensitive state. These results have led to the identification of several intriguing resistance models, either resulting from an innate property of a biofilm lifestyle (Fig. 1) or an effect induced by the antimicrobial stress itself (Fig. 2). It is tempting to speculate that any one of these models alone (such as persister formation) can fully explain biofilm resistance. However, as we currently understand them, none of these phenomena can adequately account for every aspect of the biofilm resistance phenotype. Further, these models share common features and themes, such as decreased metabolic activity seen in anaerobic microcolony environments and with persisters. Also intriguing is the possibility that the biofilm matrix might slow the progress of an antibiotic through the microcolony such that the bacteria can sufficiently activate expression of protective factors in response to the biocide.

The overlap between these resistance paradigms has led some researchers to propose a layered model of biofilm resistance, wherein the outer layers of the microcolonies provide a first-line defense by inhibiting the diffusion of antimicrobial agents, bacteria deeper within the biofilm can be further protected by altered metabolic states, and development of persisters throughout enhances bacterial survival (Stewart and Costerton 2001). Throughout, innately expressed as well as antibiotic-induced genes might provide additional protection. In short, biofilm antibiotic resistance results from an overlapping mixture of innate and induced microbial activities, intricately woven together with redundant form and function.

Our understanding of the molecular details of resistance mechanisms of bacterial biofilms is still in its infancy. While many genetic factors have been identified, many more questions remain. Even for many known resistance genes, it is uncertain how they interact with each other to establish a resistance phenotype. Much greater knowledge of genetic responses to antimicrobial agents will facilitate the production of new and better drugs to eradicate biofilms. Manipulation of regulatory and expression networks holds much promise for future treatment of biofilm infections. Indeed, quorum-sensing inhibitors have demonstrated an exceptional ability to disrupt biofilm structure and act synergistically with a number of antibiotics.

On the other hand, enhancing and nurturing the impervious nature of beneficial biofilms may lead to improvement, for example, in the production of biologically derived chemicals and bioremediation. In either case, elucidation of the genetic mechanisms of innate and induced biofilm resistance holds the key to solving this great mystery.

References

Anderl JN, Franklin MJ, Stewart PS (2000) Role of antibiotic penetration limitation in *Klebsiella pneumoniae* biofilm resistance to ampicillin and ciprofloxacin. Antimicrob Agents Chemother 44:1818–1824

Anderl JN, Zahller J, Roe F, Stewart PS (2003) Role of nutrient limitation and stationary-phase existence in *Klebsiella pneumoniae* biofilm resistance to ampicillin and ciprofloxacin. Antimicrob Agents Chemother 47:1251–1256

Bagge N, Hentzer M, Andersen JB, Ciofu O, Givskov M, Hoiby N (2004a) Dynamics and spatial distribution of beta-lactamase expression in *Pseudomonas aeruginosa* biofilms. Antimicrob Agents Chemother 48:1168–1174

Bagge N, Schuster M, Hentzer M, Ciofu O, Givskov M, Greenberg EP, Hoiby N (2004b) *Pseudomonas aeruginosa* biofilms exposed to imipenem exhibit changes in global gene expression and beta-lactamase and alginate production. Antimicrob Agents Chemother 48:1175–1187

Balaban N, Giacometti A, Cirioni O, Gov Y, Ghiselli R, Mocchegiani F, Viticchi C, Del Prete MS, Saba V, Scalise G, Dell'Acqua G (2003a) Use of the quorum-sensing inhibitor RNAIII-inhibiting peptide to prevent biofilm formation in vivo by drug-resistant *Staphylococcus epidermidis*. J Infect Dis 187:625–630

Balaban N, Gov Y, Bitler A, Boelaert JR (2003b) Prevention of *Staphylococcus aureus* biofilm on dialysis catheters and adherence to human cells. Kidney Int 63:340–345

Balaban NQ, Merrin J, Chait R, Kowalik L, Leibler S (2004) Bacterial persistence as a phenotypic switch. Science 305:1622–1625

Borriello G, Werner E, Roe F, Kim AM, Ehrlich GD, Stewart PS (2004) Oxygen limitation contributes to antibiotic tolerance of *Pseudomonas aeruginosa* in biofilms. Antimicrob Agents Chemother 48:2659–2664

Borriello G, Richards L, Ehrlich GD, Stewart PS (2006) Arginine or nitrate enhances antibiotic susceptibility of *Pseudomonas aeruginosa* in biofilms. Antimicrob Agents Chemother 50:382–384

Brooun A, Liu S, Lewis K (2000) A dose-response study of antibiotic resistance in *Pseudomonas aeruginosa* biofilms. Antimicrob Agents Chemother 44:640–646

Brown MR, Allison DG, Gilbert P (1988) Resistance of bacterial biofilms to antibiotics: a growth-rate related effect? J Antimicrob Chemother 22:777–780

Campanac C, Pineau L, Payard A, Baziard-Mouysset G, Roques C (2002) Interactions between biocide cationic agents and bacterial biofilms. Antimicrob Agents Chemother 46:1469–1474

Chan C, Burrows LL, Deber CM (2005) Alginate as an auxiliary bacterial membrane: binding of membrane-active peptides by polysaccharides. J Pept Res 65:343–351

Chernish RN, Aaron SD (2003) Approach to resistant Gram-negative bacterial pulmonary infections in patients with cystic fibrosis. Curr Opin Pulm Med 9:509–515

Costerton JW, Stewart PS, Greenberg EP (1999) Bacterial biofilms: a common cause of persistent infections. Science 284:1318–1322

Donaldson SH, Bennett WD, Zeman KL, Knowles MR, Tarran R, Boucher RC (2006) Mucus clearance and lung function in cystic fibrosis with hypertonic saline. N Engl J Med 354:241–250

Donlan RM, Costerton JW (2002) Biofilms: survival mechanisms of clinically relevant microorganisms. Clin Microbiol Rev 15:167–193

Dunne WM Jr (2002) Bacterial adhesion: seen any good biofilms lately? Clin Microbiol Rev 15:155–166

Elkins JG, Hassett DJ, Stewart PS, Schweizer HP, McDermott TR (1999) Protective role of catalase in *Pseudomonas aeruginosa* biofilm resistance to hydrogen peroxide. Appl Environ Microbiol 65:4594–4600

Elkins MR, Robinson M, Rose BR, Harbour C, Moriarty CP, Marks GB, Belousova EG, Xuan W, Bye PT (2006) A controlled trial of long-term inhaled hypertonic saline in patients with cystic fibrosis. N Engl J Med 354:229–240

Field TR, White A, Elborn JS, Tunney MM (2005) Effect of oxygen limitation on the in vitro antimicrobial susceptibility of clinical isolates of *Pseudomonas aeruginosa* grown planktonically and as biofilms. Eur J Clin Microbiol Infect Dis 24:677–687

Gibson RL, Burns JL, Ramsey BW (2003) Pathophysiology and management of pulmonary infections in cystic fibrosis. Am J Respir Crit Care Med 168:918–951

Gillis RJ, White KG, Choi KH, Wagner VE, Schweizer HP, Iglewski BH (2005) Molecular basis of azithromycin-resistant *Pseudomonas aeruginosa* biofilms. Antimicrob Agents Chemother 49:3858–3867

Harrison JJ, Ceri H, Roper NJ, Badry EA, Sproule KM, Turner RJ (2005) Persister cells mediate tolerance to metal oxyanions in *Escherichia coli*. Microbiology 151:3181–3195

Hassett DJ, Cuppoletti J, Trapnell B, Lymar SV, Rowe JJ, Yoon SS, Hilliard GM, Parvatiyar K, Kamani MC, Wozniak DJ, Hwang SH, McDermott TR, Ochsner UA (2002) Anaerobic metabolism and quorum sensing by *Pseudomonas aeruginosa* biofilms in chronically infected cystic fibrosis airways: rethinking antibiotic treatment strategies and drug targets. Adv Drug Deliv Rev 54:1425–1443

Hentzer M, Riedel K, Rasmussen TB, Heydorn A, Andersen JB, Parsek MR, Rice SA, Eberl L, Molin S, Hoiby N, Kjelleberg S, Givskov M (2002) Inhibition of quorum sensing in *Pseudomonas aeruginosa* biofilm bacteria by a halogenated furanone compound. Microbiology 148:87–102

Hoffman LR, D'Argenio DA, MacCoss MJ, Zhang Z, Jones RA, Miller SI (2005) Aminoglycoside antibiotics induce bacterial biofilm formation. Nature 436:1171–1175

Hoiby N, Frederiksen B, Pressler T (2005) Eradication of early *Pseudomonas aeruginosa* infection. J Cyst Fibros 4 [Suppl 2]:49–54

Jabra-Rizk MA, Meiller TF, James CE, Shirtliff ME (2006) Effect of farnesol on *Staphylococcus aureus* biofilm formation and antimicrobial susceptibility. Antimicrob Agents Chemother 50:1463–1469

Kaldalu N, Mei R, Lewis K (2004) Killing by ampicillin and ofloxacin induces overlapping changes in *Escherichia coli* transcription profile. Antimicrob Agents Chemother 48:890–896

Kaplan D, Christiaen D, Arad SM (1987) Chelating properties of extracellular polysaccharides from *Chlorella* spp. Appl Environ Microbiol 53:2953–2956

Keren I, Kaldalu N, Spoering A, Wang Y, Lewis K (2004a) Persister cells and tolerance to antimicrobials. FEMS Microbiol Lett 230:13–18

Keren I, Shah D, Spoering A, Kaldalu N, Lewis K (2004b) Specialized persister cells and the mechanism of multidrug tolerance in *Escherichia coli*. J Bacteriol 186:8172–8180

Kunduru MR, Pometto AL 3rd (1996) Continuous ethanol production by *Zymomonas mobilis* and *Saccharomyces cerevisiae* in biofilm reactors. J Ind Microbiol 16:249–256

Lendenmann U, Spain JC, Smets BF (1998) Simultaneous biodegradation of 2,4-dinitrotoluene and 2,6-dinitrotoluene in an aerobic fluidized-bed biofilm reactor. Environ Sci Technol 32:82–87

Lewis K (2005) Persister cells and the riddle of biofilm survival. Biochemistry (Mosc) 70:267–274

Li XZ, Webb JS, Kjelleberg S, Rosche B (2006) Enhanced benzaldehyde tolerance in *Zymomonas mobilis* biofilms and the potential of biofilm applications in fine-chemical production. Appl Environ Microbiol 72:1639–1644

Luke AK, Burton SG (2001) A novel application for *Neurospora crassa*: Progress from batch culture to a membrane bioreactor for the bioremediation of phenols. Enzyme Microb Technol 29:348–356

Mah TF, O'Toole GA (2001) Mechanisms of biofilm resistance to antimicrobial agents. Trends Microbiol 9:34–39

Mah TF, Pitts B, Pellock B, Walker GC, Stewart PS, O'Toole GA (2003) A genetic basis for *Pseudomonas aeruginosa* biofilm antibiotic resistance. Nature 426:306–310

Manefield M, Harris L, Rice SA, de Nys R, Kjelleberg S (2000) Inhibition of luminescence and virulence in the black tiger prawn (*Penaeus monodon*) pathogen *Vibrio harveyi* by intercellular signal antagonists. Appl Environ Microbiol 66:2079–2084

McLean RJ, Beauchemin D, Clapham L, Beveridge TJ (1990) Metal-binding characteristics of the gamma-glutamyl capsular polymer of *Bacillus licheniformis* ATCC 9945. Appl Environ Microbiol 56:3671–3677

Mittelman MW, Geesey GG (1985) Copper-binding characteristics of exopolymers from a freshwater-sediment bacterium. Appl Environ Microbiol 49:846–851

Morikawa M (2006) Beneficial biofilm formation by industrial bacteria *Bacillus subtilis* and related species. J Biosci Bioeng 101:1–8

Moyed HS, Bertrand KP (1983) hipA, a newly recognized gene of *Escherichia coli* K-12 that affects frequency of persistence after inhibition of murein synthesis. J Bacteriol 155:768–775

Nichols WW, Dorrington SM, Slack MP, Walmsley HL (1988) Inhibition of tobramycin diffusion by binding to alginate. Antimicrob Agents Chemother 32:518–523

Nicolella C, van Loosdrecht MC, Heijnen JJ (2000) Wastewater treatment with particulate biofilm reactors. J Biotechnol 80:1–33

Pages JM, Masi M, Barbe J (2005) Inhibitors of efflux pumps in Gram-negative bacteria. Trends Mol Med 11:382–389

Patel R (2005) Biofilms and antimicrobial resistance. Clin Orthop Relat Res:41–47

Pedersen K, Christensen SK, Gerdes K (2002) Rapid induction and reversal of a bacteriostatic condition by controlled expression of toxins and antitoxins. Mol Microbiol 45:501–510

Rachid S, Ohlsen K, Witte W, Hacker J, Ziebuhr W (2000) Effect of subinhibitory antibiotic concentrations on polysaccharide intercellular adhesin expression in biofilm-forming *Staphylococcus epidermidis*. Antimicrob Agents Chemother 44:3357–3363

Rasmussen TB, Manefield M, Andersen JB, Eberl L, Anthoni U, Christophersen C, Steinberg P, Kjelleberg S, Givskov M (2000) How *Delisea pulchra* furanones affect quorum sensing and swarming motility in *Serratia liquefaciens* MG1. Microbiology 146:3237–3244

Rasmussen TB, Skindersoe ME, Bjarnsholt T, Phipps RK, Christensen KB, Jensen PO, Andersen JB, Koch B, Larsen TO, Hentzer M, Eberl L, Hoiby N, Givskov M (2005) Identity and effects of quorum-sensing inhibitors produced by *Penicillium* species. Microbiology 151:1325–1340

Rasmussen TB, Givskov M (2006) Quorum-sensing inhibitors as anti-pathogenic drugs. Int J Med Microbiol 296:149–161

Ren D, Sims JJ, Wood TK (2002) Inhibition of biofilm formation and swarming of *Bacillus subtilis* by (5Z)-4-bromo-5-(bromomethylene)-3-butyl-2(5H)-furanone. Lett Appl Microbiol 34:293–299

Ren D, Bedzyk LA, Setlow P, England DF, Kjelleberg S, Thomas SM, Ye RW, Wood TK (2004) Differential gene expression to investigate the effect of (5Z)-4-bromo- 5-(bromomethylene)-3-butyl-2(5H)-furanone on *Bacillus subtilis*. Appl Environ Microbiol 70:4941–4949

Ren D, Zuo R, Gonzalez Barrios AF, Bedzyk LA, Eldridge GR, Pasmore ME, Wood TK (2005) Differential gene expression for investigation of *Escherichia coli* biofilm inhibition by plant extract ursolic acid. Appl Environ Microbiol 71:4022–4034

Roberts ME, Stewart PS (2005) Modelling protection from antimicrobial agents in biofilms through the formation of persister cells. Microbiology 151:75–80

Sailer FC, Meberg BM, Young KD (2003) beta-Lactam induction of colanic acid gene expression in *Escherichia coli*. FEMS Microbiol Lett 226:245–249

Schembri MA, Kjaergaard K, Klemm P (2003) Global gene expression in *Escherichia coli* biofilms. Mol Microbiol 48:253–267

Shah D, Zhang Z, Khodursky A, Kaldalu N, Kurg K, Lewis K (2006) Persisters: a distinct physiological state of *E. coli*. BMC Microbiol 6:53–61

Singh P, Cameotra SS (2004) Enhancement of metal bioremediation by use of microbial surfactants. Biochem Biophys Res Commun 319:291–297

Smith AW (2005) Biofilms and antibiotic therapy: is there a role for combating bacterial resistance by the use of novel drug delivery systems? Adv Drug Deliv Rev 57:1539–1550

Spoering AL, Lewis K (2001) Biofilms and planktonic cells of *Pseudomonas aeruginosa* have similar resistance to killing by antimicrobials. J Bacteriol 183:6746–6751

Spoering AL, Vulic M, Lewis K (2006) GlpD and PlsB participate in persister cell formation in *Escherichia coli*. J Bacteriol 188:5136–5144

Steinberg PD, Schneider R, Kjelleberg S (1997) Chemical defenses of seaweeds against microbial colonization. Biodegradation 8:211–220

Stewart PS, Costerton JW (2001) Antibiotic resistance of bacteria in biofilms. Lancet 358:135–138

Stone G, Wood P, Dixon L, Keyhan M, Matin A (2002) Tetracycline rapidly reaches all the constituent cells of uropathogenic *Escherichia coli* biofilms. Antimicrob Agents Chemother 46:2458–2461

Szomolay B, Klapper I, Dockery J, Stewart PS (2005) Adaptive responses to antimicrobial agents in biofilms. Environ Microbiol 7:1186–1191

Teitzel GM, Parsek MR (2003) Heavy metal resistance of biofilm and planktonic *Pseudomonas aeruginosa*. Appl Environ Microbiol 69:2313–2320

Walters MC 3rd, Roe F, Bugnicourt A, Franklin MJ, Stewart PS (2003) Contributions of antibiotic penetration, oxygen limitation, and low metabolic activity to tolerance of Pseudomonas aeruginosa biofilms to ciprofloxacin and tobramycin. Antimicrob Agents Chemother 47:317–323

Whiteley M, Bangera MG, Bumgarner RE, Parsek MR, Teitzel GM, Lory S, Greenberg EP (2001) Gene expression in *Pseudomonas aeruginosa* biofilms. Nature 413:860–864

Wozniak DJ, Keyser R (2004) Effects of subinhibitory concentrations of macrolide antibiotics on *Pseudomonas aeruginosa*. Chest 125:62S-69S; quiz 69S

Xu KD, Stewart PS, Xia F, Huang CT, McFeters GA (1998) Spatial physiological heterogeneity in *Pseudomonas aeruginosa* biofilm is determined by oxygen availability. Appl Environ Microbiol 64:4035–4039

Zahller J, Stewart PS (2002) Transmission electron microscopic study of antibiotic action on *Klebsiella pneumoniae* biofilm. Antimicrob Agents Chemother 46:2679–2683

Zhang S, Norrlow O, Wawrzynczyk J, Dey ES (2004) Poly(3-hydroxybutyrate) biosynthesis in the biofilm of *Alcaligenes eutrophus*, using glucose enzymatically released from pulp fiber sludge. Appl Environ Microbiol 70:6776–6782

Zheng Z, Stewart PS (2002) Penetration of rifampin through *Staphylococcus epidermidis* biofilms. Antimicrob Agents Chemother 46:900–903

Multidrug Tolerance of Biofilms and Persister Cells

K. Lewis

Abstract Bacterial populations produce a small number of dormant persister cells that exhibit multidrug tolerance. All resistance mechanisms do essentially the same thing: prevent the antibiotic from hitting a target. By contrast, tolerance apparently works by shutting down the targets. Bactericidal antibiotics kill bacteria by corrupting their targets, rather than merely inhibiting them. Shutting down the targets then protects from killing. The number of persisters in a growing population of bacteria rises at mid-log and reaches a maximum of approximately 1% at stationary state. Similarly, slow-growing biofilms produce substantial numbers of persisters. The ability of a biofilm to limit the access of the immune system components, and the ability of persisters to sustain an antibiotic attack could then account for the recalcitrance of such infections in vivo and for their relapsing nature. Isolation of *Escherichia coli* persisters by lysing a population or by sorting GFP-expressing cells with diminished translation allowed to obtain a gene expression profile.

K. Lewis
Antimicrobial Discovery Center and Department of Biology, Northeastern University,
360 Huntington Avenue, Boston, MA, 02459 USA
k.lewis@neu.edu

T. Romeo (ed.), *Bacterial Biofilms.*
Current Topics in Microbiology and Immunology 322.
© Springer-Verlag Berlin Heidelberg 2008

The profile indicated downregulated biosynthetic pathways, consistent with their dormant nature, and indicated overexpression of toxin/antitoxin (TA) modules. Stochastic overexpression of toxins that inhibit essential functions such as translation may contribute to persister formation. Ectopic expression of RelE, MazF, and HipA toxins produced multidrug tolerant cells. Apart from TA modules, *glpD* and *plsB* were identified as potential persister genes by overexpression cloning of a genomic library and selection for antibiotic tolerance. Yeast *Candida albicans* forms recalcitrant biofilm infections that are tolerant to antibiotics, similarly to bacterial biofilms. *C. albicans* biofilms produce multidrug tolerant persisters that are not mutants, but rather phenotypic variants of the wild type. Unlike bacterial persisters, however, *C. albicans* persisters were only observed in a biofilm, but not in a planktonic stationary population. Identification of persister genes opens the way to a rational design of anti-biofilm therapy. Combination of a conventional antibiotic with a compound inhibiting persister formation or maintenance may produce an effective therapeutic. Other approaches to the problem include sterile-surface materials, prodrug antibiotics, and cyclical application of conventional antimicrobials.

1 Biofilms and Persisters

According to the CDC, 65% of all infections in developed countries are caused by biofilms, bacterial communities that settle on a surface and are covered by an exopolymer matrix (Hall-Stoodley et al. 2004). These include common diseases such as childhood middle ear infection and gingivitis; infections of all known indwelling devices such as catheters, orthopedic prostheses, and heart valves; and the incurable disease of cystic fibrosis. Biofilms are produced by most if not all pathogens. *Pseudomonas aeruginosa*, causing an incurable infection in cystic fibrosis patients (Singh et al. 2000), and *Staphylococcus aureus* and *Staphylococcus epidermidis*, infecting indwelling devices (Mack et al. 2004), are probably the best-known biofilm-producing organisms. Biofilm infections are highly recalcitrant to antibiotic treatment. However, planktonic cells derived from these biofilms are in most cases fully susceptible to antibiotics. Importantly, biofilms do not actually grow in the presence of elevated levels of antibiotics, meaning they do not exhibit increased resistance as compared to planktonic cells (Lewis 2001b). But if biofilms are not resistant, how do they resist being killed? Biofilm resistance to killing has been one of the more elusive problems in microbiology, but the analysis of a simple dose-response experiment provided an unexpected insight into the puzzle (Brooun et al. 2000; Lewis 2001b; Spoering and Lewis 2001).

Most of the cells in a biofilm are actually highly susceptible to a bactericidal agent such as a fluoroquinolone antibiotic or metal oxyanions that can kill both rapidly dividing and slow-growing or nongrowing cells (Spoering and Lewis 2001; Harrison et al. 2005a, 2005b) (Fig. 1). This is important, since cells in the biofilm are slow growing, and many are probably in stationary state. The experiment also revealed a small subpopulation of cells that remain alive irrespective of the

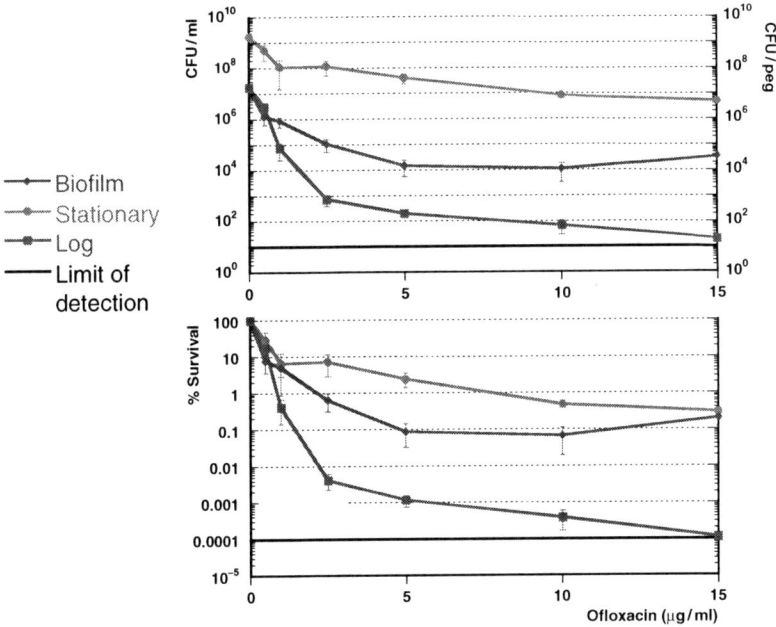

Fig. 1 Killing of logarithmic-phase, stationary-phase, and biofilm cultures of *P. aeruginosa* by ofloxacin. *Upper panel*, the concentration of cells for planktonic cultures is given in the left y axis, and the amount of cells released from a peg carrying a biofilm is given in the right y axis. The estimate of the concentration of cells in the biofilm is 10^{11}/ml of biofilm. *Lower panel*, results of the upper panel are recalculated as % survival. The limit of detection is indicated by the solid horizontal line

concentration of the antibiotic. Note that the cell concentration in log, stationary, and biofilm cultures is obviously different. Taking into account the thickness of the biofilm on the peg (*), its dimensions and the total number of cells per peg, we estimate the density of cells in the biofilm to be 10^{11}, substantially higher than in the planktonic cultures. In each case, a persister plateau is produced, meaning that starting at a certain level, further increase in the antibiotic concentration does not result in additional killing. This obviates such possible problems as diminishing the concentration of the available antibiotic by cell binding.

The level of these surviving persisters was even greater in the nongrowing stationary population. In a test tube, a stationary culture appears more tolerant than the biofilm. However, this situation is likely reversed in vivo. Antibiotic treatment will eliminate the bulk of both biofilm and planktonic cells, leaving intact persisters. At this point, the similarity with an in vitro experiment probably ends. The immune system will be able to mop up remaining planktonic persisters, just as it eliminates nongrowing cells of a population treated with a bacteriostatic antibiotic (Fig. 2). However, the biofilm matrix protects against immune cells (Leid et al. 2002; Jesaitis et al. 2003; Vuong et al. 2004), and its persisters will survive. After

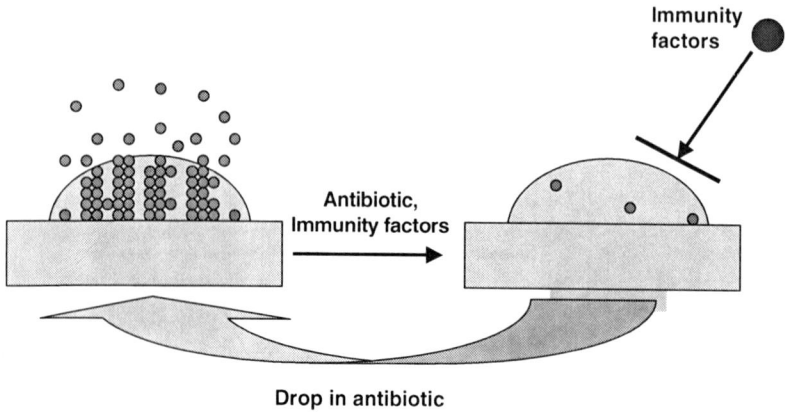

Drop in antibiotic

Fig. 2 Model of biofilm resistance based on persister survival. An initial treatment with antibiotic kills planktonic cells and the majority of biofilm cells. The immune system kills planktonic persisters, but the biofilm persister cells are protected from host defenses by the exopolysaccharide matrix. After the antibiotic concentration drops, persisters resurrect the biofilm and the infection relapses

antibiotic concentration drops, persisters will repopulate the biofilm, which will shed off new planktonic cells, producing the relapsing biofilm infection (Lewis 2001b). The problem of biofilm resistance to "everything" largely defaults to understanding persisters.

Persisters were described by Joseph Bigger in 1944 in one of the first studies of penicillin action (Bigger 1944). Bigger discovered that penicillin lysed a growing population of *Staphylococcus*, but plating this transparent solution on nutrient medium produced surviving colonies. In order to test whether these were mutants, the colonies were grown, treated with penicillin, and the new population again produced a small number of persisters surviving lysis. This experiment was repeated recently with *Escherichia coli* and several different antibiotics and produced similar results (Keren et al. 2004a; Wiuff et al. 2005). By the time the mechanism of biofilm resistance to killing was being investigated, Bigger's work was all but forgotten, a curiosity known to few microbiologists. Harris Moyed picked up the problem in the 1980s and undertook a targeted search for persister genes (Moyed and Bertrand 1983; Moyed and Broderick 1986; Scherrer and Moyed 1988; Black et al. 1991, 1994). He reasoned that treating a population of *E. coli* with ampicillin would select for mutants with increased production of persisters. After ampicillin application, cells were allowed to recover, and the enrichment process was repeated. This is different from the conventional selection for resistant mutants, where cells that can grow in the presence of antibiotic are favored. After testing for mutants that had the same MIC to ampicillin (thus not resistant), but survived better, several strains were obtained, and one of them was used to map the mutation to a *hipBA* locus. The mutant appeared to carry a mutation in the *hipA* gene, and this *hipA7* allelic strain was found to make 1% persisters in exponential cultures, about 1,000 times more than the wild type. Deletion of *hipBA* had no apparent effect on persister formation, suggesting that *hipA7* mutant (Korch et al. 2003) produced a pleiotropic artifact. Another possibility is that *hipA* is part of a redundant set of genes, and knocking out any single one does not produce a phenotype. Like Bigger's work before him, the studies of Moyed were largely forgotten.

The finding of persisters in biofilms rekindled an interest in this curiosity, which appears to be responsible for a major part of recalcitrant human infectious diseases. We will summarize here what we currently know about the biology of persisters. But first, let us consider the difference between resistance of regular cells and drug tolerance of persisters. This will provide a useful framework for the subsequent discussion of persisters and their properties.

1.1 Multidrug Resistance and Multidrug Tolerance, Two Mechanistically Distinct Menaces

Numerous mechanisms of drug resistance have been described, and in most cases we have a fairly good understanding of these processes at the molecular level. The main types of resistance are target modification by mutation; target modification by specialized enzymatic changes; target substitution, such as expressing an alternative target; antibiotic modification; antibiotic efflux; and restricted antibiotic permeation (Lewis et al. 2002; Levy and Marshall 2004). It is interesting to note that all theoretically logical possibilities of antibiotic resistance seem to have been realized in nature. Importantly, all of these mechanisms do essentially the same thing: prevent the antibiotic from binding to the target (Fig. 3). Each of these resistance mechanisms allows cells to grow at an elevated level of antibiotic. It is important to note that bactericidal antibiotics kill the cell not by merely inhibiting

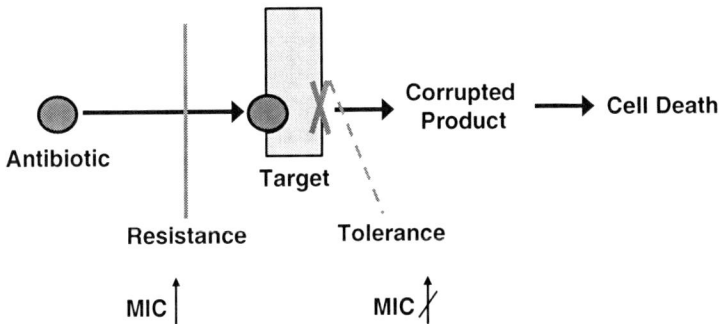

Fig. 3 Antibiotic resistance vs tolerance. Resistance mechanisms prevent an antibiotic from binding to the target, which leads to an increase in the MIC. Bactericidal antibiotics act by corrupting the target, producing a toxic product that kills the cell. Tolerance occurs when a persister protein blocks the target, preventing formation of a toxic product

the target, but by corrupting its function in a manner that creates a toxic product. Aminoglycoside antibiotics kill the cell by interrupting translation, which creates misfolded toxic peptides (Davis et al. 1986). β-Lactam antibiotics such as penicillin inhibit peptidoglycan synthesis, which activates, by a largely unknown mechanism, autolysin enzymes present in the cell wall (Bayles 2000). This leads to digestion of the peptidoglycan by autolysins and cell death. Fluoroquinolones inhibit the ligase step of the DNA gyrase and topoisomerase, without affecting the preceding nicking activity. As a result, the enzyme converts into an endonuclease (Hooper 2002).

We think that tolerance acts by a preventative blocking of the antibiotic targets (Fig. 3). If persisters are dormant and have little cell wall synthesis, translation or topoisomerase activity, then the antibiotics will bind to their targets, but will be unable to corrupt them. In this manner, tolerance does provide resistance to killing from everything, but at a price of nonproliferation.

The simplest way to form a dormant persister could be through fluctuations in the levels of a potentially large number of proteins that upon overproduction become toxic to the cell and stop growth. This possibility has been elegantly demonstrated by expressing in *E. coli* proteins that are known to inhibit growth when overproduced (Vazquez-Laslop et al. 2006). Cells overexpressing the chaperone DinJ or the *Salmonella typhimurium* PmrC, an enzyme that transfers phosphoethanolamine to lipid A, stopped growing and became highly tolerant to ampicillin and ciprofloxacin. However, it does not seem that this simple mechanism is primarily responsible for making persisters in the wild type, at least not in *E. coli*. Examination of the rate of *E. coli* persister formation over time showed very few of them in early exponential state, followed by a sharp rise in mid-exponential, and reaching about 1% in nongrowing stationary state. In order to learn whether persisters in early exponential were leftovers from stationary state, or formed de novo, the culture was kept at this stage by repeated reinoculation (Keren et al. 2004a). After four reinoculations from early exponential to early exponential, persisters completely disappeared. This simple experiment essentially rules out nonspecific mechanisms of persister formation. Indeed, mistakes are unavoidable and should happen at early exponential state as well. This experiment also indicates that persisters are preformed, rather than being produced in response to antibiotics.

Persister rise at mid-exponential state has been observed in several species, but its nature remains unknown. Quorum sensing does not seem to play a role, since addition of spent medium to early exponential cultures of *E. coli* or *P. aeruginosa* did not appreciably increase the persister level (K. Lewis, unpublished data). Whatever the mechanism of the rise, the dynamics of persister formation provide an interesting insight into the strategy of persistence. Persisters are cells that temporarily forfeit propagation in favor of survival. Their strategy is distinct from the well-studied stress responses, when the entire population expresses resistance proteins (such as heat shock or SOS) in response to a nonlethal dose of a deleterious factor. Persisters are able to survive a dose of antibiotic that kills regular cells. However, persisters only become prominent in a fairly dense cell population. This strongly suggests that persisters are essentially altruistic cells, which ensure survival of a kin population in the presence of a lethal factor. In early exponential phase, there are essentially no kin (no means for a cell to distinguish between few vs no neighbors), and no one to benefit from an act of altruism. The highest level of persisters reached in a stationary population suggests that their main function is actually ensuring survival of this nongrowing population. But since persisters are also nongrowing, why does the entire stationary population not enter into this protected state? The benefit of being a regular cell apparently stems from the ability to rapidly resume growth, which may be more problematic for a dormant persister. Since the majority of cells in an *E. coli* or *P. aeruginosa* stationary population are regular cells, this suggests that the optimal individual strategy is not to enter into persistence, again suggesting that the persister state is an altruistic behavior benefiting the kin.

The stationary state experiment also provides an important distinction between merely not growing and persistence. Some antibiotics, like β-lactams and aminoglycosides, strongly depend on cell growth for their action, and a nongrowing stationary population is indeed tolerant to these compounds. Fluoroquinolones and mitomycin C, for example, can kill nongrowing cells, and treatment with these compounds reveals a small subpopulation of tolerant persisters in the stationary state.

1.2   In Search of the Mechanism of Persister Formation

Identification of the mechanism of persister formation presents a formidable challenge due to an apparent redundancy of persister genes. Thus, attempts to identify persister genes by screening transposon insertion libraries for either increased or decreased survival to antibiotics were not successful (Hu and Coates 2005; Spoering et al. 2006). In a recent report, phoU was identified as a putative persister gene using a similar approach. However, the *phoU* mutant had a decreased MIC to a number of antibiotics. This suggests that *phoU*::Tn is a pleiotropic mutation. It is interesting to compare this experience with identifying genes controlling another function that produces dormant cells: sporulation. It is easy to obtain specific *spo* mutants specifically lacking the ability to make spores from a knockout library (*), and this is indeed how most *spo* genes were identified. Genes controlling tolerance resemble in this regard those coding for multidrug resistance pumps (MDRs). In *P. aeruginosa*, for example, there are at least 13 RND family MDRs, but knockout out of most of them does not produce a discernible phenotype (*). At the same time, overexpressing any single MDR produces multidrug resistance. We therefore reasoned that persister genes may be identified by screening or selecting a library cloned into an expression vector for gain of function. In this case, even a weak contributor to a multigene function can be identified when overexpressed. However, this approach is problematic as well, since overproduction of many proteins leads to misfolded toxic products that can stop cell growth and will create an artifact emulating a dormant state, as discussed above. It appears that standard approaches of molecular genetics are poorly suited to search for persister genes, which probably explains the slow pace of discovery in this area.

Another barrier to discovery has been a lack of approaches to isolate persister cells. The first method to isolate persisters was recently reported, based on simply sedimenting surviving cells from a culture lysed by ampicillin (Keren et al. 2004b). This method has its limitations: it requires a rapidly growing culture for ampicillin to lyse it, and the fraction of persisters in such a population is small, approximately 10^{-5}. In *E. coli*, this necessitated the use of a *hipA7* strain overproducing persisters. In addition, these persisters are exposed to an antibiotic. These limitations notwithstanding, enough cells were collected to obtain a gene expression profile. The profile showed downregulation of proteins involved in energy production and nonessential functions such as flagellar synthesis, suggesting that persisters are dormant cells. This is consistent with the finding that persisters formed by a *hipA7* (high persistence) strain of *E. coli* are nongrowing (or slow growing) cells (Balaban et al. 2004). The profile also pointed to proteins that may be responsible for dormancy: RMF, a stationary state inhibitor of translation (Yoshida et al. 2002), SulA, an inhibitor of septation (Walker 1996), and toxin-antitoxin (TA) module elements RelBE, DinJ, and MazEF (Christensen and Gerdes 2003; Christensen et al. 2003). Homologs of TA modules are found on plasmids where they constitute a maintenance mechanism (Hayes 2003). Typically, the toxin is a protein that inhibits an important cellular function such as translation or replication, and forms an inactive complex with the antitoxin. The toxin is stable, while the antitoxin is degradable. If a daughter cell does not receive a plasmid after segregation, the antitoxin level decreases due to proteolysis, leaving a toxin that either kills the cell or inhibits propagation. TA modules are also commonly found on bacterial chromosomes, but their role is largely unknown. MazEF was proposed to serve as a programmed cell death mechanism (Sat et al. 2001). However, it was reported recently that MazF and an unrelated toxin RelE do not actually kill cells, but induce stasis by inhibiting translation, a condition that can be reversed by expression of corresponding antitoxins (Pedersen et al. 2002; Christensen et al. 2003).

Expression of RelE, a toxin that causes reversible stasis by inhibiting cleaving mRNA and inhibiting translation, strongly increased tolerance to antibiotics (Keren et al. 2004b). Expression of a toxin HipA increased tolerance as well (Falla and Chopra 1998; Correia et al. 2006; Korch and Hill 2006; Vazquez-Laslop et al. 2006). Interestingly, a bioinformatics analysis indicates that HipA is a member of the Tor family of kinases, which have been extensively studied in eukaryotes (Schmelzle and Hall 2000), but have not been previously identified in bacteria. HipA is indeed a kinase, it autophosphorylates on ser150, and site-directed mutagenesis replacing it, or other conserved amino acids in the catalytic and Mg^{2+}-binding sites abolishes its ability to stop cell growth and confer drug tolerance (Correia et al. 2006). Knowing that HipA is a kinase provides an additional tool to search for the target, which is yet to be identified.

Deletion of potential candidates of persister genes noted above does not produce a discernible phenotype affecting persister production, possibly due to the high degree of redundancy of these elements. In *E. coli*, there are at least ten toxin-antitoxin (TA) modules, and more than 60 in *Mycobacterium tuberculosis* (Gerdes et al. 2005).

Several independent lines of evidence point to persister dormancy: lack of growth in the presence of antibiotics (by contrast to resistant mutants), downregulation of biosynthetic pathways, and an elegant demonstration of slow growth or no growth in persisters formed by the *E. coli hipA7* strain (Balaban et al. 2004). In the latter study, cells were placed in troughs of a multichannel chip that restricts mobility and makes it possible to simultaneously videotape growth and division of many individual cells in the channels. The device also made it possible to flush the medium, and application of ampicillin caused lysis of cells. However, cells that did not lyse were those that had little growth preceding the application of ampicillin.

Based on these data, we reasoned that dormancy may be used to physically sort naïve persister cells from a wild type population (Shah et al. 2006). Dormancy implies low levels of translation, which can then enable differential sorting based on expression of a detectable protein. In *E. coli* ASV, a degradable GFP is inserted into the chromosome in the λ attachment site and expressed from the ribosomal *rrnB*P1 promoter, the activity of which is proportional to the rate of cell growth (Fig. 4). The half-life of degradable GFP is less than 1 h, and it should be effectively cleared from dormant cells. This would then enable sorting of dim persister cells. A logarithmically growing population of *E. coli* ASV was sorted with a high-speed cell-sorter using forward light scatter, which detects particles based on size. This enabled detection of cells irrespective of their level of fluorescence. Sorting by fluorescence showed that the population consisted of two strikingly different types of cells: a bright majority and a small subpopulation of cells with no detectable fluorescence (Fig. 4). Fluorescent microscopy confirmed that the sorted bright cells were indeed bright green, while the dim ones had no detectable fluorescence. The dim cells were also smaller than the fluorescent cells, and in this regard resembled stationary state cells. Sorted dim cells were exposed to a high level of ofloxacin that rapidly kills both growing and nongrowing normal cells, but has no effect on persisters. The majority of this subpopulation survived, as compared to a drastic drop in viability of the sorted bright cells. This experiment showed that the sorted dim cells are dormant persisters.

The sorting method provides a general approach to obtaining naïve persisters from a wild type population of any species. These cells can then be used to obtain an expression profile and to study a variety of functions by biochemical methods. However, there is still room for improvement: sorting is relatively slow, costly, and results in a limited amount of material that precludes proteome analysis, for example. A rapid method for obtaining large quantities of persisters has yet to be developed.

Knowing that persisters are dormant cells bolsters the case for TA module involvement in persister formation. Indeed, TAs seem to be ideally suited for the task. Reversible action of toxins such as RelE and MazF, inhibition of important cellular functions by toxins capable of creating a dormant state, and the presence of TA modules in the chromosomes of all known free-living bacteria makes them attractive candidates for persister genes.

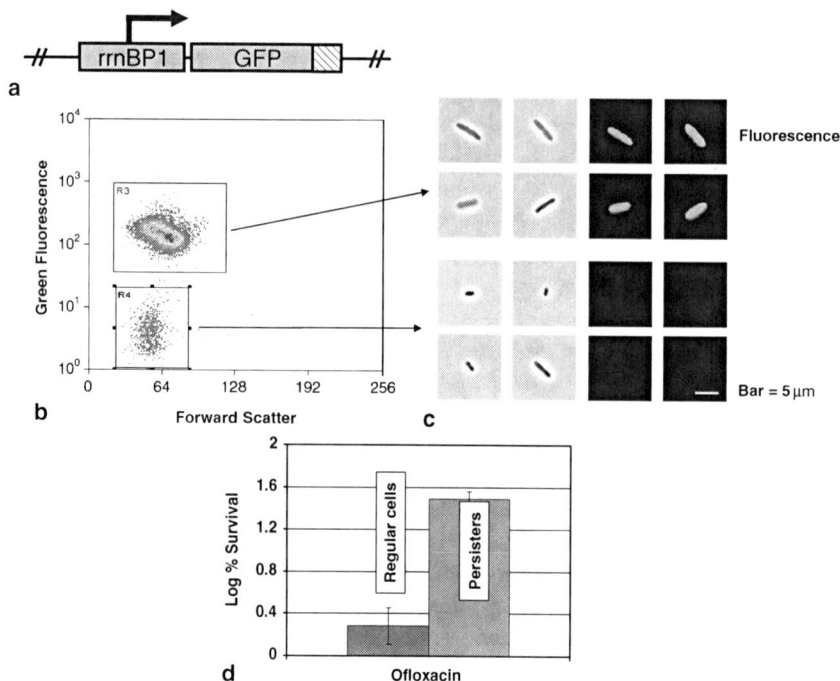

Fig. 4 Sorting of *E. coli* persister cells from a growing population. **a** Graphical representation of the reporter. An unstable variant of GFP was placed downstream of a ribosomal promoter, *rrn*BP1. **b** *E. coli* ASV cells containing the degradable rrnBP1GFP reporter cassette were grown in LB medium to mid-exponential phase (~1×10⁸ cells/ml) at 37°C with aeration and sorted. Two populations were detected using forward light-scatter, one that fluoresced brightly (R3), and another that did not (R4). **c** The sorted populations were visualized by epifluorescent microscopy (bar, 5 μm). **d** Cells were sorted as described in (**a-c**). Once sorted, both populations were treated with ofloxacin (5 μg/ml) for 3 h, diluted and spotted onto LB agar plates for colony counts. Cells carrying a degradable GFP under the control of a ribosomal promoter whose activity is proportional to the rate of growth were sorted, and the dim and bright cells were then exposed to an antibiotic to test their tolerance

1.3 A Generalized Hunt for Persister Genes

Expression cloning could in principle reveal even minor participants that contribute to persister formation. The problem with this approach, however, is that overexpression of many proteins causes nonspecific toxicity that stops cell growth and appears as natural dormancy. In order to make this approach work, we decided to use mild overexpression by cloning an *E. coli* library into a low-copy vector, using native promoters for expression, and introducing a growth step between rounds of selection by a bactericidal antibiotic (Spoering et al. 2006). In this manner, we could select against any cells that grew considerably slower than the wild type. This procedure enriched the population in cells with elevated persister production. A clone with consistent persister overproduction was sequenced and appeared to carry the *glpD* gene coding for glycerol-3-phosphate dehydrogenase. A knockout of *glpD* caused a modest decrease in persister levels in a stationary (but not exponential) culture. Interestingly, there is a second glycerol-3-phosphate dehydrogenase in *E. coli*, GlpA and a double *glpD,glpA* knockout had a considerably stronger phenotype as compared to *glpD*. This explains why we (and others) missed *glpD* in a screen of a knockout library for persister genes, and reinforces the idea of persisters being controlled by functionally redundant genes.

The opposing behavior of an overexpression vs a knockout confirms participation of GlpD in persister formation. The mechanism by which GlpD affects persistence is currently under investigation. Testing mutants affected in proteins within the GlpD metabolic network indicated another interesting component that affects persistence: PlsB, a well-conserved bacterial protein. The PlsB enzyme uses G3P to produce 1-acyl-G3P and is essential, preventing straightforward analysis of a null mutant. However, a strain with a mutant PlsB enzyme with a higher Km has been described (*plsB26*; Heath and Rock 1999), and was used to examine persister formation. The *plsB26* strain grew normally and showed the same MIC to ampicillin and ofloxacin as an isogenic wild type control. Production of persisters by the *plsB26* strain in a stationary culture was 100-1,000 times lower than the wild type, suggesting that PlsB may be a useful target for anti-persister therapy. PlsB is an abundant housekeeping protein, and its level did not change in the transcriptomes of purified persisters. It seems that PlsB, rather than being involved in persister formation, is a persister maintenance gene. It appears that in order to survive, persisters depend on the ability to maintain their membrane integrity, which requires PlsB-dependent phospholipid synthesis.

The application of expression cloning seems promising and suggests that it will yield additional persister genes and should work for any bacterial species.

The multitude of proteins that can induce multidrug tolerance is reminiscent of the many MDR pumps responsible for multidrug resistance. It appears that microbial populations have evolved two complementary and highly redundant strategies to protect themselves from antimicrobials: multidrug efflux, and when this fails, multidrug tolerance of persister cells.

1.4   Persisters and Stochastic Phenomena

Persisters make up a small subpopulation, and in mid-log phase *E. coli* produces as little as 10^{-5} surviving tolerant cells. Given that all of the cells in a population are genetically identical kin, it seems that persisters have to be produced by a stochastic process (Lewis 2000; Keren et al. 2004b; Lewis et al. 2005). Indeed, what would be the alternative? Fluctuations in the levels of a small number of dedicated proteins are probably responsible for persister formation. The absence of persisters in an early exponential population could then be due to low overall levels of persister proteins at this stage. Note that persister proteins HipA and GlpD only have an effect in stationary state, suggesting that their levels do not undergo large fluctuations during exponential growth. Indeed, both HipA and GlpD are capable of producing persisters in an exponential population upon artificial overexpression, and the lack of a phenotype for a deletion mutant at this growth phase strongly suggests that they are not reaching high enough levels in rapidly growing cells. It seems that two processes control persister formation: a stochastic fluctuation in the level of persister proteins and a controlled, regulated mean level of expression of these proteins, which is dependent on the density of the population. The fluctuation happening around an elevated base will then allow persister proteins to reach a level of expression sufficient to produce persisters in a dense population.

Stochastic phenomena have been described for a large number of functions, both in bacteria (Avery 2006; Dubnau and Losick 2006) and in eukaryotes (Kaern et al. 2005). In bacteria, stochastic processes are responsible for determining the part of the population that enters sporulation in *Bacillus subtilis* (Chung et al. 1994), spontaneous SOS induction in *E. coli* (McCool et al. 2004), or cannibalism in *B. subtilis* (Gonzalez-Pastor et al. 2003). The most visible case of bacterial decision making is chemotaxis, which relies on a trial-and-error random walk (Berg and Brown 1972). Counterclockwise rotation of a bacterial flagellum produces a bundle that propels the cell forward, but periodically a switch changes it to counterclockwise, the bundle falls apart, and the cell tumbles before the next run resumes in a new direction. The probability of a run is increased if there is a temporal increase in the concentration of an attractant. Apart from this stochastic noise within a given cell, individual cells vary dramatically in their run/tumble probability ratio (Spudich and Koshland 1976). Apparently, random changes in the expression of chemotaxis proteins produce another level of stochastic variation of individuality in a population. Two particular proteins, CheZ and CheY, of the many elements involved in the chemotaxis signal transduction appeared to be responsible for this noise generation (Korobkova et al. 2004).

The reason bacteria use stochastic processes in decision making is probably similar to our flipping a coin when a problem does not have a unique logical solution. The bacterial cell is too small to have a significant measurable difference between concentrations of a chemoeffector along its length, and trial-end-error runs solve the problem - a temporal increase in the concentration of an attractant increases the duration of the run. And when a population needs to produce a small number of specialized survivor cells, fluctuations in the level of persister proteins will induce a dormant state.

There is another level of noise in persister formation that we do not currently understand. This is the distinct variation in persister levels among populations grown under seemingly identical conditions in parallel test tubes or wells of a microtiter plate ((Wiuff et al. 2005); K. Lewis, unpublished data). It seems premature to speculate about the source of this surprising variation, but its adaptive value may be significant; just as variability of cells within a population increases the chances of kin survival, variability of persister levels among populations will be similarly adaptive.

1.5   Is Multidrug Tolerance Transmissible?

Drug resistance is highly transmissible. Virtually every mechanism of resistance we know of can be present on a plasmid or a transposon. Transmissibility enables migration of resistance genes from unknown soil microorganisms into human pathogens, which is largely responsible for the increasing failure of once effective antibiotic therapies. One may suggest by simple analogy that multidrug tolerance may be transmissible as well. However, MDT mechanisms are universally present in bacteria, since all tested species make persisters, and there does not seem to be a need for transmission. At the same time, several observations point to possible transmission of drug tolerance.

As we have noted in a previous section, TA modules were first discovered as a plasmid maintenance mechanism (Gerdes et al. 1986). The toxins may be grouped into two categories: those that rapidly kill the daughter cell that did not receive a plasmid (membrane-acting proteins that cause leaks) and those that cause reversible stasis, such as translational inhibitors RelE and MazF (Gerdes et al. 2005). Interestingly, the killer toxins improve plasmid maintenance 100- to 1,000-fold, while stasis toxins improve plasmid maintenance by only four- to tenfold. The obvious question is why has an ineffective maintenance mechanism been recruited, while an effective one is available and may easily be borrowed from another plasmid by horizontal transmission. It may be that the stasis toxins justify their presence by increasing the host's multidrug tolerance.

An interesting case of possible transmissible MDT is the presence of a HipBA homolog on an R391 transposon-like IncJ element found in Gram-negative bacteria (Boltner et al. 2002). The element adds an additional copy of the specialized MDT gene *hipA*, which may increase the level of persister production.

Finally, there is an unusual case of transmissible resistance that appears more like tolerance. The plasmid-coded Qnr proteins confer relatively low (but clinically significant) resistance to fluoroquinolones in Gram-negative species (Tran and Jacoby 2002; Vetting et al. 2006). The crystal structure of a Qnr homolog, the chromosomally coded MfpA of *M. tuberculosis* has been reported. Remarkably, MfpA is a structural DNA mimic that binds DNA gyrase, the target of fluoroquinolones (Hegde et al. 2005). The pentapeptide repeat in the Qnr family proteins produces a helical structure peculiarly resembling the double helix of DNA. MfpA binds to and sequesters the gyrase, inhibiting the enzyme and preventing the gyrase/fluoroquinolone complex from nicking the DNA. It would be interesting to see how much tolerance Qnr and MfpA add to the cell: how many more persisters will survive treatment with a fluoroquinolone?

1.6   Persisters in Yeast Biofilms

Eukaryotic yeasts have a lifestyle that is very similar to that of prokaryotic micro-organisms. Not surprisingly, analogous adaptations evolved in these two groups in response to similar environmental challenges through convergent evolution. For example, yeasts form biofilms that, similarly to bacterial biofilms, are responsible for highly recalcitrant infections (Kumamoto and Vinces 2005).

The focus of yeast biofilm research has been on *C. albicans*, an important human pathogen that causes oral thrush, relapsing vaginosis, and is a leading cause of morbidity and mortality in immunocompromised individuals. The biofilm forms when single cells attach to a surface and grow into microcolonies, which then merge and produce a complex 3D structure that is held together by hyphae and an exopolymer matrix (Chandra et al. 2001). The biofilm contains a mixture of yeast, hyphae, and pseudohyphae. Similarly to bacteria, yeast biofilm exopolymer matrix restricts penetration of immune system components (Hoyle et al. 1990; von Eiff et al. 1999), but does not appreciably hinder diffusion of antifungal drugs (Baillie and Douglas 2000; Samaranayake et al. 2005).

Genes encoding multidrug resistance (MDR) pumps MDR1, CDR1, and CDR2 are upregulated upon attachment of *C. albicans* cells to a surface, and this accounts for the resistance of young biofilms to azole antibiotics (Mukherjee et al. 2003). However, the high level of drug resistance of mature biofilms (≥48 h) was not affected by deletion of all three of these genes either singly or in combination, including an *mdr1Δ cdr1Δ cdr2Δ* triple mutant (Ramage et al. 2002; Mukherjee et al. 2003; Kumamoto 2005). Decreased ergosterol content (Mukherjee et al. 2003; Kumamoto 2005) and a diminished level of ergosterol biosynthetic gene expression (Garcia-Sanchez et al. 2004) have been reported in mature *Candida* biofilms and may contribute to drug resistance. Indeed, azoles act by inhibiting ergosterol biosynthesis, and amphotericin B binds to ergosterol. However, ergosterol is unlikely to be involved in the action of echinocandins that inhibit the synthesis of cell wall B-glucan (Datry and Bart-Delabesse 2006), or chlorhexidine, a membrane-active antiseptic that is very effective against bacteria that lack sterols. In essence, the mechanism of *C. albicans* biofilm antifungal resistance remains largely unknown.

We recently examined biofilm resistance of *C. albicans*, following the same approaches used previously for bacteria. A dose-dependent experiment with two highly microbicidal agents, amphotericin B and chlorhexidine (the only compounds that can kill nongrowing yeast cells) showed complete elimination of cells in an exponential and stationary planktonic populations. However, a distinctly biphasic killing was observed in a mature biofilm, indicating the presence of persisters (LaFleur et al. 2006). Similarly to bacteria, yeast persisters are not mutants: upon reinoculation, surviving cells reproduced the original wild type population with a new fraction of persister cells. Staining with fluorescein, which specifically binds to dead yeast cells, showed live persisters within a yeast biofilm treated with amphotericin. These rare live cells were either yeast or pseudohyphae and were morphologically unremarkable. Sorting this stained population showed that dim cells form colonies, while bright ones do not. This method, similarly to the approach described for bacteria, opens the way for obtaining a gene expression profile of yeast persisters.

Quite unexpectedly, *C. albicans* persisters were only apparent in a biofilm culture, and not in a nongrowing stationary population. Both biofilm and planktonic stationary cultures produce a mix of cell types, including yeast, pseudohyphae, and hyphae. The specific production of persisters in a biofilm is distinctly different from what is observed in bacteria, where a stationary planktonic culture actually makes more persisters than a biofilm (Spoering and Lewis 2001). This probably suggests that the biofilm, and not the planktonic population is the survival mode of yeast life, and that is where specialized survivor cells are produced.

Dependence of persister production on biofilm formation suggested that these two forms of yeast populations may share part of the same developmental program. A number of genes involved in yeast biofilm formation have been identified, and mutants deleted in these elements were tested for persister production (LaFleur et al. 2006). Surprisingly, all tested mutants appeared to produce normal levels of persisters. Among the tested mutants was the flo8 strain, which does not make hyphae, and its biofilm consists of a simple layer of yeast cells attached to the surface (Cao et al. 2006). This indicates that attachment is sufficient for persister formation.

A surface contact-dependent Mck1p kinase that affects biofilm formation and invasiveness in *C. albicans* was recently described (Kumamoto 2005), but a strain deleted in mck1 was able to produce normal levels of persisters as well. Our analysis of biofilm mutants suggests that known genes are not involved in persister formation. Persister isolation based on cell sorting opens the possibility of obtaining their transcription profile, which is likely to point to persister genes.

2   In Search of Therapy: Persister Eradication

Given the prominent role of tolerance in infectious disease, the need for compounds that could eradicate persisters is obvious. Another important factor to consider is the potential causality between tolerance and the rise of resistance. A lengthy, lingering infection that is not eradicated due to its tolerance is likely to provide a fertile ground for producing resistant mutants, or for acquisition of resistance determinants through horizontal transmission from other species. A mathematical model predicts that tolerance substantially increases the danger of resistance (Levin and Rozen 2006). This observation provides an additional incentive for developing compounds that are able to sterilize, rather than merely suppress an infection.

Before we describe possible approaches to eradicating persisters, let us put the problem into perspective and consider the state of affairs in the general field of anti-infective drug discovery. This has been described as a crisis due to an innovation gap (Walsh 2003b). Forty years separate the discovery of the last major class, fluoroquinolones, from the recently introduced linezolid, a narrow-spectrum protein synthesis inhibitor acting against Gram-positive bacteria (Zurenko et al. 1997). During this time, pathogens did not take a break, but rapidly acquired resistance to all known antibiotics. There have been several recent successes in addition to linezolid, which include daptomycin (Cubicin, Cubist), a natural product lipopeptide, ramoplanin, a natural product depsipeptide (Genome Therapeutics, currently Oscient), and dalbavancin (Vicuron, currently Pfizer), a molecule related to vancomycin, all acting against Gram-positive species. The recently approved tigecycline (Tigacyl, Wyeth) is a fairly broad-spectrum compound loosely based on tetracycline that is not subject to existing resistances toward the parent antibiotic.

The crisis nonetheless looms, caused by a conspicuous lack of novel classes of broad-spectrum antibiotics in the face of growing drug resistance (Dougherty et al. 2002; Boggs and Miller 2004; Bush 2004; Clardy and Walsh 2004; Projan and Shlaes 2004; Schmid 2004; Silver 2006). The reasons for the crisis are well understood. Bacteria and fungi were the source of natural products, including broad-spectrum compounds, that fueled the golden era of antibiotic discovery during the 1940s and 1950s; Schatz et al. 1944; Demain and Fang 2000; Walsh 2003a). However, culturable microorganisms on which anti-infective drug discovery has been based make up only 0.01%-1% of the total diversity in the environment (Osburne et al. 2000), and this limited resource became overmined.

Synthetic approaches were unable to replace natural products in developing new anti-infective therapies. While it is certainly possible to find synthetic inhibitors against defined bacterial protein targets by screening compound libraries in vitro, attempts to convert such hits into broad-spectrum leads using medicinal chemistry have not been successful. The main obstacle to development of broad-spectrum antibiotics is the penetration barrier of the cell envelope of Gram-negative bacteria. The barrier is multilayered and consists of an inner membrane that restricts the passage of hydrophilic compounds, an outer membrane that restricts penetration of amphipathic compounds (which essentially all drugs are), and MDR pumps that efflux amphipathic substances across the outer membrane. In short, the envelope is designed in a way to restrict penetration of *all* molecules, while nutrients enter through porins and specialized transporters (Lewis and Lomovskaya 2002; Li and Nikaido 2004). We do not have a rational approach to impart a drug candidate with an ability to cross this barrier, and trial-and-error synthesis does not produce permeating compounds with a measurable degree of probability. As a result, most companies have focused on development of narrow-spectrum antibiotics, or on compounds that are modifications of existing classes. We find ourselves increasingly vulnerable to multidrug-resistant Gram-negative pathogens (Meyer 2005).

Some encouraging developments in solving the antibiotic crisis are worth mentioning. Developing MDR inhibitors is an attractive strategy to produce broad-spectrum dual-compound therapies (Markham and Neyfakh 1996; Hsieh et al. 1998; Markham et al. 1999). Plants use a dual antimicrobial/MDR inhibitor combination to combat their microbial pathogens (Stermitz et al. 2000a). For example, berberine, a weak antimicrobial, is strongly potentiated by 5′-methoxyhydnocarpin, an MDR inhibitor that has no activity on its own. However, presently described plant MDR inhibitors are limited to Gram-positive bacteria, which may explain why there are so few plant pathogens among this group of microorganisms. A synthetic MDR inhibitor active against RND MDRs of Gram-negative species is being developed by Mpex Pharmaceuticals and has entered clinical trials (Lomovskaya and Bostian 2006). NovoBiotic Pharmaceuticals is taking a different approach, based on the assumption that additional classes of broad-spectrum compounds are harbored by the 99% of unculturable microorganisms. Unculturable microorganisms are grown in their natural environment in a diffusion chamber (Kaeberlein et al. 2002) and are then used as a source of secondary metabolites for drug discovery.

An anti-persister drug based on traditional approaches could be produced by combining a conventional antibiotic, such as a fluoroquinolone, and an inhibitor of an essential persister protein. However, it is not clear that inhibition of a single persister protein will be sufficient to have a desired effect. As mentioned above, a mutant in an essential PlsB protein causes *E. coli* persisters to drop 1-2 log (Spoering et al. 2006), but whether this will be clinically sufficient is not clear.

The requirements for an anti-persister compound are similar to a broad-spectrum antibiotic: it has to penetrate well into cells of both Gram-positive and Gram-negative species. From the above analysis, we can see that the probability of this happening with traditional discovery approaches is 1 drug or less per 40 years. Unlike a conventional antibiotic, however, anti-persister therapy will face an additional hurdle. FDA only requires testing against rapidly growing bacteria, and market conditions are excellent for a new conventional broad-spectrum antibiotic. Why then commit resources to a considerably more challenging anti-persister dual therapy, even if a suitable target will be identified?

In trying to combat persisters, we may have encountered the ultimate adversary. Indeed, persisters evolved through billions of years to perform a single task: survival. During this time, they have encountered all possible harmful compounds, and the inability of any currently employed antibiotics to eliminate persisters provides a sobering view of the magnitude of the challenge. A general recipe in dealing with persisters is then to devise approaches that are not borrowed from and probably do not exist in nature.

2.1   Pulse-Dosing with a Conventional Antibiotic

A disarmingly simple approach to sterilize an infection was first proposed by Bigger in 1944 (Bigger 1944). The idea is to kill all regular cells with a high dose of an antibiotic, then allow the antibiotic levels to drop, which will cause remaining persisters to wake up and convert into regular cells. If a second antibiotic application is administered just after persisters start to grow, an essentially complete sterilization may be achieved. This approach works very well in a test tube, and a *P. aeruginosa* biofilm can be essentially sterilized with two consecutive applications of a fluoroquinolone (K. Lewis, unpublished data). Pulse-dosing probably does not exist in nature, leaving persisters unprepared to handle this simple regimen.

Perhaps understandably, this approach has not been received with enthusiasm by specialists in clinical microbiology. The goal of established therapies is to maintain the plasma level of an antibiotic at its maximum, which will discourage resistance development. Most importantly, an optimal pulse-dosing regimen will probably vary from patient to patient. However, it seems that patients may have taken solving the problem of intractable persistent infections into their own hands. Individuals who suffer from persistent infections that require a lengthy therapy often do get cured, but why a year-long regimen is better than a week-long one is unclear. An accidental perfect oscillating dosing may very well be responsible for persister eradication in such cases. The patients probably experiment with the dosing by being absent-minded, which sooner or later produces the perfect administration regimen. Curing of persistent infections may therefore result from patient noncompliance. Analyzing how persistent infections are cured may shed some useful light on the likelihood of developing a rational regimen for sterilizing pulse-dosing.

2.1.1 Sterile-Surface Materials

Antiseptics can kill persister cells, but they are obviously toxic and largely unsuitable for systemic applications. Polymeric materials impregnated with antimicrobials have been developed and introduced both as consumer goods and to prevent biofilm formation on catheters. Leaching of the antimicrobial leads to obvious problems: loss of efficacy over time, toxicity in the case of antiseptics, and creation of excellent conditions for developing resistance to a slowly releasing antibiotic.

The ideal approach would be to create a permanently sterile material by covalently attaching an antimicrobial compound to the surface. There is an obvious problem with this approach: once attached to the surface, an antimicrobial molecule loses its mobility and is unable to attack the pathogen. We reasoned that this problem may be solved by linking the antimicrobial to a long, flexible polymeric chain anchored covalently to the surface of a material (Tiller et al. 2001). Quaternary ammonium compounds (QACs) seemed a good choice for the antimicrobial moiety because their target is primarily the microbial membrane (Denyer and Hugo 1991) and they accumulate in the cell driven by the membrane potential (Severina et al. 2001). Attaching a long chain of poly(4-vinyl-N-alkylpyridinium bromide) to an amino glass slide produced a material that remained largely sterile (Tiller et al. 2001). Conceptually similar immobilized polymers poly [2-(dimethylamino)ethyl methacrylate] (Lee et al. 2004) and N-alkyl-polyethylenimine rapidly depolarized and killed *S. aureus* or *E. coli* cells coming in contact with the surface, with no indication of surviving persisters (Milovic et al. 2005). Importantly, in order to be effective, the sterile-surface polymers have to be long enough to penetrate across the cell envelope (Morgan et al. 2000; Lin et al. 2003), while their shorter versions were ineffective (Lin et al. 2003).

The only known mechanism of resistance to hydrophobic cations is efflux by MDRs (Hsieh et al. 1998). The pathogens are forced to actively accumulate these compounds by electrophoresis across the charged membrane, making these substances especially dangerous (Severina et al. 2001). MDRs probably evolved to counter this threat, and then broadened their spectrum to include other amphipathic compounds as well (Lewis 2001a). However, MDRs evolved to extrude small molecules, not large polymers. Indeed, the activity of a surface modified with *N*-hexyl-PVP was similar against a panel of *S. aureus* strains that consisted of a mutant deleted in the major NorA MDR, a wild type, and a wild-type strain carrying additionally a QacA pump on a natural transmissible plasmid (Lin et al. 2002). Similarly, the soluble analog poly(vinyl-*N*-methylpyridinium iodide) showed the same minimal inhibitory concentrations with all three strains (hexyl-PVP itself is insoluble in water), while the soluble monomer *N*-hexylpyridinium expectedly had the highest activity against the knockout mutant and the lowest against the strain expressing both the NorA and the QacA MDRs. No resistant mutants were found after repeated exposure of bacteria to a surface modified with *N*-alkyl-PEI (Milovic et al. 2005).

It is interesting to consider how MDR resistance is countered in nature. Plants produce QAC-type berberine alkaloids that are actively accumulated by bacteria (Severina et al. 2001), can damage the membrane, and intercalate into the DNA (Jennings and Ridler 1983). From this perspective, berberine appears to be a perfect anti-infective: neither of its targets can be mutated. The only way for bacteria to resist such a compound is by pumping it out via an MDR (Hsieh et al. 1998). Plants, however, in addition to berberine, also make methoxyhydnocarpin, an MDR inhibitor that disables the resistance mechanism of Gram-positive pathogens and acts in synergy with the antimicrobial (Stermitz et al. 2000a). Bacteria, especially Gram-negative species, probably responded by evolving a vast diversity of MDRs, and it is likely that the current state of affairs in this competition is a stalemate between plants producing antimicrobials/MDR inhibitors (Guz et al. 2001; Stermitz et al. 2002; Tegos et al. 2002; Morel et al. 2003; Stermitz et al. 2003; Belofsky et al. 2004, 2006) and bacteria with their large arsenal of MDR pumps. But unlike natural antimicrobials, macromolecular polymers made of amphipathic cations probably do not exist in nature, which would explain why pathogens lack protection from sterile surface materials (Lewis and Klibanov 2005). The attractive properties of cationic polymers are likely to lead to the development of commercial products with sterile surfaces. These materials, however, do not address the need for systemic sterilizing antibiotics.

2.2 Sterilizing Antibiotics

Unlike antiseptics, known target-specific antibiotics do not sterilize an infection. Indeed, persister tolerance is aimed at preventing death from an otherwise bactericidal antibiotic that can only corrupt active targets. Antiseptics that can kill persisters do not rely on specific targets, damage the cell membrane or DNA/proteins in general, and are obviously toxic. Given this set of facts, developing a single-molecule sterilizing antibiotic does not appear feasible. However, let us consider a perfect antibiotic from first principles. This compound is benign, but a bacterial enzyme converts it into a reactive antiseptic in the cytoplasm. The active molecule does not leave the cytoplasm (because of increased polarity) and attaches covalently to many targets, killing the cell. Irreversible binding to the targets creates a sink that will allow the compound to avoid MDR efflux.

Several existing antimicrobials closely match the properties of this idealized prodrug antibiotic. These are isoniazid, pyrazinamide, ethionamide, and metronidazole. The first three are anti-Mtb drugs, while metronidazole is a broad-spectrum compound acting against anaerobic bacteria. All four compounds convert into active antiseptic-type molecules inside the cell that bind covalently to their targets. It seems to be no accident that prodrug antibiotics make up the core of anti-Mtb drug arsenal. As we have mentioned, *M. tuberculosis* probably forms the most intransigent persisters, and excellent bactericidal properties are a critical feature for any antituberculosis antibiotic. Targets have been identified for isoniazid and ethionamide (Vilcheze et al. 2005), suggesting a relatively limited reactivity for these compounds. The targets are most likely relatively preferred molecules, the top ones on a list that may include many if not most bacterial proteins and DNA. At the same time, the existence of preferred targets indicates that the prodrug products are not that reactive, and there is considerable room for developing better sterilizing antibiotics based on the same principle.

3 Conclusions

Over half a century has passed since the discovery of drug tolerant persisters, but their study is still an emerging field. The presence of persisters in biofilms provides an important incentive to understand their nature. Recent advances in isolating persisters, determining their transcriptome, and finding candidate persister genes are hopeful indications that the pace of progress in understanding this elusive problem is picking up. Formidable obstacles remain, due to difficulty in isolating sufficient amounts of persister cells, the apparent redundancy of persister genes, and the temporary phenotype of these cells. The mechanism of drug tolerance appears to be mechanistically distinct from resistance and is based on shutting down antibiotic targets. Persisters are specialized cells that have evolved to survive all possible natural threats and in confronting persisters, we may have met our ultimate adversary. The challenge is to develop approaches they were unlikely to encounter in nature, such as combination therapies, prodrugs which are activated inside the bacterial cell, sterile-surface materials, and pulse-dosing.

Acknowledgements Work described in this chapter was supported by NIH grant GM061162.

References

Avery SV (2006) Microbial cell individuality and the underlying sources of heterogeneity. Nat Rev Microbiol 4:577-587

Baillie GS, Douglas LJ (2000) Matrix polymers of *Candida* biofilms and their possible role in biofilm resistance to antifungal agents. J Antimicrob Chemother 46:397-403

Balaban NQ, Merrin J, Chait R, Kowalik L, Leibler S (2004) Bacterial persistence as a phenotypic switch. Science 305:1622-1625

Bayles KW (2000) The bactericidal action of penicillin: new clues to an unsolved mystery. Trends Microbiol 8:274-278

Belofsky G, Percivill D, Lewis K, Tegos GP, Ekart J (2004) Phenolic metabolites of Dalea versicolor that enhance antibiotic activity against model pathogenic bacteria. J Nat Prod 67:481-484

Belofsky G, Carreno R, Lewis K, Ball A, Casadei G, Tegos GP (2006) Metabolites of the smoke tree, *Dalea spinosa*, potentiate antibiotic activity against multidrug-resistant *Staphylococcus aureus*. J Nat Prod 69:261-264

Berg HC, Brown DA (1972) Chemotaxis in *Escherichia coli* analysed by three-dimensional tracking. Nature 239:500-504

Bigger JW (1944) Treatment of staphylococcal infections with penicillin. Lancet ii:497-500

Black DS, Kelly AJ, Mardis MJ, Moyed HS (1991) Structure and organization of hip, an operon that affects lethality due to inhibition of peptidoglycan or DNA synthesis. J Bacteriol 173:5732-5739

Black DS, Irwin B, Moyed HS (1994) Autoregulation of hip, an operon that affects lethality due to inhibition of peptidoglycan or DNA synthesis. J Bacteriol 176:4081-4091

Boggs AF, Miller GH (2004) Antibacterial drug discovery: is small pharma the solution? Clin Microbiol Infect 10 [Suppl 4]:32-36

Boltner D, MacMahon C, Pembroke JT, Strike P, Osborn AM (2002) R391: a conjugative integrating mosaic comprised of phage, plasmid, and transposon elements. J Bacteriol 184:5158-5169

Brooun A, Liu S, Lewis K (2000) A dose-response study of antibiotic resistance in *Pseudomonas aeruginosa* biofilms. Antimicrob Agents Chemother 44:640-646

Bush K (2004) Antibacterial drug discovery in the 21st century. Clin Microbiol Infect 10 [Suppl 4]: 10-17

Cao F, Lane S, Raniga PP, Lu Y, Zhou Z, Ramon K, Chen J, Liu H (2006) The Flo8 transcription factor is essential for hyphal development and virulence *Candida albicans*. Mol Biol Cell 17:295-307

Chandra J, Kuhn DM, Mukherjee PK, Hoyer LL, McCormick T, Ghannoum MA (2001) Biofilm formation by the fungal pathogen *Candida albicans*: development, architecture, and drug resistance. J Bacteriol 183:5385-5394

Christensen SK, Gerdes K (2003) RelE toxins from bacteria and Archaea cleave mRNAs on translating ribosomes, which are rescued by tmRNA. Mol Microbiol 48:1389-1400

Christensen SK, Pedersen K, Hansen FG, Gerdes K (2003) Toxin-antitoxin loci as stress-response-elements: ChpAK/MazF and ChpBK cleave translated RNAs and are counteracted by tmRNA. J Mol Biol 332:809-819

Chung JD, Stephanopoulos G, Ireton K, Grossman AD (1994) Gene expression in single cells of *Bacillus subtilis*: evidence that a threshold mechanism controls the initiation of sporulation. J Bacteriol 176:1977-1984

Clardy J, Walsh C (2004) Lessons from natural molecules. Nature 432:829-837

Correia FF, D'Onofrio A, Rejtar T, Li L, Karger BL, Makarova K, Koonin EV, Lewis K (2006) Kinase activity of overexpressed HipA is required for growth arrest and multidrug tolerance in *Escherichia coli*. J Bacteriol 188:8360-8367

Datry A, Bart-Delabesse E (2006) Caspofungin: mode of action and therapeutic applications. Rev Med Interne 27:32-39

Davis BD, Chen LL, Tai PC (1986) Misread protein creates membrane channels: an essential step in the bactericidal action of aminoglycosides. Proc Natl Acad Sci U S A 83:6164-6168

Demain AL, Fang A (2000) The natural functions of secondary metabolites. In: Fiechter IA (ed) History of modern biotechnology. Springer, Berlin New York Heidelberg, pp 1-39

Denyer SP, Hugo WB (1991) Mechanisms of action of chemical biocides: their study and exploitation. Society for Applied Bacteriology

Dougherty TJ, Barrett JF, Pucci MJ (2002) Microbial genomics and novel antibiotic discovery: new technology to search for new drugs. Curr Pharm Des 8:1119-1135

Dubnau D, Losick R (2006) Bistability in bacteria. Mol Microbiol 61:564-572

Falla TJ, Chopra I (1998) Joint tolerance to beta-lactam and fluoroquinolone antibiotics in *Escherichia coli* results from overexpression of hipA. Antimicrob Agents Chemother 42:3282-3284

Garcia-Sanchez S, Aubert S, Iraqui I, Janbon G, Ghigo JM, d'Enfert C (2004) *Candida albicans* biofilms: a developmental state associated with specific and stable gene expression patterns. Eukaryot Cell 3:536-545

Gerdes K, Rasmussen PB, Molin S (1986) Unique type of plasmid maintenance function: postsegregational killing of plasmid-free cells. Proc Natl Acad Sci U S A 83:3116-3120

Gerdes K, Christensen SK, Lobner-Olesen A (2005) Prokaryotic toxin-antitoxin stress response loci. Nat Rev Microbiol 3:371-382

Gonzalez-Pastor JE, Hobbs EC, Losick R (2003) Cannibalism by sporulating bacteria. Science 301:510-513

Guz NR, Stermitz FR, Johnson JB, Beeson TD, Wilen S, Hsiang J-F, Lewis K (2001) Flavonolignan and flavone inhibitors of a *Staphylococcus aureus* multidrug resistance (MDR) pump. Structure-activity relationships. J Med Chem 44:261-268

Hall-Stoodley L, Costerton JW, Stoodley P (2004) Bacterial biofilms: from the natural environment to infectious diseases. Nat Rev Microbiol 2:95-108

Harrison JJ, Ceri H, Roper NJ, Badry EA, Sproule KM, Turner RJ (2005a) Persister cells mediate tolerance to metal oxyanions in *Escherichia coli*. Microbiology 151:3181-3195

Harrison JJ, Turner RJ, Ceri H (2005b) Persister cells, the biofilm matrix and tolerance to metal cations in biofilm and planktonic *Pseudomonas aeruginosa*. Environ Microbiol 7:981-994

Hayes F (2003) Toxins-antitoxins: plasmid maintenance, programmed cell death, and cell cycle arrest. Science 301:1496-1499

Heath RJ, Rock CO (1999) A missense mutation accounts for the defect in the glycerol-3-phosphate acyltransferase expressed in the plsB26 mutant. J Bacteriol 181:1944-1946

Hegde SS, Vetting MW, Roderick SL, Mitchenall LA, Maxwell A, Takiff HE, Blanchard JS (2005) A fluoroquinolone resistance protein from *Mycobacterium tuberculosis* that mimics DNA. Science 308:1480-1483

Hooper DC (2002) Target modification as a mechanism of antimicrobial resistance. In: Lewis K, Salyers A, Taber H and Wax R (eds) Bacterial resistance to antimicrobials: mechanisms genetics medical practice and public health. Marcell Dekker, New York, pp 161-192

Hoyle BD, Jass J, Costerton JW (1990) The biofilm glycocalyx as a resistance factor. J Antimicrob Chemother 26:1-5

Hsieh PC, Siegel SA, Rogers B, Davis D, Lewis K (1998) Bacteria lacking a multidrug pump: a sensitive tool for drug discovery. Proc Natl Acad Sci U S A 95:6602-6606

Hu Y, Coates AR (2005) Transposon mutagenesis identifies genes which control antimicrobial drug tolerance in stationary-phase *Escherichia coli*. FEMS Microbiol Lett 243:117-124

Jennings BR, Ridler PJ (1983) Interaction of chromosomal stains with DNA. An electrofluorescence study. Biophys Struct Mech 10:71-79

Jesaitis AJ, Franklin MJ, Berglund D, Sasaki M, Lord CI, Bleazard JB, Duffy JE, Beyenal H, Lewandowski Z (2003) Compromised host defense on *Pseudomonas aeruginosa* biofilms: characterization of neutrophil and biofilm interactions. J Immunol 171:4329-4339

Kaeberlein T, Lewis K, Epstein SS (2002) Isolating uncultivable microorganisms in pure culture in a simulated natural environment. Science 296:1127-1129

Kaern M, Elston TC, Blake WJ, Collins JJ (2005) Stochasticity in gene expression: from theories to phenotypes. Nat Rev Genet 6:451-464

Keren I, Kaldalu N, Spoering A, Wang Y, Lewis K (2004a) Persister cells and tolerance to antimicrobials. FEMS Microbiol Lett 230:13-18

Keren I, Shah D, Spoering A, Kaldalu N, Lewis K (2004b) Specialized persister cells and the mechanism of multidrug tolerance in *Escherichia coli*. J Bacteriol 186:8172-8180

Korch SB, Hill TM (2006) Ectopic overexpression of wild-type and mutant hipA genes in *Escherichia coli*: effects on macromolecular synthesis and persister formation. J Bacteriol 188:3826-3836

Korch SB, Henderson TA, Hill TM (2003) Characterization of the hipA7 allele of *Escherichia coli* and evidence that high persistence is governed by (p)ppGpp synthesis. Mol Microbiol 50:1199-1213

Korobkova E, Emonet T, Vilar JM, Shimizu TS, Cluzel P (2004) From molecular noise to behavioural variability in a single bacterium. Nature 428:574-578

Kumamoto CA (2005) A contact-activated kinase signals *Candida albicans* invasive growth and biofilm development. Proc Natl Acad Sci U S A 102:5576-5581

Kumamoto CA, Vinces MD (2005) Alternative *Candida albicans* lifestyles: growth on surfaces. Annu Rev Microbiol 59:113-133

LaFleur MD, Kumamoto CA, Lewis K (2006) *Candida albicans* biofilms produce antifungal-tolerant persister cells. Antimicrob Agents Chemother 50:3839-3846

Lee SB, Koepsel RR, Morley SW, Matyjaszewski K, Sun Y, Russell AJ (2004) Permanent, nonleaching antibacterial surfaces. 1. Synthesis by atom transfer radical polymerization. Biomacromolecules 5:877-882

Leid JG, Shirtliff ME, Costerton JW, Stoodley AP (2002) Human leukocytes adhere to, penetrate, and respond to *Staphylococcus aureus* biofilms. Infect Immun 70:6339-6345

Levin BR, Rozen DE (2006) Non-inherited antibiotic resistance. Nat Rev Microbiol 4:556-562

Levy SB, Marshall B (2004) Antibacterial resistance worldwide: causes, challenges and responses. Nat Med 10:S122-S129

Lewis K (2000) Programmed death in bacteria. Microbiol Mol Biol Rev 64:503-514

Lewis K (2001a) In search of natural substrates and inhibitors of MDR pumps. J Mol Microbiol Biotechnol 3:247-254

Lewis K (2001b) Riddle of biofilm resistance. Antimicrob Agents Chemother 45:999-1007

Lewis K, Klibanov AM (2005) Surpassing nature: rational design of sterile-surface materials. Trends Biotechnol 23:343-348

Lewis K, Lomovskaya O (2002) Drug efflux. In: Lewis K, Salyers A, Taber H and Wax R (eds) Bacterial resistance to antimicrobials: mechanisms genetics medical practice and public health. Marcel Dekker, New York, pp 61-90

Lewis K, Salyers A, Taber H, Wax R (2002) Bacterial resistance to antimicrobials: mechanisms genetics medical practice and public health. Marcel Dekker, New York

Lewis K, Spoering A, Kaldalu N, Keren I, Shah D (2005) Persisters: specialized cells responsible for biofilm tolerance to antimicrobial agents. In: Pace J, Rupp ME and Finch RG (eds) Biofilms infection, and antimicrobial therapy. Taylor & Francis, Boca Raton, pp 241-256

Li XZ, Nikaido H (2004) Efflux-mediated drug resistance in bacteria. Drugs 64:159-204

Lin J, Tiller JC, Lee SB, Lewis K, Klibanov AM (2002) Insights into bactericidal action of surface-attached poly(vinyl-N-hexylpyridinium) chains. Biotechnol Lett 24:801-805

Lin J, Qiu S, Lewis K, Klibanov AM (2003) Mechanism of bactericidal and fungicidal activities of textiles covalently modified with alkylated polyethylenimine. Biotechnol Bioeng 83:168-172

Lomovskaya O, Bostian KA (2006) Practical applications and feasibility of efflux pump inhibitors in the clinic - a vision for applied use. Biochem Pharmacol 71:910-918

Mack D, Becker P, Chatterjee I, Dobinsky S, Knobloch JK, Peters G, Rohde H, Herrmann M (2004) Mechanisms of biofilm formation in *Staphylococcus epidermidis* and *Staphylococcus aureus*: functional molecules, regulatory circuits, and adaptive responses. Int J Med Microbiol 294:203-212

Markham PN, Neyfakh AA (1996) Inhibition of the multidrug transporter NorA prevents emergence of norfloxacin resistance in *Staphylococcus aureus*. Antimicrob Agents Chemother 40:2673-2674

Markham PN, Westhaus E, Klyachko K, Johnson ME, Neyfakh AA (1999) Multiple novel inhibitors of the NorA multidrug transporter of *Staphylococcus aureus*. Antimicrob Agents Chemother 43:2404-2408

McCool JD, Long E, Petrosino JF, Sandler HA, Rosenberg SM, Sandler SJ (2004) Measurement of SOS expression in individual *Escherichia coli* K-12 cells using fluorescence microscopy. Mol Microbiol 53:1343-1357

Meyer AL (2005) Prospects and challenges of developing new agents for tough Gram-negatives. Curr Opin Pharmacol 5:490-494

Milovic NM, Wang J, Lewis K, Klibanov AM (2005) Immobilized N-alkylated polyethylenimine avidly kills bacteria by rupturing cell membranes with no resistance developed. Biotechnol Bioeng 90:715-722

Morel C, Stermitz FR, Tegos G, Lewis K (2003) Isoflavone MDR efflux pump inhibitors from *Lupinus argenteus*. Synergism between some antibiotics and isoflavones. J Agricult Food Chem 51:5677-5679

Morgan HC, Meier JF, Merker RL (2000) Method of creating a biostatic agent using interpenetrating network polymers. US Patent No. 6,146,688

Moyed HS, Bertrand KP (1983) hipA, a newly recognized gene of *Escherichia coli* K-12 that affects frequency of persistence after inhibition of murein synthesis. J Bacteriol 155:768-775

Moyed HS, Broderick SH (1986) Molecular cloning and expression of hipA, a gene of Escherichia coli K-12 that affects frequency of persistence after inhibition of murein synthesis. J Bacteriol 166:399-403

Mukherjee PK, Chandra J, Kuhn DM, Ghannoum MA (2003) Mechanism of fluconazole resistance in *Candida albicans* biofilms: phase-specific role of efflux pumps and membrane sterols. Infect Immun 71:4333-4340

Osburne MS, Grossman TH, August PR, MacNeil IA (2000) Tapping into microbial diversity for natural products drug discovery. ASM News 66:411-417

Pedersen K, Christensen SK, Gerdes K (2002) Rapid induction and reversal of a bacteriostatic condition by controlled expression of toxins and antitoxins. Mol Microbiol 45:501-510

Projan SJ, Shlaes DM (2004) Antibacterial drug discovery: is it all downhill from here? Clin Microbiol Infect 10 [Suppl 4]:18-22

Ramage G, Bachmann S, Patterson TF, Wickes BL, Lopez-Ribot JL (2002) Investigation of multi-drug efflux pumps in relation to fluconazole resistance in *Candida albicans* biofilms. J Antimicrob Chemother 49:973-980

Samaranayake YH, Ye J, Yau JY, Cheung BP, Samaranayake LP (2005) In vitro method to study antifungal perfusion in *Candida* biofilms. J Clin Microbiol 43:818-825

Sat B, Hazan R, Fisher T, Khaner H, Glaser G, Engelberg-Kulka H (2001) Programmed cell death in *Escherichia coli*: some antibiotics can trigger mazEF lethality. J Bacteriol 183:2041-2045

Schatz A, Bugie E, Waksman SA (1944) Streptomycin, a substance exhibiting antibiotic activity against Gram-positive and Gram-negative bacteria. Proc Soc Exp Biol Med 55:66-69

Scherrer R, Moyed HS (1988) Conditional impairment of cell division and altered lethality in hipA mutants of *Escherichia coli* K-12. J Bacteriol 170:3321-3326

Schmelzle T, Hall MN (2000) TOR, a central controller of cell growth. Cell 103:253-262

Schmid MB (2004) Seeing is believing: the impact of structural genomics on antimicrobial drug discovery. Nat Rev Microbiol 2:739-746

Severina II, Muntyan MS, Lewis K, Skulachev VP (2001) Transfer of cationic antibacterial agents berberine, palmatine and benzalkonium through bimolecular planar phospholipid film and *Staphylococcus aureus* membrane. IUBMB Life Sciences 52:321-324

Shah DV, Zhang Z, Kurg K, Kaldalu N, Khodursky A, Lewis K (2006) Persisters: a distinct physiological state of E. coli. BMC Microbiol 6:53

Silver LL (2006) Antibacterial drug discovery and development - SRI's 11th Annual Summit. Antibacterial trends and current research. IDrugs 9:394-397

Singh PK, Schaefer AL, Parsek MR, Moninger TO, Welsh MJ, Greenberg EP (2000) Quorum-sensing signals indicate that cystic fibrosis lungs are infected with bacterial biofilms. Nature 407:762-764

Spoering AL, Lewis K (2001) Biofilms and planktonic cells of *Pseudomonas aeruginosa* have similar resistance to killing by antimicrobials. J Bacteriol 183:6746-6751

Spoering AL, Vulic M, Lewis K (2006) GlpD and PlsB participate in persister cell formation in *Escherichia coli*. J Bacteriol 188:5136-5144

Spudich JL, Koshland DE Jr (1976) Non-genetic individuality: chance in the single cell. Nature 262:467-471

Stermitz FR, Lorenz P, Tawara JN, Zenewicz L, Lewis K (2000a) Synergy in a medicinal plant: antimicrobial action of berberine potentiated by 5'-methoxyhydnocarpin, a multidrug pump inhibitor. Proc Natl Acad Sci U S A 97:1433-1437

Stermitz FR, Scriven LN, Tegos G, Lewis K (2002) Two flavonols from *Artemisia annua* which potentiate the activity of berberine and norfloxacin against a resistant strain of *Staphylococcus aureus*. Planta Med 68:1140-1141

Stermitz FR, Cashman KK, Halligan KM, Morel C, Tegos GP, Lewis K (2003) Polyacylated neohesperidosides from *Geranium caespitosum*: bacterial multidrug resistance pump inhibitors. Bioorg Med Chem Lett 13:1915-1918

Tegos G, Stermitz FR, Lomovskaya O, Lewis K (2002) Multidrug pump inhibitors uncover the remarkable activity of plant antimicrobials. Antimicrob Agents Chemother 46:3133-3141

Tiller JC, Liao CJ, Lewis K, Klibanov AM (2001) Designing surfaces that kill bacteria on contact. Proc Natl Acad Sci U S A 98:5981-5985

Tran JH, Jacoby GA (2002) Mechanism of plasmid-mediated quinolone resistance. Proc Natl Acad Sci U S A 99:5638-5642

Vazquez-Laslop N, Lee H, Neyfakh AA (2006) Increased persistence in *Escherichia coli* caused by controlled expression of toxins or other unrelated proteins. J Bacteriol 188:3494-3497

Vetting MW, Hegde SS, Fajardo JE, Fiser A, Roderick SL, Takiff HE, Blanchard JS (2006) Pentapeptide repeat proteins. Biochemistry 45:1-10

Vilcheze C, Weisbrod TR, Chen B, Kremer L, Hazbon MH, Wang F, Alland D, Sacchettini JC, Jacobs WR Jr (2005) Altered NADH/NAD+ ratio mediates coresistance to isoniazid and ethionamide in mycobacteria. Antimicrob Agents Chemother 49:708-720

Von Eiff C, Heilmann C, Peters G (1999) New aspects in the molecular basis of polymer-associated infections due to staphylococci. Eur J Clin Microbiol Infect Dis 18:843-846

Vuong C, Voyich JM, Fischer ER, Braughton KR, Whitney AR, DeLeo FR, Otto M (2004) Polysaccharide intercellular adhesin (PIA) protects *Staphylococcus epidermidis* against major components of the human innate immune system. Cell Microbiol 6:269-275

Walker GC (1996) The SOS response of *Escherichia coli*. In: Neidhardt FC (ed) *Escherichia coli* and Samonella. Cellular and molecular biology. ASM Press, Washington DC, pp 1400-1416

Walsh C (2003a) Antibiotics. Actions, origins, resistance. ASM Press, Washington DC

Walsh C (2003b) Where will new antibiotics come from? Nat Rev Microbiol 1:65-70

Wiuff C, Zappala RM, Regoes RR, Garner KN, Baquero F, Levin BR (2005) Phenotypic tolerance: antibiotic enrichment of noninherited resistance in bacterial populations. Antimicrob Agents Chemother 49:1483-1494

Yoshida H, Maki Y, Kato H, Fujisawa H, Izutsu K, Wada C, Wada A (2002) The ribosome modulation factor (RMF) binding site on the 100S ribosome of *Escherichia coli*. J Biochem 132:983-989

Zurenko GE, Ford CW, Hutchinson DK, Brickner SJ, Barbachyn MR (1997) Oxazolidinone antibacterial agents: development of the clinical candidates eperezolid and linezolid. Expert Opin Investig Drugs 6:151-158

Biofilms on Central Venous Catheters:
Is Eradication Possible?

R. M. Donlan

Abstract Biofilms on indwelling medical devices such as central venous catheters result in significant morbidity and mortality and have a substantial impact on healthcare delivery. Because routine systemic treatment of patients with catheter-associated bloodstream infections is often ineffective, due to the tolerance of biofilm organisms on these devices, other strategies such as the antimicrobial lock treatment (ALT) have been used. This approach involves the instillation of high concentrations of the antimicrobial agent directly into the biofilm-containing

R. M. Donlan
Division of Healthcare Quality Promotion, Centers for Disease Control and Prevention,
1600 Clifton Road, N.E., Mail Stop C16, Atlanta, GA 30333, USA
rld8@cdc.gov

T. Romeo (ed.), *Bacterial Biofilms.*
Current Topics in Microbiology and Immunology 322.

catheter for exposure (i.e., dwell) times sufficient to eradicate the biofilm. Results from human studies, animal studies, and laboratory studies using in vitro model systems have suggested that eradication of a biofilm is possible, depending on the organisms in the biofilm, biofilm age, the antimicrobial agent used, and the dwell/duration of the treatment. The most effective antimicrobial agents are those (1) that are less affected by the extracellular polymeric substance matrix of the biofilm, (2) that have a more rapid bactericidal effect, or (3) for which the mechanism of action is not dependent upon the growth rates of the organisms. Combining agents may also provide synergy. Fungal biofilms have proven to be much more difficult to treat using the ALT, though newer fungicidal drugs such as the echinocandins hold promise in this regard. However, a serious drawback with the ALT is the potential for the development of resistance. Newer treatments, incorporating agents not classified as antibiotics, appear to effectively eradicate biofilms in in vitro models and should be evaluated in animal and patient studies. Promising technologies that incorporate novel approaches such as ultrasound, bacteriophage, quorum-sensing inhibitors, or enzymes may also provide useful approaches in the future.

1 Biofilms and Infections Associated with Indwelling Medical Devices

Use of indwelling medical devices such as catheters, mechanical heart valves, and prosthetic joints for patient care is associated with increased risk of infection including bloodstream and urinary tract infections. In the United States, approximately 80,000 central venous catheter-associated (CVC-associated) bloodstream infections occur in intensive care units each year (Centers for Disease Control and Prevention 2002). These infections result in significant morbidity and mortality and a potentially large increase in the cost of caring for patients. These infections occur when microorganisms colonize the catheter and establish a biofilm. Biofilms are dynamic sessile microbial communities in which the organisms produce an extracellular polymeric matrix that results in a distinct structure. Microorganisms also exhibit slower growth rates and greater tolerance to antimicrobial agents when growing in a biofilm (Donlan and Costerton 2002). The process of biofilm formation is complex, dependent upon multiple factors including properties of the substratum, presence of conditioning film, hydrodynamics, physical and chemical properties of the liquid in contact with the device surface, and properties of the colonizing microbial cells (Donlan 2002). Biofilm-associated microorganisms may elicit disease processes by detachment of individual cells or aggregates from the device surface, by production of endotoxins, or by providing a niche for the development of antibiotic-resistant organisms (Donlan and Costerton 2002). Biofilm-associated organisms also exhibit tolerance to antimicrobial agents (Stewart et al. 2004). The implications of tolerance are that treatment of device-associated infections with systemic antimicrobial agents is generally ineffective. For example, a study by Marre et al. (1997) evaluated whether treatment with vancomycin, gentamicin, or other antimicrobial agents could successfully resolve catheter-related bacteremia in hemodialysis patients. The treatment was successful in only 32% of patients, suggesting that systemic antimicrobial treatment is not an effective method for controlling biofilms on catheters.

2 Central Venous Catheter Biofilms

Central venous catheters (CVCs) are inserted for administration of fluids, blood products, medications, nutritional solutions, hemodynamic monitoring, and to provide vascular access for dialysis (Flowers et al. 1989) (Fig. 1). These devices have been shown to pose a greater risk of device-related infection than any other indwelling medical device (Maki 1994; Klevins et al. 2005). Microorganisms may colonize both external and luminal surfaces of CVCs (Raad et al. 1993). Biofilms may form within 3 days of catheter insertion (Anaissie et al. 1995). It has been reported (Raad et al. 1993) that catheters in place for less than 10 days are more heavily colonized on the external surfaces while long-term catheters (up to 30 days) tend to be more heavily colonized within the lumen. Organisms that colonize CVCs may originate from the skin at the insertion site and migrate along the catheter's external surface or from the catheter hub, due to handling by healthcare workers, in which case the organisms travel up the lumen (Elliott et al. 1997; Raad

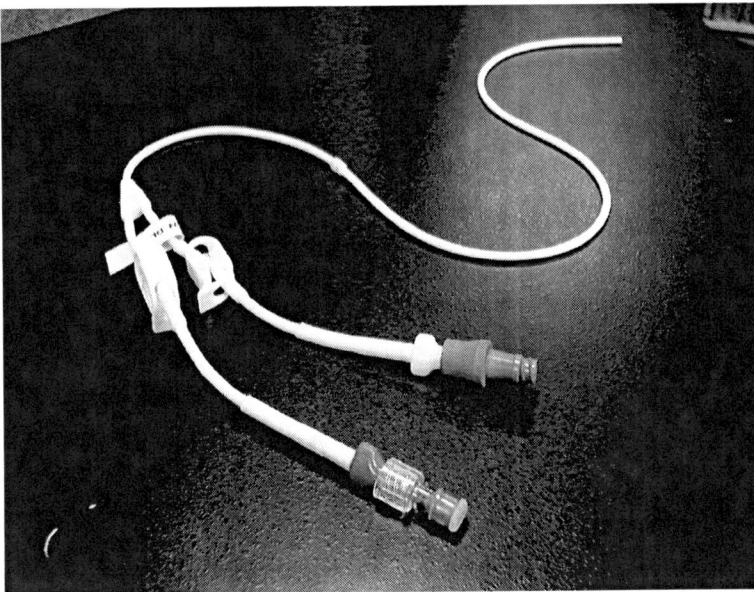

Fig. 1 Central venous catheter

Table 1 Microorganisms that have been isolated from biofilms on explanted central venous catheters[a]

Gram–positive bacteria	Gram–negative bacteria	Other organisms
Corynebacterium spp.	Acinetobacter spp.	Candida spp.
Enterococcus spp.	Acinetobacter calcoaceticus	Candida albicans
Enterococcus faecalis	Acinetobacter anitratus	Candida tropicalis
Enterococcus faecium	Enterobacter cloacae	Mycobacterium chelonei
Staphylococcus spp.	Enterobacter aerogenes	
Staphylococcus aureus	Escherichia coli	
Methicillin Resistant –Staphylococcus aureus	Klebsiella pneumoniae	
Staphylococcus epidermidis	Klebsiella oxytoca	
Coagulase–negative staphylococci	Pseudomonas aeruginosa	
Streptococcus spp. (alpha hemolytic)	Pseudomonas putida	
Streptococcus spp. (viridans streptococci)	Proteus spp.	
Streptococcus pneumoniae	Providencia spp.	
	Serratia marcescens	

[a] List of organisms is based upon the following references: Aufwerber et al. 1991; Anaissie et al. 1995; Darouiche et al. 1999; Maki et al. 1997; Kamal et al. 1991; Kowalewska-Grochowska et al. 1991; Flowers et al. 1989; Brun-Buisson et al. 1987; Raad et al. 1993; Zuffrey et al. 1988; Moreno et al. 2006.

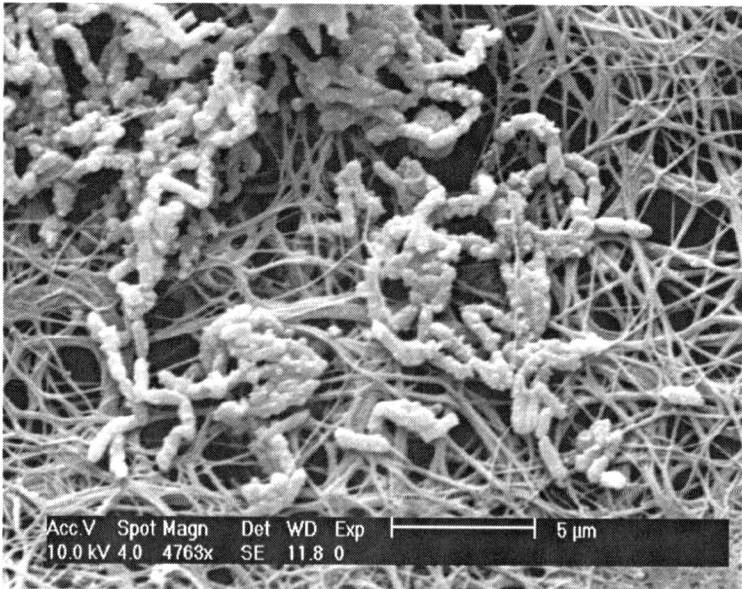

Fig. 2 Scanning electron microscopic image of the luminal surface of an explanted catheter containing a biofilm of *Alcaligenes xylosoxidans* in a fibrin-like matrix. Photo by Janice Carr

1998). Organisms may also originate through hematogenous seeding of the device from another nidus of infection in the catheterized patient (Anaissie et al. 1995). A listing of microorganisms that have been recovered from biofilms on CVCs removed from patients is shown in Table 1. For several of the studies represented by these data, multiple organisms were isolated from the same catheter biofilm. Platelets, plasma, and tissue proteins such as fibronectin, fibrin, and laminin will also adsorb to the CVC surface after it is exposed to the bloodstream (Raad 1998) and these materials will alter the surface characteristics and affect microbial attachment (Murga et al. 2001; Rupp et al. 1999). Figure 2 shows a biofilm containing bacilli embedded in a fibrin-like matrix on the surface of an explanted catheter. These organisms were also isolated from separate segments of the same catheter on microbiological media and identified as a pure culture of *Alcaligenes xylosoxidans* (R.M. Donlan, unpublished data).

3 Biofilm Eradication

In general, from an operational point of view, the presence of biofilms on a substratum is synonymous with the presence of irreversibly attached microbial cells on that surface, though a strict definition of biofilms will also include the extracellular polymeric substance (EPS) and other materials entrapped in this matrix (Donlan and Costerton 2002). Because methods for detecting and quantifying the EPS component of the biofilm are limited, it is generally not quantified or even described when biofilms are measured. It is possible that residual EPS on a surface could affect cell attachment kinetics or in some way impact device usage. However, for the purposes of this paper, biofilm eradication is defined as the removal of all detectable microbial cells from a substratum. The definition is of course method-dependent, since different methods exhibit different recovery efficiency and therefore a different limit of detection.

4 Studies Evaluating the Use of Antimicrobial Lock Treatments Incorporating Antimicrobial Drugs to Treat Catheter-Associated Biofilms

If systemic antimicrobial treatment has only limited efficacy against catheter-associated biofilms, are there other strategies that might be more effective? One treatment approach for intravascular catheters is the antimicrobial lock, in which the indwelling colonized catheter is instilled in situ with a high concentration of an antimicrobial agent for a dwell time sufficient to eradicate the biofilms. By design, the volume of antimicrobial agent solution is sufficient to fill the lumen but not spill into circulation. The antimicrobial lock technique (ALT) was first reported by Messing et al. (1988) in a study in which 11 patients with silicone tunneled CVCs who had experienced multiple episodes of catheter-related bloodstream infections (CRBSI) with Gram-positive and Gram-negative organisms were treated using the ALT containing 1.5 mg ml^{-1} amikacin, 0.2 mg ml^{-1}minocycline, or 1.0 mg ml^{-1} vancomycin. These concentrations would provide levels in the milligram per milliliter range in the catheter lumen. These concentrations were considered too low to treat systemic infections, even if 1 ml of the 2-ml ALT were to reach blood circulation (which the authors considered to be the extreme case). The ALT was applied for 12 h each day for approximately 2 weeks. In order to assess whether supplemental systemic antimicrobial agents would improve the efficacy of the ALT, patients were treated with the ALT alone or with the ALT plus short-term systemic antimicrobial agents. The type and dosing of systemic antimicrobial agents selected depended upon the organism isolated from blood cultures. Treatment was considered ineffective if catheters were removed due to CRBSI. In this study, one catheter in each treatment group (2/22 cases) was removed; 20/22 were successfully treated. In the two treatment failures, recurrent fever was observed and both in-line and catheter tip cultures were negative for the bacteria originally isolated but positive for fungi. There was no significant difference between groups (with or without systemic antimicrobial agents), suggesting that use of systemic antimicrobial agents in conjunction with the ALT did not improve the treatment.

4.1 Case Studies

Results of several other evaluations of the ALT in catheterized patients are shown in Tables 2, 3, 4, 5, 6. The antimicrobial agents chosen for the ALT were based upon the results of broth microdilution testing of blood culture isolates. In each table, organisms listed were isolated from individual patients prior to the antimicrobial lock treatment. In some cases multiple organisms were isolated from the same patient. And in some studies, catheters from multiple patients were colonized by the same organisms. With the exception of the study by Messing et al. (1988), the basis for the choice of drug concentration was not given. Berrington and Gould (2001) suggested that bactericidal rather than bacteriostatic agents be used, and at the highest practical concentration, keeping in mind that the agents may diffuse into circulation and cause adverse affects to the patient. Mermel et al. (2001) suggested that ALTs should contain 1-5 mg ml^{-1} in a sufficient volume to fill the lumen of the catheter. The basis for the use of these high concentrations makes sense in light of the fact that biofilm-associated organisms are extremely tolerant to antimicrobial agents (Stewart et al. 2004). The choice of dwell time appears be driven more by practical considerations, such as the time interval between periods of use of the device rather than upon pharmacokinetics. However, Benoit et al. (1995) measured vancomycin and gentamicin concentrations in intraluminal fluids and showed that levels were maintained above 2.5 mg ml^{-1} during an 8- to 12-h dwell time. In vitro (Anthony and Rubin 1999) and patient (Haimi-Cohen et al. 2001) studies have

Table 2 Case studies depicting the efficacy of the antimicrobial lock treatment incorporating vancomycin for catheter–associated infections

Ref.[a]	Organism	Treatment	Outcome
1	*Enterococcus* spp. (1)[b]	Concen: 0.1 mg ml⁻¹ Dwell: 48–72 h Duration: 14 days	Symptoms resolve, patient remains well 12 weeks post–ALT
	CoNS[c] (1)	Concen: 0.1 mg ml⁻¹ Dwell: 48–72 h Duration: 14 days	Symptoms resolve, patient remains well 12 weeks post–ALT
	CoNS (1)	Concen: 0.1 mg ml⁻¹ Dwell: 48–72 h Duration: 14 days	Symptoms resolve, patient remains well 4 weeks post–ALT
2	*S. epidermidis* (5)	Concen: 2.0 mg ml⁻¹ Dwell: 12 h Duration: 10–14 days	Catheter blood culture negative, no infection 4 weeks post–ALT
	S. epidermidis (1)	Concen: 2.0 mg ml⁻¹ Dwell: 12 h Duration: 10–14 days	Catheter blood cultures positive post–ALT=failure
3	*S. epidermidis* (1)	Concen: 83 mg ml⁻¹ Dwell: 12 h Duration: 5–17 days	Catheter blood culture negative 8 weeks post–ALT, symptoms resolve
	Corynebacterium spp. (1)	Concen: 33 mg ml⁻¹ Dwell: 12 h Duration: 5–17 days	Catheter blood culture negative 8 weeks post–ALT, symptoms resolve
4	*S. epidermidis* (1)	Concen: 5 mg ml⁻¹ Dwell: 8–12 h Duration: 13 days	Catheter blood culture negative 12 weeks post–ALT, absence of fever
5	*S. epidermidis* (4) *S. aureus* (2) *E. faecalis* (1) *Corynebacterium* spp. (1)	Concen: 0.1 mg ml⁻¹ Dwell: 4 h Duration: 15 days Included 5% heparin	Catheter blood cultures negative post–ALT, absence of fever

[a] 1, Bailey et al. 2002; 2, Johnson et al. 1994; 3, Krzywda et al. 1995; 4, Benoit et al. 1995; 5, Capdevila et al. 1993
[b] Number of separate catheters colonized by the same organism
[c] Coagulase–negative staphylococci

shown that several antimicrobial agents used in ALTs will maintain activity through much longer dwell times (7–14 days). Several observations regarding the case studies of Messing et al. (1988) and studies presented in Tables 2–6 can be made. In most cases, the vancomycin ALT against Gram-positive organisms was successful; the two treatment failures were against coagulase-negative staphylococci (Bailey et al. 2002) and *Staphylococcus epidermidis* (Johnson et al. 1994). Other agents that were effective against *S. epidermidis* were cefazolin, erythromycin, clindamycin, and in most cases nafcillin. For Gram-negative organisms, amikacin, mezlocillin, and ciprofloxacin were effective while gentamicin and the cephalosporin ALTs occasionally failed or required prolonged duration to succeed. *Candida*

Table 3 Case studies depicting the efficacy of the antimicrobial lock treatment incorporating the aminoglycosides gentamicin or amikacin for catheter–associated infections

Ref.[a]	Organism	Treatment	Outcome
1	*Escherichia coli, Klebsiella* spp.	Gentamicin Concen: 0.02 mg ml[-1] Dwell: 48–72 h Duration: 14 days	Clinical relapse within 2 weeks post–ALT, catheter removed
	Pseudomonas aeruginosa	Included 5000 U ml[-1] heparin	Clinical relapse within 2 weeks post–ALT, catheter removed
	Acinetobacter baumannii, CoNS[b]		Clinical relapse within 2 weeks post–ALT, catheter removed
	Citrobacter spp.		Clinical relapse within 2 weeks post–ALT
	Pseudomonas aeruginosa		Symptoms resolved, patient well for 12 weeks post–ALT
2	*Enterobacter cloacae*	Amikacin Concen: 2.0 mg ml[-1] Dwell: 12 h Duration: 10–14 days	Catheter blood culture negative, no infection 4 weeks post–ALT
3	*Staphylococcus epidermidis*	Gentamicin Concen: 13 mg ml[-1] Dwell: 12 h Duration: 5–17 days	Catheter blood culture negative 8 weeks post–ALT, symptoms resolve
4	*Chryseomonas luteola* *Klebsiella pneumoniae* *Citrobacter diversus*	Gentamicin Concen: 5 mg ml[-1] Dwell: 8–12 h Duration: 7 days	Catheter blood culture negative 12 weeks post–ALT, absence of fever
	Escherichia coli	Gentamicin Concen: 5 mg ml[-1] Dwell: 8–12 h Duration: 8 days	Catheter blood culture negative 12 weeks post–ALT, absence of fever
5		Amikacin Concen: 50 mg ml[-1] Dwell: 8 h	Catheter blood culture negative 6 weeks post–ALT for all
	Corynebacterium spp., *Veillonella* spp.	Duration: 14 days	
	Klebsiella pneumoniae	Duration: 14 days	
	Klebsiella pneumoniae	Duration: 7 days	
	Acinetobacter spp.	Duration: 9 days	
	Escherichia coli, *Citrobacter freundii,* *Enterococcus faecalis*	Duration: 10 days	
	Citrobacter freundii, *Enterobacter* spp.	Duration: 7 days	

[a] 1, Bailey et al. 2002; 2, Johnson et al. 1994; 3, Krzywda et al. 1995; 4, Benoit et al. 1995; 5, Rao et al. 1992
[b] Coagulase–negative staphylococci

Table 4 Case studies depicting the efficacy of the antimicrobial lock treatment incorporating selected β–lactams or cephalosporins for catheter–associated infections

Ref.[a]	Organism	Treatment	Outcome
1	*Enterobacter cloacae*	Mezlocillin Concen: 2.0 mg ml⁻¹ Dwell: 12 h Duration: 10–14 days	Catheter blood culture negative, no infection 4 weeks post–ALT
	Enterococcus faecalis	Ampicillin + Gentamicin Concen. each agent: 2.0 mg ml⁻¹ Dwell: 12 h Duration: 10–14 days	Catheter blood culture negative, no infection 4 weeks post–ALT
2	*Staphylococcus epidermidis*	Cefazolin Concen: 67 mg ml⁻¹ Dwell: 12 h Duration: 5–17 days	Catheter blood culture negative 8 weeks post–ALT, symptoms resolve
	Staphylococcus epidermidis (2)[b]	Nafcillin Concen: 167 mg ml⁻¹ Dwell: 12 h Duration: 5–17 days	Catheter blood culture negative 8 weeks post–ALT, symptoms resolve
	Staphylococcus epidermidis	Nafcillin Concen: 167 mg ml⁻¹ Dwell: 12 h Duration: 5–17 d	Catheter blood culture positive post–ALT, CRBSI not resolved
	Staphylococcus epidermidis (3)	Nafcillin Concen: 83 mg ml⁻¹ Dwell: 12 h Duration: 5–17 days	Catheter blood culture negative 8 weeks post–ALT, symptoms resolve
	Moraxella osloensis	Ceftriaxone Concen: 167 mg ml⁻¹ Dwell: 12 h Duration: 5–17 days	Catheter blood culture negative 8 weeks post–ALT, symptoms resolve
	Klebsiella pneumoniae	Ceftazidime Concen: 300 mg ml⁻¹ Dwell: 12 h Duration: 5–17 days	Catheter blood culture negative 8 weeks post–ALT, symptoms resolve
	Escherichia coli, *Klebsiella pneumoniae,* *Streptococcus intermedius,* *Streptococcus salivarius,*	Ceftazidime Concen: 167 mg ml⁻¹ Dwell: 12 h Duration: 5–17 days	Catheter blood culture negative 8 weeks post–ALT, symptoms resolve
	Staphylococcus epidermidis *Klebsiella oxytoca* *Klebsiella pneumoniae*	Ceftriaxone Concen: 83 mg ml⁻¹ Dwell: 12 h Duration: 5–17 days	Catheter blood culture positive post–ALT, CRBSI not resolved

(continued)

Table 4 (continued)

Ref.[a] Organism	Treatment	Outcome
Klebsiella pneumoniae	Ceftriaxone Concen: 83 mg ml^{-1} Dwell: 12 h Duration: 5–17 days	Catheter blood culture positive post–ALT, CRBSI not resolved
Klebsiella pneumoniae	Ceftazidime Concen: 83 mg ml^{-1} Dwell: 12 h Duration: 5–17 days	Catheter blood culture positive post–ALT, CRBSI not resolved

[a] 1, Johnson et al. 1994; 2, Krzywda et al. 1995
[b] Number of separate catheters colonized by the same organism

Table 5 Case studies depicting the efficacy of the antimicrobial lock treatment incorporating minocycline, clindamycin, erythromycin, or ciprofloxacin for catheter–associated infections

Ref[a]	Organism	Treatment	Outcome
1	Staphylococcus epidermidis	Erythromycin Concen: 67 mg ml^{-1} Dwell: 12 h Duration: 5–17 days	Catheter blood culture negative 8 weeks post–ALT, symptoms resolve
	Staphylococcus epidermidis	Clindamycin Concen: 100 mg ml^{-1} Dwell: 12 h Duration: 5–17 days	Catheter blood culture negative 8 weeks post–ALT, symptoms resolve
2	Pseudomonas aeruginosa	Ciprofloxacin Concen: 0.1 mg ml^{-1} Dwell: 4 h Duration: 15 days	Catheter blood cultures negative post–ALT, absence of fever

[a] 1, Krzywda et al. 1995; 2,Capdevila et al. 1993

biofilms were also very tolerant to an amphotericin ALT. In each study, the ALT for fungi contained amphotericin B, though drug concentrations varied from 0.33-2.5 mg ml^{-1}.

4.2 Animal Model Studies

Several studies have investigated the efficacy of the ALT using in vivo model systems. Capdevila et al. (2001) implanted silicone catheters into the inferior cava vein of New Zealand White rabbits and challenged them with two different *Staphylococcus aureus* cultures. Biofilms that formed after 18 h were treated with 2.5 mg ml^{-1} vancomycin, 1.0 mg ml^{-1} ciprofloxacin, 2500 IU ml^{-1} sodium heparin,

Table 6 Case studies depicting the efficacy of the antimicrobial lock treatment incorporating amphotericin B for catheter–associated infections

Ref[a]	Organism	Treatment	Outcome

1	*Candida parapsilosis*	Concen: 2.5 mg ml^{-1}
		Dwell: 48–72 h
		Duration: 14 days
		    removed
			Clinical relapse within
			    2 weeks post–ALT,
			    catheter

2	*Candida albicans*	Concen: 2.0 mg ml^{-1}
		Dwell: 12 h
		Duration: 10–14 days
		    post–ALT
			Catheter blood culture
			    negative, no infection
			    4 weeks

3	*Candida parapsilosis (2)*[b]	Concen: 0.33 mg ml^{-1}
		Dwell: 12 h
		Duration: 5–17 days
			Catheter blood culture
			    positive post–ALT,
			    CRBSI not resolved

	Candida tropicalis	Concen: 0.33 mg ml^{-1}
		Dwell: 12 h
		Duration: 5–17 days
		    not resolved
			Catheter blood
			    culture positive
			    post–ALT, CRBSI

	Candida tropicalis,
	Klebsiella pneumoniae
		Concen: 0.33 mg ml^{-1}
		+ 83 mg ml^{-1}
		Ceftazidime
		Dwell: 12 h
		Duration: 5–17 days
			Catheter blood post– ALT,
			    culture positive CRBSI
			    not resolved

[a] 1, Bailey et al. 2002; 2, Johnson et al. 1994; 3, Kryzwda et al. 1995
[b] Number of separate catheters colonized by the same organism

or a combination of vancomycin and heparin at the levels given above for a 24-h dwell time. Ciprofloxacin was more effective than vancomycin, though neither antimicrobial agent was able to completely eradicate the biofilms. Heparin alone was completely ineffective, and heparin in combination with either ciprofloxacin or vancomycin did not improve efficacy. A study by Giacometti et al. (2005) found that neither vancomycin nor ciprofloxacin were highly effective against catheter-associated biofilms of *S. aureus* in a rat model, though quinupristin-dalfopristin provided a 4-log biofilm plate count reduction compared to the untreated control under the same conditions. These results are not encouraging since none of the agents evaluated in these studies were capable of eradicating biofilms on implanted devices. Another approach is to combine agents with different modes of action. Kandemir et al. (2005) implanted silicone drains into the medullary canal of rats that were then inoculated with *Pseudomonas aeruginosa* and allowed biofilms to form on these inserts for 14 days. The treatment groups included the control, a group treated subcutaneously every 24 h with 1,500 mg kg^{-1} ceftazidime, and a group treated subcutaneously every 24 h with 1,500 mg kg^{-1} ceftazidime in combination with 100 mg kg^{-1} clarithromycin, which was administered orally every 12 h. After treatment for 20 days, biofilm levels on the implanted drains and explanted bone material were determined. Ceftazidime treatment had no significant effect on biofilm plate count levels compared to the untreated control group, but the addition of clarithromycin significantly reduced biofilms. In another study (Yamasaki et al. 2001), 60 mg kg^{-1} roxithromycin was given to mice containing 48-h *S. aureus* tissue biofilms by intramuscular (im) injection every 12 h alone or in combination with 200 mg kg^{-1} imipenem (by im injection) for 5 or 8 days. Treatment with either antimicrobial agent alone was ineffective in reducing biofilm plate count levels, but the combination significantly reduced, but did not eradicate viable bacteria isolated from tissue biofilms. In summary, none of these approaches were shown to eradicate biofilms in the animal models in which they were tested, though certain antimicrobial agents or combinations of agents may significantly reduce biofilm levels.

4.3 Studies Using In Vitro Model Systems: Bacterial Biofilms

There is a large number of published papers reporting the effect of antimicrobial agents against biofilms of clinically relevant organisms. Table 7 presents the results of five such studies that specifically investigated, in in vitro model systems, the ability of various agents as ALT to eradicate biofilms. These studies show that biofilms of Gram-positive and Gram-negative bacteria can be eradicated when high concentrations of the agent are used for extended dwell/duration periods. Vancomycin and teicoplanin were most effective against biofilms of *Staphylococcus* spp. when it was used for an extended duration or in combination with other antimicrobial agents. The requirement for prolonged contact times can be expected in light of the substantially slower growth rates of biofilm-associated bacteria and that the glycopeptide antimicrobial agents act by inhibiting cell wall synthesis (Stratton 2005). This is reflected by the extremely high biofilm-minimal inhibitory concentration values in the published literature (Ceri et al. 1999; Sandoe et al. 2006). The efficacy of the glycopeptides has also been shown to be substantially lowered by the slime (EPS) of *S. epidermidis* (Souli and Giamarellou 1998). It would be expected that an extended contact period would be required to completely inactivate biofilm organisms with glycopeptides. Nafcillin, a narrow-spectrum β-lactamase-resistant β-lactam that inhibits cell-wall synthesis, was also effective against Gram-positive bacteria in one of two studies (Lee et al. 2006). Other agents that appeared particularly effective against Gram-positive bacteria were linezolid, rifampin, and ciprofloxacin. Linezolid is an oxazolidinone that inhibits protein synthesis of most Gram-positive bacteria, including strains that are resistant to the β-lactams and glycopeptides (Shinabarger et al. 1997). Others have demonstrated the efficacy of linezolid (El-Azizi et al. 2005; Giacometti et al. 2005), rifampin (a RNA polymerase inhibitor) (Monzon et al. 2002; Peck et al. 2003), and ciprofloxacin (an inhibitor of DNA synthesis) (Giacometti et al. 2005). The basis for the efficacy of these agents against biofilms of Gram-positive bacteria may be their rapid mode of action, ability to penetrate the biofilm extracellular polymeric substance (EPS) matrix (Souli and Giamarellou, 1998), or the fact that their activity is not growth rate-dependent.

Table 7 Evaluation of the ability of antimicrobial agents to eradicate biofilms in in vitro model systems

Ref[a]	Treatment conditions	Outcome
1	*S. epidermidis* MA1 PVC catheter Vancomycin (0.45 mg/ml) Exposure: 1 h every 8 h	Mean log CFU catheter⁻¹ reduction after 3 days =3.19, biofilms eradicated on 0 catheters
	Vancomycin treatment supplemented with: Netilmicin (0.025 mg/ml) Exposure: 1 h every 8 h	Mean log CFU catheter⁻¹ reduction after 3 days=4.98, biofilms eradicated on 3/5 catheters
	Vancomycin treatment supplemented with: Rifampin (0.15 mg/ml) Exposure: 90 min every 12 h	Mean log CFU catheter⁻¹ reduction after 3 days=4.83, biofilms eradicated on 2/5 catheters
	Vancomycin treatment supplemented with: Fosfomycin (0.50 mg/ml) Exposure: 4 h every 6 h	Mean log CFU catheter⁻¹ days=4.92, biofilms 3 eradicated on 3/5 catheters
	Vancomycin (0.45 mg/ml) Exposure: continuous	Mean log CFU catheter⁻¹ reduction after 3 days=3.29, biofilms eradicated on 0 catheters
	Vancomycin (5 mg/ml) Exposure: every 12 h	Mean log CFU catheter⁻¹ reduction after 3 days=5.75 (complete eradication on all catheters)
2	*E. aerogenes* (clinical strain) Silicone catheter Aztreonam (83.3 mg/ml) Exposure: 12 h	0 CFU cm⁻¹ after 4 day exposure
	S. epidermidis M187sp11 Silicone catheter Ceftriaxone (83.3 mg/ml) Exposure: 12 h	0 CFU cm⁻¹ after 7–day exposure
	K. pneumoniae (clinical strain) Gentamicin (13.3 mg/ml) Exposure: 12 h	0 CFU cm⁻¹ after 1–day exposure
	S. aureus ATCC29213 Nafcillin (83.3 mg/ml) Exposure: 12 h	0 CFU cm⁻¹ after 1–day exposure
	S. aureus ATCC 29213 Vancomycin (83.3 mg/ml) Exposure: 12 h	0 CFU cm⁻¹ after 7–day exposure
	C. albicans ATCC90028 Amphotericin B (1.0 mg/ml) Exposure: 12 h	0 CFU cm⁻¹ after 7–day exposure
	C. tropicalis (clinical strain) Amphotericin B (1.0 mg/ml) Exposure: 12 h	0 CFU cm⁻¹ after 7–day exposure
	C. albicans ATCC 90028 Fluconazole (2.0 mg/ml) Exposure: 12 h	0 CFU cm⁻¹ after 7–day exposure
	C. tropicalis (clinical strain) Fluconazole (2.0 mg/ml) Exposure: 12 h	20 CFU cm⁻¹ after 7–day exposure

(continued)

Table 7 (continued)

Ref[a]	Treatment conditions	Outcome
3	*S. epidermidis* ATCC 35984 Polyurethane coupons Linezolid (2.0 mg/ml) Exposure: continuous	0 CFU coupon[-1] after 3–day exposure
	Vancomycin (10 mg/ml) Exposure: continuous	0 CFU coupon[-1] after 10–day exposure
	Gentamicin (10 mg/ml) Exposure: continuous	6.81 Log CFU coupon[-1] after 10–day exposure
	Eperezolid (4 mg/ml) Exposure: continuous	0 CFU coupon[-1] after 7–day exposure
4	*S. epidermidis* P15 Catheter segments Vancomycin (0.004 mg/ml) Amikacin (0.016 mg/ml) Exposure: continuous	0 CFU catheter segment[-1] after 24–h exposure
	Vancomycin (0.004 mg/ml) Rifampin (0.001 mg/ml) Exposure: continuous	0 CFU catheter segment[-1] after 24 h exposure
	Teicoplanin (0.064 mg/ml) Amikacin (0.016 mg/ml) Exposure: continuous	0 CFU catheter segment[-1] after 24–h exposure
	Teicoplanin (0.064 mg/ml) Rifampin (0.001 mg/ml) Exposure: continuous	0 CFU catheter segment[-1] after 24–h exposure
5	*S. epidermidis* ATCC 35984 Polyurethane sheets Vancomycin (5.0 mg/ml) Exposure: continuous	0 CFU sheet[-1] after 5–day exposure
	Teicoplanin (5 mg/ml) Exposure: continuous	0 CFU sheet[-1] after 7–day exposure
	Ciprofloxacin (5 mg/ml) Exposure: continuous	0 CFU sheet[-1] after 1–day exposure
	Rifampin (5 mg/ml) Exposure: continuous	0 CFU sheet[-1] after 1–day exposure
	Cefazolin (5 mg/ml) Exposure: continuous	>2 Log CFU sheet[-1] after 14–day exposure (Not eradicated)
	Gentamicin (5 mg/ml) Exposure: continuous	>7 Log CFU sheet[-1] after 14–day exposure (Not eradicated)
	Nafcillin (5 mg/ml) Exposure: continuous	0 CFU sheet[-1] after 14–day exposure
	Erythromycin (5 mg/ml) Exposure: continuous	>8 Log CFU sheet[-1] after 14–day exposure (not eradicated)

[a] 1, Gaillard et al. 1990; 2, Andris et al. 1998; 3, Curtin et al. 2003; 4, Pascual et al. 1994; 5, Lee et al. 2006

Studies documenting the ability of antimicrobial agents to eradicate biofilms of Gram-negative bacteria are few. However, as shown in Table 7, the gentamicin ALT was very effective against *Klebsiella pneumoniae*, eradicating biofilm organisms within 1 day. This is interesting in light of the fact that several studies have suggested aminoglycoside antimicrobials, such as gentamicin, are ineffective against biofilms of Gram-negative bacteria due to binding of the agent by the biofilm EPS matrix (Gordon et al. 1988; Hoyle et al. 1992), slow growth rates of bacteria in biofilms (Nichols et al. 1989), or diminished oxygen levels in biofilm cell clusters (Tresse et al. 1995). Others have found some, but not all, of the aminoglycosides to be effective against biofilms. For example, Ceri et al. (1999) and Domingue et al. (1994) showed that tobramycin was highly effective against biofilms of *P. aeruginosa* while gentamicin and amikacin were generally ineffective. It would be interesting to evaluate antimicrobial locks containing one or more of the aminoglycosides against biofilms of other organisms. An advantage of gentamicin is that it is fast-acting (Stratton 2005), and this property combined with the very high concentration apparently allows it to overwhelm the cells in the biofilm. There is also evidence that the fluoroquinolone antimicrobial agents may be effective against biofilm organisms. Ali-Abdi et al. (2006) found that biofilms of *P. aeruginosa* grown on catheter pieces for 6 days could be eradicated by 4 µg ml^{-1} (16X MIC) of ciprofloxacin and 32 µg/ml (64X MIC) of ofloxacin. They also provided data showing more rapid diffusion of these antimicrobial agents through 1% alginate, the primary component of the *P. aeruginosa* EPS matrix, suggesting that at least part of the reason for the efficacy was due to greater penetration through the biofilm matrix.

Combining agents with differing modes of action may also be effective in eradicating or reducing biofilm counts. For example, Sandoe et al. (2006) showed that the addition of 10 µg ml^{-1} gentamicin to linezolid reduced the minimal biofilm inhibitory concentration (MBIC) for *Enterococcus faecalis* and *E. faecium* isolates; addition of gentamicin to vancomycin also reduced the MBIC for *E. faecium*. Peck et al. (2003) found that erythromycin and rifampin were synergistic with vancomycin in reducing (but not eradicating) biofilms of *S. epidermidis*. Pascual et al. (1994) showed that vancomycin and teicoplanin were effective against biofilms of *S. epidermidis* only when combined with either amikacin or rifampin. Rifampin is highly effective against staphylococci, including β-lactamase-producing strains and methicillin-resistant strains (Kucers et al. 1997). However, resistant mutants of staphylococci easily arise in vivo and in vitro in the presence of rifampin, though resistance can usually be prevented by combining other antimicrobial agents with rifampin. Gagnon et al. (1991) found that cell-wall-active agents such as vancomycin, β-lactams, and cephalosporins demonstrated synergy with rifampin against biofilms of *S. epidermidis*. Rifamipin had a much higher rate of action against biofilms in vitro than several other antimicrobial agents, including cephalosporins, vancomycin, gentamicin, and ciprofloxacin, but rifampin-resistant survivors also developed with time (Richards et al. 1991). However, combining the cell-wall-active agents (vancomycin, cefazolin) with rifampin effectively killed the surviving rifampin-resistant organisms. The authors surmised that the molecular characteristics of rifampin allowed more rapid uptake by the biofilm cells, with resistant survivors being inactivated by the slower acting cell-wall-active agents. Souli and Giamarellou (1998) also showed that the slime of *S. epidermidis* did not influence the activity of rifampin but substantially reduced the activity of most antimicrobial agents tested. Simon and Simon (1990), in an in vitro study of *S. epidermidis* growing on Teflon vein catheters, also found that rifampin combined with teicoplanin or vancomycin completely inactivated biofilms of all the strains tested.

Several studies have reported on the ability of the macrolide antimicrobial agents to provide synergy with other agents. Yasuda et al. (1994) showed that clarithromycin reduced the EPS and hexose concentrations (a measure of the EPS) of *S. epidermidis* biofilms and enhanced the diffusion of ofloxacin or cefotiam through biofilms of this organism. This proposed mechanism of synergy (of the macrolides with other agents) is also consistent with the results of studies in animal models (Kandemir et al. 2005; Yamasaki et al. 2001).

4.3.1 Effect of Biofilm Age

One of the problems with comparing different in vitro studies, even for the same organism, is that conditions used for growing the biofilms may differ significantly. One of the more important variables influencing biofilm susceptibility is the age of the biofilm (Anwar et al. 1992; Chuard et al. 1993; Amorena et al. 1999). A study by Monzon et al. (2002) showed that increasing age of *S. epidermidis* biofilms significantly reduced the efficacy of cephalothin, clindamycin, erythromycin, vancomycin, and teicoplanin; the activity of rifampin and tetracycline were either unaffected or only minimally so. The increasing amounts of EPS produced as the biofilm ages would result in nutrient and oxygen gradients, affecting cell metabolism and growth rates. These factors in turn would impact the activity of the antimicrobial agents.

4.4   In Vivo and In Vitro Studies Investigating Fungal Biofilms

An increasing proportion of device-related infections, in particular device-related bloodstream and urinary tract infections, are caused by *Candida* spp. Mortality rates due to *Candida* infections on vascular catheters may be as high as 26%-38% (Kojic and Darouiche 2004). Because removal of the catheter is normally required to resolve these infections (Kojic and Darouiche 2004), special consideration should be given to the effect of antifungal agents on *Candida* biofilms. Schinabeck et al. (2004) used a rabbit model to examine the effect of either a liposomal amphotericin B or fluconazole lock on biofilms of *Candida albicans* on implanted catheters. The concentration and dwell times for these locks were 10 mg ml^{-1} and 8 h in both cases for 7 days. Results showed that the liposomal amphotericin B lock completely eradicated the biofilms (2-3 log reduction, $p<0.001$). By comparison, there was no significant difference between the fluconazole-treated and untreated catheters. Results from a study by Andris et al. (1998), shown in Table 7, showed that amphotericin B was more effective than fluconazole against *Candida tropicalis*, but equally effective (albeit at half the dosage) against *C. albicans* in an in vitro model system. Others have evaluated the effect of various antifungal drugs against *Candida* spp. Ramage et al. (2002) found that fluconazole at all doses tested was ineffective against biofilms of different strains of *C. albicans*, while amphotericin B provided more than 95% reduction in viable cells at the three highest concentrations tested, though it was ineffective at levels that can be used therapeutically (0.125 µg ml^{-1} and 0.5 µg ml^{-1}). Kuhn et al. (2002) found that the lipid formulation of amphotericin B (liposomal amphotericin B) was more effective against *Candida* spp. biofilms than the non-lipid formulation. Ramage et al. (2002) found that caspofungin was the most effective antimicrobial agent tested: 0.125 and 0.5 µg ml^{-1} (levels achievable in humans) provided greater than 99% reduction (i.e., >2 log reduction) in viable cells after 48 h. Kuhn et al. (2002) also found the echinocandins, caspofungin, and micafungin to be effective against biofilms of *C. albicans* and *Candida parapsilosis*. In in vitro model studies, Bachmann et al. (2003) showed that caspofungin was synergistic with fluconazole against *C. albicans* biofilms. It can be expected that the polyenes, such as amphotericin B, which are fungicidal and bind to membrane sterols resulting in membrane permeability and cell death (Georgopapadakou and Walsh 1994), would be more effective against *Candida* biofilms than the azoles (fluconazole), which are fungicidal and act by increasing membrane permeability and membrane-bound enzyme activity (Kucers et al. 1997).

The echinocandins are also fungicidal, but act by inhibiting the synthesis of 1,3-β-D-glucan, a cell-wall component (Graybill 2001). The results of these in vitro studies suggest that the echinocandins and in some cases liposomal amphotericin B could be effective against biofilms of *Candida* spp., though as the studies discussed suggest, total eradication of fungal biofilms may be elusive.

4.5   Can the ALT Eradicate Biofilms?

In reviewing the results of case studies, when high concentrations of the antimicrobial agent (usually in the milligram per milliliter range) were used for dwell times of 12 h or more, most studies showed that within 14 days of treatment bloodstream infections associated with catheter biofilms of different bacteria were treated successfully. However, for most of the studies shown in Tables 2-6 (Krzywda et al. 1995; Messing et al. 1988; Bailey et al. 2002; Johnson et al. 1994; Benoit et al. 1995), treatment efficacy was based upon negative blood cultures (collected through the CVC) following treatment and/or absence of clinical symptoms in the patient. None of these studies provided definitive evidence for the presence or absence of biofilms on the catheter surface. However, Capdevila et al. (1993) and Rao et al. (1992) did examine catheter tips for biofilms using the roll tip method (with its questionable accuracy [Donlan 2001]) and found that negative blood cultures and absence of clinical symptoms predicted biofilm eradication, though the number of samples in each of these studies was relatively small. Therefore, the evidence from case studies (and studies using animal models) supporting the ability of the antimicrobial lock treatment to eradicate biofilms is weak. On the other hand, studies using in vitro models suggest that antimicrobial agents can eradicate bacterial biofilms, when used in sufficiently high concentrations for extended periods of time. The agents that were most effective against bacterial biofilms of Gram-positive organisms were vancomycin, teicoplanin, linezolid, rifampin, and ciprofloxacin. Gentamicin was effective against biofilms of Gram-negative organisms. ALTs that contain combinations of antimicrobial agents with differing modes of action may also be effective. For example, the macrolides can degrade the biofilm EPS and enhance diffusion of other supplemental bactericidal agents through this matrix. Fungal biofilms appear to be much more difficult to eradicate, though the use of liposomal amphotericin B or the echinocandins appear promising.

One concern with the use of antimicrobial agents at such high concentrations is the possibility of toxicity to the patient resulting from the diffusion of the lock solution into systemic circulation or if the lock is inadvertently flushed into circulation. Berrington and Gould (2001) suggested highest practical concentrations for antimicrobial agents in the ALT, including vancomycin (50 mg ml^{-1}), gentamicin (40 mg ml^{-1}), and teicoplanin (133 mg ml^{-1}). These levels are generally well above effective concentrations used in most of the in vitro model system studies discussed, so this would suggest that drug concentrations used in these studies would be relevant in patient treatment. Conditions impacting biofilm growth and susceptibility of biofilm organisms to the antimicrobial agent would be substantially different in an animal or human than in an in vitro model, supporting the case for more rigorous testing conditions in order to provide more clinically relevant data. For example, inclusion of serum proteins, use of mixed species biofilms, and growth of the biofilms for extended time periods would provide a more clinically relevant biofilm model system. This in turn could provide stronger evidence for the efficacy of the ALT against biofilms of different microorganisms.

Another concern with the use of antimicrobial agents in antimicrobial locks is the potential for the development of resistance. The increased use of antimicrobial agents in the healthcare setting, especially in intensive care units, has selected for resistant organisms. For example, the increase in vancomycin-resistant enterococci (VRE) in healthcare institutions has been linked in part to the increase in vancomycin use, which in turn provided selective pressure leading to the spread of VRE (Tenover 2001; Schwalbe et al. 1987; Kaplan et al. 1988; Sieradzki et al. 2003). Proportions of resistant staphylococci, enterococci, and pseudomonads are highest in organisms isolated from patients in intensive care units where use of antimicrobial agents is greatest (Archibald et al. 1997). In light of the concern for the development and spread of organisms resistant to vancomycin, the CDC has recommended that the use of ALTs containing vancomycin should be discouraged (Centers for Disease Control and Prevention 1995) or used only in special circumstances, such as the treatment of patients with long-term cuffed or tunneled catheters or ports who have had a history of multiple CRBSI (Centers for Disease Control and Prevention 2002).

As noted by Berrington and Gould (2001), until a systematic study comparing the use of the ALT alone, ALT plus systemic antimicrobial agents, and systemic antimicrobial agents alone for the treatment of catheter-related infections is conducted, the results of such studies evaluating the ALT should be viewed cautiously.

5 Studies Investigating the Use of Novel Catheter Lock Solutions for the Eradication of Biofilms

Other less conventional approaches, using agents that are not classified as antimicrobial agents, have been evaluated using the ALT approach. For example, tetrasodium EDTA (ethylene diamine tetraacetic acid) or disodium EDTA used alone or in combination with minocycline have been used effectively against bacterial and fungal biofilms. EDTA has antimicrobial properties (Raad et al. 1997; Root et al. 1988) against bacteria and fungi, and may also destabilize the biofilm structure (Turakhia et al. 1983). Percival et al. (2005) and Kite et al. (2004) showed that 40 mg ml^{-1} of tetrasodium EDTA could eradicate biofilms in an in vitro model and on explanted hemodialysis catheters, respectively. In the in vitro model system, the treatment eradicated biofilms after a 21-h dwell time against biofilms of *S. epidermidis*, *P. aeruginosa*, *K. pneumoniae*, and *E. coli* grown for 48 h; after a 25-h dwell time, biofilms of MRSA and *C. albicans* were also eradicated. Raad et al. (2003) tested 18-h biofilms of *S. epidermidis*, *S. aureus*, and *C. albicans* against combinations of minocycline and disodium EDTA and found that 0.1 mg ml^{-1} minocycline plus 30 mg ml^{-1} EDTA significantly reduced (but did not eradicate) biofilms of each organism. Biofilms on explanted catheter tips were also substantially reduced (tenfold or more) by a combination of 3 mg ml^{-1} minocycline and 30 mg ml^{-1} EDTA. This treatment approach was also effective in the treatment of CRBSI in three different patient studies, as evidenced by remission of symptoms and negative catheter tip cultures (Raad et al. 1997). A novel lock treatment containing taurolidine (2 H-1,2,4-thiadiazine-4,4'-methylenebis(tetrahydro-1,1,1'-tetroxide) eradicated 72-h biofilms of *S. aureus*, *S. epidermidis*, and *E. faecalis* in an in vitro model when they were exposed to 5,000 U ml^{-1} for 24 h (Shah et al. 2002). Taurolidine is a derivative of the amino acid taurine, which inhibits and kills a broad range of microorganisms. Its proposed mechanism of action is based on the interaction of methylol derivatives with components of the bacterial cell wall, resulting in cell damage (Torres-Viera et al. 2000). Metcalf et al. (2004) instilled 70% ethanol in a Hickman catheter and combined this treatment with intravenous amoxicillin to resolve an *E. coli* bloodstream infection in a patient. The catheter was locked with ethanol between total parenteral nutrition infusions for a period of 3 days, and remained free of infection for more than 3 years, when the study was completed. Assuming patient compatibility, the next step would be to evaluate these treatments in animal models and in patient studies, using guidelines that have been suggested for the traditional antimicrobial lock treatments.

6 New Technologies

The scientific and patent literature is filled with novel concepts that have been reported to control or eradicate microbial biofilms. Several that appear promising follow.

6.1 Ultrasound as an Adjuvant for Antibiotics

Ultrasound has been shown to enhance the efficacy of antimicrobial agents against planktonic and biofilm Gram-negative bacteria, and has been termed the bioacoustic effect (Johnson et al. 1998). Johnson et al. (1998) demonstrated that 20 µg ml^{-1} or 30 µg ml^{-1} gentamicin could eradicate biofilms of *E. coli* in an in vitro model when combined with 70-kHz ultrasound, whereas biofilms treated with either ultrasound alone or gentamicin alone were unaffected. Carmen et al. (2005) demonstrated that ultrasound could act synergistically with gentamicin to reduce biofilms of *E. coli* in an animal model. Biofilms were grown for 24 h on polyethylene disks that were then implanted in New Zealand White female rabbits. Gentamicin (8 mg kg^{-1}) was injected subcutaneously into the rabbits every 24 h, and ultrasound was applied using a 28.5-kHz ultrasound transducer at an intensity of 500 mW cm$^{2-1}$. Disks were removed from the rabbits after 72 h of exposure to the treatment and processed to recover and quantify the biofilms. The results show that ultrasound plus gentamicin was significantly more effective than gentamicin alone for controlling *E. coli* biofilms. The mechanism for this synergy is not known, but may be the result of perturbation of the cell membrane or to a stress response by the organisms (Rediske et al. 1999).

6.2 Bacteriophage

Bacteriophage can also be instilled into catheters as a lock treatment to eradicate biofilms. Doolittle et al. (1995) reported that Phage T4 significantly reduced biofilms of *E. coli* in an in vitro model system. Biofilms were grown for 28 h in a Modified Robbins Device prior to the addition of phage. Viable biofilm cell counts were reduced by 6 logs within 5 h of treatment. Phage numbers increased initially during the first 5 h then decreased as the number of surviving biofilm cells diminished. Hanlon et al. (2001) investigated the effect of Phage F116 on biofilms of *P. aeruginosa* in microtiter plates and showed that intact biofilms were more tolerant to phage attack than suspended biofilm cells. They also found that an increase in biofilm age did not appear to significantly decrease susceptibility, as has been observed during the treatment of biofilms with antimicrobial agents. Phage treatment was effective on biofilms grown for 20 days prior to treatment. Other published studies have also demonstrated the efficacy of phage lock treatments against biofilms of different bacteria (Hughes et al. 1998; Sillankova et al. 2004). For this treatment approach to work, organisms isolated from the colonized device would need to be screened against a bank of phages to determine the specific phage strain with greatest lytic ability. This strain could be grown to a high titer then instilled into the indwelling catheter as a lock treatment.

6.3　Quorum-Sensing Inhibitors

Bacteria such as *P. aeruginosa* have been shown to regulate biofilm formation by means of quorum-sensing agents. These are small diffusible molecules produced by the organisms themselves that regulate virulence and important biofilm properties such as architecture and susceptibility to antimicrobial agents and the host immune system. It has been recognized that quorum-sensing inhibitors could provide a viable approach for the control of clinically relevant bacteria (Rice et al. 2005). In *P. aeruginosa*, quorum sensing is mediated by the N-acylated homoserine lactones (AHLs) (Davies et al. 1998). Crude extracts of garlic have been shown to specifically inhibit quorum-sensing gene expression in *P. aeruginosa* (Rasmussen et al. 2005). Using an in vitro model, Rasmussen et al. also demonstrated that *P. aeruginosa* PAO1 biofilms grown in the presence of garlic extract were substantially more susceptible to tobramycin treatment than were untreated or garlic extract-only-treated biofilms. Bjarnsholt et al. (2005) showed that polymorphonuclear leukocytes (PMNs) were activated in the presence of biofilms of *P. aeruginosa* grown for 3 days in media containing 2% garlic extract, resulting in extensive grazing and phagocytosis of the biofilm. Quorum-sensing-deficient *P. aeruginosa* mutants also exhibited PMN activation and phagocytosis, supporting the role of garlic extract as a quorum-sensing inhibitor. Studies in an animal model also showed that treatment could stimulate the immune response and clear the introduced bacteria. It is still unclear from this work how established biofilms of this organism would respond to this treatment.

Balaban et al. (2003) described a heptapeptide termed RNA III-inhibiting peptide (RIP) that was shown to inhibit biofilm formation and toxin formation by *S. aureus*. *S. aureus* continually secretes a 33-kDa protein called ribonucleic acid III (RNA III)-activating peptide (RAP). When RAP achieves a threshold concentration, it induces phosphorylation of its target protein (termed TRAP), a 21-kDa protein which in turn results in increased cell adhesion and toxin production. RIP will down-regulate TRAP phosphorylation, reducing cell adhesion and suppressing toxin production. *S. epidermidis* incubated in the presence of 5-µg RIP exhibited reduced adhesion to HaCat human skin keratinocytes and polystyrene surfaces. Mutant TRAP strains of *S. aureus* (simulating cells that had been inhibited with RIP) also produced significantly less biofilm in flow cells and in membrane colony biofilm systems (Balaban et al. 2005). These authors also found that biofilm formation by several *S. aureus* and *S. epidermidis* strains on Dacron grafts implanted into rats was significantly reduced when these grafts were first soaked in a 20-mg/l RIP solution and this treatment was combined with parenteral administration of RIP. These results suggest that RIP might also be capable of eradicating biofilms on CVCs.

6.4 Enzymes That Degrade the Biofilm EPS

The biofilm structural matrix, termed EPS, is composed of polymers, primarily polysaccharide in nature. Alginate, the EPS of *P. aeruginosa*, can be enzymatically depolymerized by alginate lyases. Hatch and Shiller (1998) showed that alginate retarded the diffusion of aminoglycosides and inhibited their antimicrobial activity. However, addition of alginate lyase allowed greater penetration of gentamicin and tobramycin through alginate and greater activity of these agents against *P. aeruginosa*. This would suggest that alginate lyase might enhance the effectiveness of antimicrobial agents in the treatment of biofilms. In a subsequent study, Alkawash et al. (2006) treated biofilms of two different mucoid *P. aeruginosa* strains in an in vitro model with 64 μg ml^{-1} gentamicin with and without 20 U ml^{-1} alginate lyase and found that the enzyme significantly improved the efficacy of the antimicrobial agent. After 120 h, the combination treatment had eradicated the biofilms of each strain, whereas between 6 and 7 log CFU ml^{-1} of the biofilm organisms that were treated with gentamicin alone still survived after this exposure period. Enzymes targeting other polymers comprising the bacterial EPS might also be effective in this regard.

7 Evaluating Biofilm Eradication Strategies

A systematic approach to assess biofilm eradication treatment strategies might include the following suggestions.

- Develop an in vitro model that reasonably simulates the indwelling catheter biofilm with respect to substratum, properties of the growth medium, biofilm age and cell density, and presence of serum proteins, and that uses bloodstream isolates of clinically relevant organisms. Murga et al. (2001) and Curtin and Donlan (2006) can be consulted for examples of in vitro model systems for growing and testing biofilms on indwelling medical devices. The Drip Flow Reactor (Curtin and Donlan 2006) can be modified and used for growing and testing biofilms on the lumens of central venous catheters (Figs. 3, 4). Other screening approaches incorporating the MBEC Device (Ceri et al. 1999) or the CDC Biofilm Reactor (Donlan et al. 2004) can provide higher throughput testing but under less relevant conditions.
- Validate results obtained in the in vitro model under more rigorous conditions by using either explanted biofilms (as done by Kite et al. 2004) and/or an animal model. Treatments that appear effective in in vitro models often do not show the same level of effectiveness in animal models, due to such complicating factors as the response of the host immune system and presence of serum proteins.
- Ascertain that the treatment can be tolerated by the patient and is compatible with the normal-use regimen of the device.

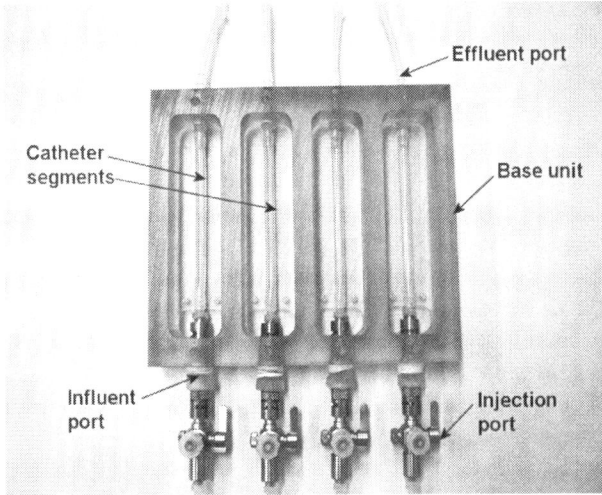

Fig. 3 Modified Drip Flow Reactor

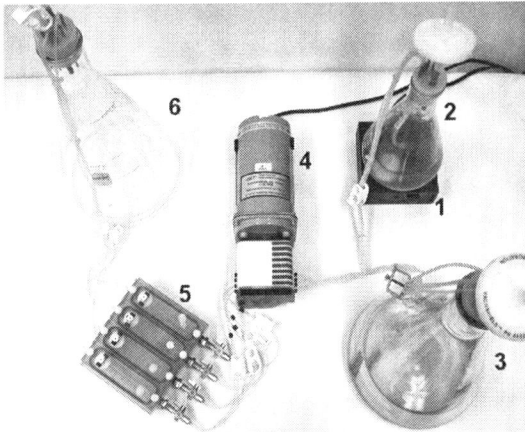

Fig. 4 In vitro model system for growing and testing biofilms on central venous catheters. *1*, Mixing plate; *2*, bacterial inoculum; *3*, sterile medium reservoir; *4*, peristaltic pump; *5*, Modified Drip Flow Reactor containing catheter sections; *6*, waste reservoir. Bacterial inoculum is pumped through catheter sections for approximately 2 h to provide initial bacterial attachment, followed by continuous irrigation with sterile medium to allow biofilm growth

- Assure that catheter biofilms are recovered and quantified when conducting clinical studies to evaluate the treatment. Resolution of patient symptoms may not predict eradication of the biofilm from the catheter. Biofilm recovery and detection methods should also be validated (Donlan et al. 2001).

8   Conclusions

Infections associated with the use of central venous catheters place a burden on healthcare providers and are a significant source of morbidity and mortality among populations. These infections are commonly associated with microorganisms that have colonized the device and formed microbial biofilms. The tolerance toward antimicrobial agents exhibited by biofilms has led to the use of treatment strategies that directly target the catheter surface, with the intent that elevated concentrations of the antimicrobial agent for extended dwell times will kill and eradicate the biofilms and avoid the alternative - removing the catheter from the patient. Antimicrobial lock treatments have met with mixed success in laboratory and clinical studies. All antimicrobial agents are not equal with respect to biofilm eradication. Consideration should be given to the mechanism of action of the agent, its interaction with the biofilm EPS, and the effect of such biofilm-specific characteristics as local oxygen concentration, biofilm age, and cellular growth rate. Systematic studies in larger patient populations are needed. However, the potential for the spread of antimicrobial resistance, particularly in very sick patients, argues in favor of newer approaches that are highly effective in eradicating biofilms while avoiding or greatly minimizing the use of antimicrobial agents. Clearly, this presents a challenge that will require the scientific community to work closely with the regulatory community, healthcare industry, and infectious disease clinicians.

Acknowledgements Use of trade names is for identification only and does not constitute endorsement by the Public Health Service or by the U.S. Department of Health and Human Services. The author would like to thank Janice Carr for the scanning electron microscopic image used in Fig. 2.

References

Ali-Abdi A, Mohammadi-Mehr M, Agha Alaei YA (2006) Bactericidal activity of various antibiotics against biofilm-producing *Pseudmonas aeruginosa*. Int J Antimicrob Agents 27:196-200

Alkawash MA, Soothill JS, Schiller NL (2006) Alginate lyase enhances antibiotic killing of mucoid *Pseudomonas aeruginosa* in biofilms. APMIS 114:131-138

Amorena B, Gracia E, Monzon M, Leiva J, Oteiza C, Perez M, Alabart J-L, Hernandez-Yago J (1999) Antibiotic susceptibility assay for *Staphylococcus aureus* in biofilms developed in vitro. J Antimicrob Chemother 44:43-55

Anaissie E, Samonis G, Kontoyiannis D, Costerton J, Sabharwal U, Bodey G, Raad I (1995) Role of catheter colonization and infrequent hematogenous seeding in catheter-related infections. Eur J Clin Microbiol Infect Dis 14:135-137

Andris DA, Krzywda EA, Edmiston CE, Krepel CJ, Gohr CM (1998) Elimination of intraluminal colonization by antibiotic lock in silicone vascular catheters. Nutrition 14:427-432

Anthony TU, Rubin LG (1999) Stability of antibiotics used for antibiotic-lock treatment of infections of implantable venous devices (ports). Antimicrob Agents Chemother 43:2074-2076

Anwar H, Strap JL, Chen K, Costerton JW (1992) Dynamic interactions of biofilms of mucoid *Pseudomonas aeruginosa* with tobramycin and pipercillin. Antimicrob Agents Chemother 36:1208-1214

Archibald L, Phillips L, Monnet D, McGowan JE Jr, Tenover F, Gaynes R (1997) Antimicrobial resistance in isolates from inpatients and outpatients in the United States: increasing importance of the intensive care unit. Clin Infect Dis 24:211-215

Aufwerber E, Ringertz S, Ransjo U (1991) Routine semiquantitative cultures and central venous catheter-related bacteremia. APMIS 99:627-630

Bachmann SP, Ramage G, VandeWalle K, Patterson TF, Wickes BL, Lopez-Ribot JL (2003) Antifungal combinations against *Candida albicans* biofilms in vitro. Antimicrob Agents Chemother 47:3657-3659

Bailey E, Berry N, Cheesbrough JS (2002) Antimicrobial lock therapy for catheter-related bacteraemia among patients on maintenance haemodialysis. J Antimicrob Chemother 50:611-618

Balaban N, Giacometti A, Cirioni O, Gov Y, Ghiselli R, Mocchegiani F, Viticchi C, Simona Del Prete M, Saba V, Scalise G, Dell'Acqua G (2003) Use of the quorum-sensing inhibitor RNAIII-Inhibiting peptide to prevent biofilm formation in vivo by drug-resistant *Staphylococcus epidermidis*. J Infect Dis 187:625-630

Balaban N, Stoodley P, Fux CA, Wilson S, Costerton JW, Dell'Acqua G (2005) Prevention of staphylococcal biofilm-associated infections by the quorum sensing inhibitor RIP. Clin Orthop Rel Res 437:48-54

Benoit J-L, Carandang G, Sitrin M, Annow PM (1995) Intraluminal antibiotic treatment of central venous catheter infections in patients receiving parenteral nutrition at home. Clin Infect Dis 21:1286-1288

Berrington A, Gould FK (2001) Use of antibiotic locks to treat colonized central venous catheters. J Antimicrob Chemother 48:597-603

Bjarnsholt T, Jensen PO, Rasmussen TB, Christophersen L, Calum H, Hentzer M, Hougen H-P, Rygaard J, Moser C, Eberl L, Hoiby N, Givskov M (2005) Garlic blocks quorum sensing and promotes rapid clearing of pulmonary *Pseudomonas aeruginosa* infections. Microbiology 151:3873-3880

Brun-Buisson C, Abrouk F, Legrand P, Huet Y, Larabi S, Rapin M (1987) Diagnosis of central venous catheter-related sepsis. Critical level of quantitative tip culture. Arch Intern Med 147:873-877

Capdevila JA, Segarra A, Planes AM, Ramirez-Arellano M, Pahissa A, Piera L, Martinez-Vazquez JM (1993) Successful treatment of haemodialysis catheter-related sepsis without catheter removal. Nephrol Dial Transplant 8:231-234

Capdevila JA, Gavalda J, Fortea J, Lopez P, Martin MT, Gomis X, Pahissa A (2001) Lack of antimicrobial activity of sodium heparin for treating experimental catheter-related infection due to *Staphylococcus aureus* using the antibiotic-lock technique. Clin Microbiol Infect 7:206-212

Carmen JC, Roeder BL, Nelson JL, Robison RL, Robison RA, Schaalje GB, Pitt WG (2005) Treatment of biofilm infections on implants with low-frequency ultrasound and antibiotics. Am J Infect Control 33:78-82

Centers for Disease Control and Prevention (1995) Recommendations for preventing the spread of vancomycin resistance: recommendations of the Hospital Infection Control Practices Advisory Committee (HICPAC). MMWR 44 (No. RR-12):1-13

Centers for Disease Control and Prevention (2002) Guidelines for prevention of intravascular catheter-related infections. MMWR 51 (No. RR-10):1-32

Ceri H, Olson ME, Stremick C, Read RR, Morck D, Buret A (1999) The Calgary Biofilm Device: new technology for rapid determination of antibiotic susceptibilities of bacterial biofilms. J Clin Microbiol 37:1771-1776

Chuard C, Vaudaux P, Waldvogel FA, Lew DP (1993) Susceptibility of *Staphylococcus aureus* growing on fibronectin-coated surfaces to bactericidal antibiotics. Antimicrob Agents Chemother 37:625-632

Curtin J, Cormican M, Fleming G, Keelehan J, Colleran E (2003) Linezolid compared with eperezolid, vancomycin, and gentamicin in an in vitro model of antimicrobial lock therapy for

Staphylococcus epidermidis central venous catheter-related biofilm infections. Antimicrob Agents Chemother 47:3145-3148

Curtin JJ, Donlan RM (2006) Using bacteriophages to reduce formation of catheter-associated biofilms by *Staphylococcus epidermidis*. Antimicrob Agents Chemother 50:1268-1275

Darouiche RO, Raad II, Heard SO, Thornby JI, Wenker OC, Gabrielli A, Berg J, Khardori N, Hanna H, Hachem R, Harris R, Mayhall G (1999) A comparison of two antimicrobial-impregnated central venous catheters. N Engl J Med 340:1-8

Davies DG, Parsek MR, Pearson JP, Iglewski BH, Costerton JW, Greenberg EP (1998) The involvement of cell-to-cell signals in the development of a bacterial biofilm. Science 280:295-298

Domingue G, Ellis B, Dasgupta M, Costerton JW (1994) Testing antimicrobial susceptibilities of adherent bacteria by a method that incorporates guidelines of the National Committee for Clinical Laboratory Standards. J Clin Microbiol 32:2564-2568

Donlan RM (2001) Biofilm formation: a clinically relevant microbiological process. Clin Infect Dis 33:1387-1392

Donlan RM (2002) Biofilms: microbial life on surfaces. Emer Infect Dis 8:881-890

Donlan RM, Costerton JW (2002) Biofilms: survival mechanisms of clinically relevant microorganisms. Clin Microbiol Rev 15:167-193

Donlan RM, Murga R, Bell M, Toscano CM, Carr JH, Novicki TJ, Zuckerman C, Corey LC, Miller JM (2001) Protocol for detection of biofilms on needleless connectors attached to central venous catheters. J Clin Microbiol 39:750-753

Donlan RM, Piede JA, Heyes CD, Sanii L, Murga R, Edmonds P, El-Sayed I, El-Sayed MA (2004) Model system for growing and quantifying *Streptococcus pneumoniae* biofilms in situ and in real time. Appl Environ Microbiol 70:4980-4988

Doolittle MM, Cooney JJ, Caldwell DE (1995) Lytic infection of *Escherichia coli* biofilms by bacteriophage T4. Can J Microbiol 41:12-18

El-Azzi M, Rao S, Kanchanapoom T, Kardori N (2005) In vitro activity of vancomycin, quinupristin/dalfopristin, and linezolid against intact and disrupted biofilms of staphylococci. Ann Clin Microb Antimicrob 4:1-9

Elliott TSJ, Moss HA, Tebbs SE, Wilson IC, Bonser RS, Graham TR, Burke LP, Faroqui MH (1997) Novel approach to investigate a source of microbial contamination of central venous catheters. Eur J Clin Microbiol Infect Dis 16:210-213

Flowers RH, Schwenzer KJ, Kopel RF, Fisch MJ, Tucker SI, Farr BM (1989) Efficacy of an attachable subcutaneous cuff for the prevention of intravascular catheter-related infection. JAMA 261:878-883

Gagnon RF, Richards GK, Wiesenfeld L (1991) *Staphylococcus epidermidis* biofilms: unexpected outcome of double and triple antibiotic combinations with rifampin. ASAIO Trans 37:M158-M160

Gaillard J-L, Merlino R, Pajot N, Goulet O, Fauchere J-L, Ricour C, Vernon M (1990) Conventional and nonconventional modes of vancomycin administration to decontaminate the internal surface of catheters colonized with coagulase-negative staphylococci. JPEN J Paren Enteral Nutrit 14:593-597

Georgopapadakou NH, Walsh TJ (1994) Human mycoses: drugs and targets for emerging pathogens. Science 264:371-373

Giacometti A, Cirioni O, Ghisselli R, Orlando F, Mocchegiani F, Silvestri C, Licci A, De Fusco M, Provinciali M, Saba V, Scalise G (2005) Comparative efficacies of quinupristin-dalfopristin, linezolid, vancomycin, and ciprofloxacin in treatment, using the antibiotic-lock technique, of experimental catheter-related infection due to *Staphylococcus aureus*. Antimicrob Agents Chemother 49:4042-4045

Gordon CA, Hodges NA, Marriott C (1988) Antibiotic interaction and diffusion through alginate and exopolysaccharide of cystic fibrosis-derived *Pseudomonas aeruginosa*. J Antimicrob Chemother 22:667-674

Graybill JR (2001) The echinocandins, first novel class of antifungals in two decades: will they live up to their promise? Int J Clin Pract 55:633-638

Haimi-Cohen Y, Husain N, Meenan J, Karayaicin G, Lehrer M, Rubin LG (2001) Vancomycin and ceftazidime bioactivities persist for at least 2 weeks in the lumen in ports: simplifying treatment of port-associated bloodstream infections by using the antibiotic lock technique. Antimicrob Agents Chemother 45:1565-1567

Hanlon GW, Denyer SP, Olliff CJ, Ibrahim LJ (2001) Reduction of exopolysaccharide viscosity as an aid to bacteriophage penetration through *Pseudomonas aeruginosa* biofilms. Appl Environ Microbiol 67:2746-2753

Hatch RA, Schiller NL (1998) Alginate lyase promotes diffusion of aminoglycosides through the extracellular polysaccharide of mucoid *Pseudomonas aeruginosa*. Antimicrob Agents Chemother 42:974-977

Hoyle BD, Wong CKW, Costerton JW (1992) Disparate efficacy of tobramycin on Ca^{+2}-, Mg^{+2}-, and HEPES-treated *Pseudomonas aeruginosa* biofilms. Can J Microbiol 38:1214-1218

Hughes KA, Sutherland IW, Jones MV (1998) Biofilm susceptibility to bacteriophage attack: the role of phage-borne polysaccharide depolymerase. Microbiol 144:3039-3047

Johnson DC, Johnson FL, Goldman S (1994) Preliminary results treating persistent central venous catheter infections with the antibiotic lock technique in pediatric patients. Pediatr Infect Dis J 13:930-931

Johnson LL, Peterson RV, Pitt WG (1998) Treatment of bacterial biofilms on polymeric biomaterials using antibiotics and ultrasound. J Biomater Sci Polymer Edn 9:1177-1185

Kamal GD, Pfaller MA, Rempe, LE, Jebson PJR (1991) Reduced intravascular catheter infection by antibiotic bonding. A prospective, randomized, controlled trial. JAMA 265:2364-2368

Kandemir O, Oztuna V, Milcan A, Bayramoglu A, Celik, HH, Bayarslan C, Kaya A (2005) Clarithromycin destroys biofilms and enhances bactericidal agents in the treatment of *Pseudomonas aeruginosa* osteomyelitis. Clin Othop Rel Res 430:171-175

Kaplan AH, Gilligan PH, Facklam RR (1988) Recovery of resistant enterococci during vancomycin prophylaxis. J Clin Microbiol 26:1216-1218

Kite P, Eastwood K, Sugden S, Percival SL (2004) Use of in vivo-generated biofilms from hemodialysis catheters to test the efficacy of a novel antimicrobial catheter lock for biofilm eradication in vitro. J Clin Microbiol 42:3073-3076

Klevins RM, Tokars JI, Andrus M (2005) Nephrol News Issues 19:37-38, 43

Kowalewska-Grochowska K, Richards R, Moysa GL, Lam K, Costerton JW, King EG (1991) Guidewire catheter change in central venous catheter biofilm formation in a burn population. Chest 100:1090-1095

Kojic EM, Darouiche RO (2004) Candida infections of medical devices. Clin Microbiol Rev 17:255-267

Krzywda EA, Andris DA, Edmiston CE, Quebbeman EJ (1995) Treatment of Hickman catheter sepsis using antibiotic lock technique. Infect Control Hosp Epidemiol 16:596-598

Kucers A, Crowe SM, Grayson ML, Hoy JF (1997) The use of antibiotics: a clinical review of antibacterial, antifungal, and antiviral drugs, 5th edn, Butterworth-Heinemann, Oxford.

Kuhn DM, Chandra GT, Mukheree PK, Ghannoum MA (2002) Antifungal susceptibility of *Candida* biofilms: unique efficacy of Amphotericin B lipid formulations and Echinocandins. Antimicrob Agents Chemother 46:1773-1780

Lee J-Y, Ko KS, Peck KR, Oh WS, Song J-H (2006) In vitro evaluation of the antibiotic lock technique (ALT) for the treatment of catheter-related infections caused by staphylococci. J Antimicrob Chemother 57:1110-1115

Maki DG (1994) Infections caused by intravascular devices used for infusion therapy: pathogenesis, prevention, and management. In: Bisno AL, Waldvogel FA (eds) Infections associated with indwelling medical devices. American Society for Microbiology, Washington DC, pp 155-212

Maki DG, Stolz SM, Wheeler S, Mermel LA (1997) Prevention of central venous catheter-related bloodstream infection by use of an antiseptic-impregnated catheter. A randomized, controlled, trial. Ann Intern Med 127:257-266

Marr KA, Sexton DJ, Conlon PJ, Corey GR, Schwab SJ, Kirkland KB (1997) Catheter-related bacteremia and outcome of attempted catheter salvage in patients undergoing hemodialysis. Ann Inter Med 127:275-280

Mermel LA, Farr BM, Sheretz RJ, Raad II, O'Grady N, Harris JS, Craven DE (2001) Guidelines for the management of intravascular catheter-related infections. Clin Infect Dis 32:1249-1272

Messing B, Peitra-Cohen S, Debure A, Beliah M, Bernier J-J (1988) Antibiotic-lock technique: a new approach to optimal therapy for catheter-related sepsis in home-parenteral nutrition patients. JPEN J Paren Enteral Nutrit 12:185-189

Metcalf SCL, Chambers ST, Pithie AD (2004) Use of ethanol locks to prevent recurrent central line sepsis. J Infect 49:20-22

Monzon M, Oteiza C, Leiva J, Lamarta M, Amorena B (2002) Biofilm testing of *Staphylococcus epidermidis* clinical isolates: low performance of vancomycin in relation to other antibiotics. Diag Microbiol Infect Dis 44:319-324

Moreno CA, Rosenthal VD, Olarte N, Gomez WV, Sussmann O, Agudelo JG, Rojas C, Osorio L, Linares C, Valderrama A, Mercado PG, Bernate PH, Vergara GR, Pertuz AM, Mojica BE, Navarrete Mdel P, Romero AS, Henriquez D (2006) Device-associated infection rate and mortality in intensive care units of 9 Colombian hospitals: findings of the International Nosocomial Infection Control Consortium. Infect Control Hosp Epidemiol 27:349-356

Murga R, Miller JM, Donlan RM (2001) Biofilm formation by Gram-negative bacteria on central venous catheter connectors: effect of conditioning films in a laboratory model. J Clin Microbiol 39:2294-2297

Nichols WW, Evans MJ, Slack MPE, Walmsley HL (1989) The penetration of antibiotics into aggregates of mucoid and non-mucoid *Pseudomonas aeruginosa*. J Gen Microbiol 135:1291-1303

Pascual A, Ramirez de Arellano E, Perea EJ (1994) Activity of glycopeptides in combination with amikacin or rifampin against *Staphylococcus epidermidis* biofilms on plastic catheters. Eur J Clin Microbiol Infect Dis 13:515-517

Peck KR, Kim SW, Jung S-I, Kim Y-S, Oh WS, Lee JY, Jin JH, Kim S, Song J-H, Kobayashi H (2003) Antimicrobials as potential adjuctive agents in the treatment of biofilm infections with *Staphylococcus epidermidis*. Chemother 49:189-193

Percival SL, Kite P, Eastwood K, Murga R, Carr J, Arduino MJ, Donlan RM (2005) Tetrasodium EDTA as a novel central venous catheter lock solution against biofilm. Infect Control Hosp Epidemiol 26:515-519

Raad I (1998) Intravascular-catheter-related infections. Lancet 351:893-898

Raad I, Costerton JW, Sabharwal U, Sacilowski M, Anaissie W, Bodey GP (1993) Ultrastructural analysis of indwelling vascular catheters: a quantitative relationship between luminal colonization and duration of placement. J Infect Dis 168:400-407

Raad I, Buzaid A, Rhyne J, Hachem R, Darouiche R, Safar H, Albitar M, Sherertz RJ (1997) Minocycline and ethylenediaminetetraacetate for the prevention of recurrent vascular catheter infections. Clin Infect Dis 25:149-151

Raad I, Chatzinikolaou I, Chaiban G, Hanna H, Hachem R, Dvorak T, Cook G, Costerton W (2003) In vitro and ex vivo activities of minocycline and EDTA against microorganisms embedded in biofilm on catheter surfaces. Antimicrob Agents Chemother 47:3580-3585

Ramage G, VandeWalle K, Bachmann SP, Wickes BL, Lopez-Ribot JL (2002) In vitro pharmacodynamic properties of three antifungal agents against preformed *Candida albicans* biofilms determined by time-kill studies. Antimicrob Agents Chemother 46:3634-3636

Rao JS, O'Meara A, Harvey T, Breatnach F (1992) A new approach to the management of Broviac catheter infection. J Hosp Infect 22:109-116

Rasmussen TB, Bjarnsholt T, Skindersoe ME, Hentzer M, Kristoffersen P, Kote M, Nielsen J, Eberl L, Givskov M (2005) Screening for quorum-sensing inhibitors (QSI) by use of a novel genetic system, the QSI Selector. J Bacteriol 187:1799-1814

Rediske AM, Roeder BL, Brown MK, Nelson JL, Robison RL, Draper DO, Schaalje GB, Robison RA, Pitt WG (1999) Ultrasonic enhancement of antibiotic action on *Escherichia coli* biofilms: an in vivo model. Antimicrob Agents Chemother 43:1211-1214

Rice SA, McDougald D, Kumar N, Kjelleberg S (2005) The use of quorum-sensing blockers as therapeutic agents for the control of biofilm-associated infections. Curr Opin Investig Drugs 6:178-184

Richards GK, Gagnon RF, Prentis J (1991) Comparative rates of antibiotic action against *Staphylococcus epidermidis* biofilms. ASAIO Trans 37:M160-M162

Root JL, McIntyre R, Jacobs NJ, Daghlian CP (1988) Inhibitory effect of disodium EDTA upon the growth of *Staphylococcus epidermidis* in vitro: relation to infection prophylaxis of Hickman catheters. Antimicrob Agents Chemother 32:1627-1631.

Rupp ME, Ulphani JS, Fey PD, Mack D (1999) Characterization of *Staphylococcus epidermidis* polysaccharide intercellular adhesion/hemaglutinin in the pathogenesis of intravascular catheter-associated infection in a rate model. Infect Immun 67:2656-2659

Sandoe JAT, Wysome J, West AP, Heritage J, Wilcox MH (2006) measurement of ampicillin, vancomycin, linezolid, and gentamicin activity against enterococcal biofilms. J Antimicrob Chemother 57:767-770

Schinabeck MK, Long LA, Hossain MA, Chandra J, Mukherjee PK, Mohamed S, Ghannoum MA (2004) Rabbit model of *Candida albicans* biofilm infection: liposomal amphotericin B antifungal lock therapy. Antimicrob Agents Chemother 48:1727-1732

Schwalbe RS, Stapleton JT, Gilligan PH (1987). Emergence of vancomycin resistance in coagulase-negative staphylococci. N Eng J Med 316:927-931

Shah CB, Mittelman MW, Costerton JW, Parenteau S, Pelak M, Arsenault R, Mermel LA (2002) Antimicrobial activity of a novel catheter lock solution. Antimicrob Agents Chemother 46:1674-1679

Shinabarger DL, Marotti KR, Murray RW, Lin AH, Melchior EP, Swaney SM, Dunyak DS, Demyan WF, Buysse JM (1997) Mechanism of action of oxazolidinones: effects of linezolid and eperezolid on translation reactions. Antimicrob Agents Chemother 41:2132-2136

Sieradzki K, Leski T, Dick J, Borio L, Tomasz A (2003) Evolution of a vancomycin-intermediate *Staphylococcus aureus* strain in vivo: multiple changes in antibiotic resistance phenotypes of a single lineage of methicillin-resistant *S. aureus* under the impact of antibiotics administered for chemotherapy. J Clin Microbiol 41:1687-1693

Sillankova SR, Oliveira MJ, Vieira MJ, Sutherland IW, Azeredo J (2004) Bacteriophage Phi S1 infection of *Pseudomonas fluorescens* planktonic cells versus biofilms. Biofouling 20:133-138

Simon VC, Simon M (1990) Antibacterial activity of teicoplanin and vancomycin in combinations with rifampicin, fusidic acid or fosfomycin against staphylococci on vein catheters. Scan J Infect Dis Suppl 72:14-19

Souli M, Giamarellou H (1998) Effects of slime produced by clinical isolates of coagulase-negative staphylococci on activities of various antimicrobial agents. Antimicrob Agents Chemother 42:939-941

Stewart PS, Mukherjee PK, Ghannoum MA (2004) Biofilm antimicrobial resistance. In: Ghannoum M, O'Toole GA (eds) Microbial biofilms. ASM Press, Washington DC, pp 250-268

Stratton CW (2005) Molecular mechanisms of action for antimicrobial agents: general principles and mechanisms for selected classes of antibiotics. In: Lorian V (ed) Antibiotics in laboratory medicine. Lippincott, Williams, and Wilkins, Philadelphia, pp 532-536

Tenover FC (2001) Development and spread of bacterial resistance to antimicrobial agents: an overview. Clin Infect Dis 33 [Suppl 3]:S108-S115

Torres-Viera C, Thauvin-Eliopoulos C, Souli M, DeGirolami P, Farris MG, Wennersten CB, Sofia RD, Eliopoulos GM (2000) Activities of taurolidine in vitro and in experimental enterococcal endocarditis. Antimicrob Agents Chemother 44:1720-1724

Tresse O, Jouenne T, Junter G-A (1995) The role of oxygen limitation in the resistance of agar-entrapped, sessile-like *Escherichia coli* to aminoglycoside and B-lactam antibiotics. J Antimicrob Chemother 36:521-526

Turakhia MH, Cooksey KE, Characklis WG (1983) Influence of a calcium-specific chelant on biofilm removal. Appl Environ Microbiol 46:1236-1238

Yamasaki O, Akiyama H, Toi Y, Arata J (2001) A combination of roxithromycin and imipenem as an antimicrobial strategy against biofilm formed by *Staphylococcus aureus*. J Antimicrob Chemother 48:573-577

Yasuda H, Ajiki Y, Koga T, Yokata T (1994) Interaction between clarithromycin and biofilms formed by *Staphylococcus epidermidis*. Antimicrob Agents Chemother 38:138-141

Zufferey J, Rime R, Francioli P, Bille J (1988) Simple method for rapid diagnosis of catheter-associated infection by direct acridine orange staining of catheter tips. J Clin Microbiol 26:175-177

Role of Bacterial Biofilms in Urinary Tract Infections

J. K. Hatt and P. N. Rather(☒)

Abstract Bacterial urinary tract infections represent the most common type of nosocomial infection. In many cases, the ability of bacteria to both establish and maintain these infections is directly related to biofilm formation on indwelling devices or within the urinary tract itself. This chapter will focus on the role of biofilm formation in urinary tract infections with an emphasis on Gram-negative bacteria. The clinical implications of biofilm formation will be presented along with potential strategies for prevention. In addition, the role of specific pathogen-encoded functions in biofilm development will be discussed.

P. N. Rather

Department of Microbiology and Immunology, Emory University School of Medicine 3001 Rollins Research Center, Atlanta, GA, USA

prather@emory.edu

T. Romeo (ed.), *Bacterial Biofilms.*
Current Topics in Microbiology and Immunology 322.
© Springer-Verlag Berlin Heidelberg 2008

1   Introduction

Urinary tract infections (UTIs) represent the most commonly acquired bacterial infection. These infections are prevalent in both outpatient and hospital populations and are responsible for an estimated seven million office visits, one million emergency room visits and 100,000 hospitalizations annually (Foxman 2003). Urinary tract infections account for an estimated 25%-40% of nosocomial infections and represent the most common type of these infections (Bagshaw and Laupland 2006; Foxman 2003; Kalsi et al. 2003; Maki and Tambyah 2001; Wagenlehner and Naber 2006). The annual healthcare costs of urinary tract infections are estimated at $1.6 billion per year (Foxman 2003).

The risk of developing a urinary tract infection increases significantly with the use of indwelling devices such as catheters and urethral stents/sphincters. Indwelling catheters are the primary contributing factor in the development of these infections. The use of catheters to manage urinary incontinence in nursing home and spinal cord injury patients makes these populations especially vulnerable to these infections. Remarkably, the risk of developing catheter-associated urinary tract infections increases 5% with each day of catheterization and virtually all patients are colonized by day 30 (Maki and Tambyah 2001). The ability of various bacteria to form biofilms on catheters is well documented and will be expanded on later in this chapter. For all these indwelling devices, a number of studies support the role of biofilms in the establishment of infection (Reid et al. 1992; Choog and Whitfield 2000; Silverstein and Donatucci 2003; Trautner et al. 2004; Tenke et al. 2006).

The predominant pathogens in urinary tract infections are *Escherichia coli* (25%), followed by *Enterococci* (16%), *Pseudomonas aeruginosa* (11%), *Klebsiella pneumoniae* (8%), *Candida albicans* (8%), *Enterobacter* (5%), *Proteus mirabilis* (5%), and coagulase-negative *Staphylococci* (4%) (Emori and Gaynes 1993) (Table 1). These pathogens are typically found in the lower intestinal tract and can be introduced into the urinary tract via contaminated indwelling devices.

This chapter will focus on the role of bacterial biofilms in the development of urinary tract infections. In the first section, we will discuss the role of biofilm

Table 1 Frequency of urinary tract infections by various pathogens

Organism	Percentage of infections
E. coli	25
Enterococcus spp.	16
P. aeruginosa	11
K. pneumoniae	8
C. albicans	8
Enterobacter spp.	5
P. mirabilis	5
Coagulase–negative *Staphylococci*	4

From Emori and Gaynes 2003

formation on indwelling devices with an emphasis on crystalline biofilms formed by urease-producing organisms. Next, the role of selected, uropathogen-specific factors that contribute to biofilm formation will be discussed. In the final section, the role of biofilms on, or in, host urinary tissues will be addressed.

2 General Aspects of Biofilm Formation on Indwelling Urinary Tract Devices

A common prerequisite to initial biofilm formation on indwelling urinary devices is the formation of a conditioning film from urinary components (polysaccharides and proteins) (Reid 1999; Trautner and Darouiche 2004; Tenke et al. 2006; Donlan 2001). This conditioning film facilitates the initial attachment of microorganisms, which normally adhere poorly to uncoated surfaces. The subsequent attachment of microorganisms and biofilm development is similar to that described elsewhere in this book and will not be described further. A variety of indwelling devices are commonly used in the urological setting, including open and closed urinary catheters, urethral stents and sphincters, and penile prostheses. For each type of device, biofilm formation has been documented from infection sites (Reid 1992; Nickel et al. 1993; Stickler et al. 1998; Morris et al. 1999; Choog and Whitfield 2000; Tenke et al. 2006). Of these devices, catheters are the primary culprit in the development of urinary tract infections (Morris et al. 1999; Stickler et al. 1998; Trautner and Darouiche 2004). Early studies documented that a catheter removed from a patient with recalcitrant urosepsis, who failed antibiotic therapy, contained a thick biofilm adherent to the catheter (Nickel et al. 1985). This work was followed up by additional studies that documented extensive biofilm formation on urinary catheters by scanning electron microscopy (Ohkawa et al. 1990), which primarily included studies of catheters from patients that failed antibiotic therapy (Nickel et al. 1989). A variety of Gram-negative bacteria were colonizing the catheters including *P. aeruginosa, Enterococcus faecalis, E. coli* and *P. mirabilis*, and these same organisms were isolated from infected urine (Nickel et al. 1989). Interestingly, this early study revealed the association of crystalline deposits with *P. mirabilis* biofilms and laid the groundwork for a large number of subsequent studies that revealed the relationship between urease production and crystalline biofilms on catheters.

3 Biofilm Formation by Urease-Producing Organisms

3.1 Crystalline Biofilms

Indwelling Foley catheters are used extensively to manage urinary incontinence in elderly patients or those with bladder dysfunction, such as spinal cord injury patients. However, these devices place a patient at high risk for the development of urinary tract infections. A unique type of crystalline biofilm can form on catheters by urease producing organisms, such as members of the Proteeae (*P. mirabilis, Providencia stuartii, Morganella morganii*) and *K. pneumoniae* (Stickler et al. 1993, 1998). The production of urease by these organisms results in the cleavage of urea that occurs at concentrations of 0.4–0.5 M in urine (Li et al. 2002). The ammonia that is generated by urease activity raises the pH of the urine resulting in calcium and magnesium phosphate crystal formation within the biofilm matrix (Stickler et al. 1993, 1998). Studies by Nickel (1987) have demonstrated that biofilm development is a prerequisite for crystal formation as the matrix may act as a nucleation site for crystal development, and the higher concentration of cells in the biofilm allows for a greater localized concentration of urease. It has been proposed that crystalline biofilm formation is a multistep process as follows:

1. Introduction of a urease producing organism
2. Formation of a conditioning film on the surface of the catheter
3. Bacterial adherence to the catheter
4. Biofilm development and production of exopolysaccharides
5. Elevation of pH within the urine and biofilm by urease production
6. Crystallization of calcium and magnesium phosphate within the biofilm matrix (Morris et al. 1997)

Using a laboratory model for a catheterized bladder, Stickler and colleagues have demonstrated that with *P. mirabilis*, the development of biofilms on a catheter surface generally begins near the eye-hole with microcolonies forming at this site (Stickler et al. 2003b). Calcium and magnesium phosphate crystals begin to form and the biofilm then extends down the lumenal surface. Eventually the encrustations will block the catheter. This leads to bladder distension and urine leakage, or more serious complications such as pyelonephritis when urine from the distended bladder is refluxed into the kidney. In addition, crystalline biofilms that form on the outside of the catheter can cause irritation and trauma to the mucosa of the urethra. *P. mirabilis* appears to be the predominant organism in encrusted biofilms (Stickler et al. 1993) and appears to be the most effective organism at producing crystalline biofilms using in vitro models (Stickler et al. 1998). This is likely due to the fact that the *P. mirabilis* urease is six- to tenfold more active than other bacterial ureases (Tenke et al. 2006).

On solid surfaces, *P. mirabilis* undergoes a unique form of migration termed swarming. This process requires a complex cellular differentiation from a short vegetative rod to an elongated swarmer cell, reviewed in Rather 2005. The expression of urease in *P. mirabilis* is increased 30- to 50-fold in differentiated swarmer cells (Allison et al. 1992). However, although *P. mirabilis* is clearly capable of swarming on catheter surfaces (Sabbuba et al. 2002; Stickler et al. 1999), swarmer cell differentiation does not appear to be required for crystalline biofilm formation. Studies by Jones et al. (2005) examined a panel of *P. mirabilis* mutants and found that strains defective in swarming were actually more proficient at biofilm formation. The basis for this inverse relationship is unclear in *P. mirabilis*, but similar findings have been reported in *Salmonella typhimurium* and involve decreased surfactin production (Mireles et al. 2001).

3.2   Control of Crystalline Biofilms

At the current time, all types of urinary catheters are subject to encrustation (Morris et al. 1997). A number of strategies have been used in an attempt to decrease crystalline biofilm formation on catheters, most of which have had limited success. In artificial bladder models, catheter material composed of silicone was blocked less effectively than silver gel-coated latex catheters (Morris et al. 1997). However, all materials tested were eventually encrusted. The use of chlorhexidine as an antiseptic was not effective at reducing Gram-negative infections and selected for resistant isolates that were also multidrug-resistant (Stickler 2002). In laboratory models, the use of triclosan to disinfect the retention balloon of catheters reduced crystalline biofilms significantly and may have utility in clinical settings (Stickler et al. 2003). An additional strategy, the use of silver impregnated catheters has also given mixed results and may have limited utility (Rosch et al. 1999) The direct role of urease and alkaline urine in crystalline biofilms has prompted studies to examine the effect of acidic drinks such as cranberry juice on crystalline biofilm formation; however, studies in catheterized patients revealed no difference in patients that had ingested this liquid (Morris and Stickler 2001). A separate strategy involves decreasing urine pH by the use of urease inhibitors. Using in vitro models, the addition of acetohydroxamic acid or flurofamide restricted the increase in urine pH in the presence of *P. mirabilis* and led to reduced deposits of calcium and magnesium salts on the catheters (Morris and Stickler 1998). At the present time, the utility of these treatments in a clinical setting has not been determined, but they offer promising options for the control of crystalline biofilms.

Recently, a more general strategy to control biofilm formation on urinary catheters has been described by Burton et al. and is based on the inhibition of GlmU, an *N*-acetyl-D-glucosamine-1-phosphate acetyltransferase (Burton et al. 2006). This enzyme synthesizes the activated nucleotide sugar UDP-GlcNAc, a precursor for synthesis of β-1,6-*N*-acetyl-D-glucosamine polysaccharide adhesin that is required for biofilm formation in *E. coli* and *Staphylococcus epidermidis*. Inhibitors of GlmU, such as iodoacetamide (IDA) and *N*-ethyl maleimide (NEM) inhibited *E. coli* biofilm formation. Moreover, derivatives of NEM such as *N*-phenyl maleimide (NPM), *N,N'*-(1,2phenylene) dialeimide (oPDM), and *N*-(1-prenyl) malemide (PyrM) were effective at inhibiting biofilm formation by *P. aeruginosa*, *K. pneumoniae*, *S. epidermidis*, and *E. faecalis* (Burton et al. 2006). The addition of protamine sulfate to oPDM enhanced anti-biofilm activity and catheters coated with both compounds were virtually free of bacterial colonization (Burton et al. 2006). The promising combination of oPDM plus PS may have broad utility in the prevention of bacterial colonization of urinary catheters and this strategy may be applicable to other indwelling medical devices colonized by organisms that require the β-1,6-*N*-acetyl-D-glucosamine adhesion for biofilm formation.

3.3 Infectious Urinary Stones

Approximately 15% of urinary stones are initiated by infections (Bichler et al. 2002). These stones can form in the bladder or kidney and are often associated with abnormalities of the urinary tract or obstructions (Bichler et al. 2002; Abrahams and Stoller 2003). These stones are composed of struvite (magnesium ammonium phosphate) or apatite (calcium phosphate). The formation of these stones is strongly correlated with urease-producing bacteria and for the purposes of this chapter, these stones will be considered a free-floating crystallized biofilm. In fact, bladder and kidney stones are remarkably similar to the crystals that form on catheters (Griffith et al. 1976). The first step in production of these stones was revealed from studies by Griffin et al. who established the key role of urease in the formation of these stones (Griffith et al. 1976). These in vitro studies demonstrated that the expression of bacterial urease increased the pH of urine, leading to the formation of calcium and magnesium crystals, a precursor to stone formation. Urine that was urea-free or contained urease inhibitors did not support crystal formation. McLean et al. proposed that the next step in stone formation is the interaction of crystals with bacterial exopolysaccharides (McLean et al. 1989). The O-antigen of LPS can be acidic due to uronic acid and this acidic nature can facilitate the binding of Ca^{2+} and Mg^{2+} (Knirel et al. 2003; Clapham et al. 1990; Dumanski et al. 1994). Studies by both Torzewski and co-workers and Dumanski et al. have extended these findings and found that the sugar composition of *P. mirabilis* lipopolysaccharide can influence the crystallization process (Torzewska et al. 2003; Dumanski et al. 1994). However, a paradox of these studies is that strains with the strongest cation binding supported the poorest crystal growth. It is proposed that the weaker binding is enough for concentration, but facilitates the release for crystal growth (Dumanski et al. 1994). Stone formation is facilitated by additional bacterial growth or interactions between stones, and the stone eventually comprises a matrix with imbedded bacteria (Nickel et al. 1987; McLean et al. 1989; Li et al. 2002; Takeuchi et al. 1984). Studies by Li et al. used *P. mirabilis* cells constitutively expressing green fluorescent protein (GFP) to visualize cells within urinary stones in a mouse model of urinary tract infection. Further examination by electron microscopy revealed that stones were compact in the core and had a loose structure in the outer layers (Li et al. 2002). Within the stone matrix, vegetative cells were the predominant type with some swarmer cells evident. The majority of bacteria resided in the outer layers of the stone. It was proposed that this location allows access to nutrients, yet may confer protection from immune responses.

A second possible mechanism of urinary stone formation involves nanobacteria (Kajander and Ciftcioglu, 1998). These organisms were capable of forming carbonite apatite at pH 7.4 in the absence of urease activity. In a survey of human kidney stones, nanobacteria were found in all stones tested ($n = 30$). Therefore, stone formation can be initiated by urease-producing organisms and by urease-independent mechanisms at neutral pH involving nanobacteria.

4 *Escherichia coli* Factors Involved in UTI Biofilm Development

4.1 Type 1 pili in UPEC Pathogenesis and Intracellular Biofilm Formation

Uropathogenic *Escherichia coli* (UPEC) strains are the most common causes of urinary tract infections (UTIs) (Hooton and Stamm 1997). Moreover, UPEC biofilms are responsible for many catheter-associated and chronic UTIs (Nicolle 2005). UPEC strains can vary greatly in their ability to cause UTIs. This is most likely due to the different repertoire of virulence factors associated with each UPEC strain (Foxman et al. 1995; Johnson et al. 1998; Marrs et al. 2005). Virulence factors described for UPEC include α-hemolysin, cytotoxic necrotizing factor 1 (cnf1), lipopolysaccharide (LPS), capsule, the siderophores aerobactin and enterobactin, proteases, and a number of adhesive organelles (Johnson 1991; Oelschlaeger et al. 2002). However, no single virulence factor has been identified that is specific to or definitive of UPEC. Despite this fact and despite their presence in a majority of wild type *E. coli* strains (Hagberg et al. 1981; Langermann et al. 1997), perhaps the single most important virulence factor of UPEC is the expression of type 1 pili, which are key factors that are involved in the formation of UPEC biofilms in living tissue and on abiotic surfaces. This is due to the contribution of type 1 pili to adherence to, invasion of, and persistence within the bladder.

4.1.1 The *fim* Gene Cluster

Type 1 pili are peritrichously expressed proteinaceous cell surface structures found on many members of the *Enterobacteriaceae*, including the majority of both commensal and pathogenic strains of *E. coli* (Johnson 1991; Yamamoto et al. 1995). Type 1 pili of *E. coli* are thick rod-shaped composite structures approximately 7 nm wide and 1 μm in length (Brinton 1965). In *E. coli*, the nine genes of the *fim* gene cluster encode the structural and regulatory proteins that produce type 1 pili or fimbria. The *fimAFGH* genes are structural genes encoding the major and minor protein components of the pilus rod and tip fibrillum (Brinton 1965). The pilus is composed primarily of multiple copies of the FimA major subunit protein organized in a right-handed helical rod, the FimF adaptor protein, and the tip fibrillum composed of the minor protein component FimG and multiple copies of the mannose-specific adhesin FimH (Brinton 1965; Klemm and Christiansen 1987; Russell and Orndorff

1992; Jones et al. 1995). It has also been suggested that the FimH adhesin may be interspersed along the pilus rod (Krogfelt et al. 1990). The assembly of the pilus itself is dependent on the *fimC* and *fimD* genes that encode the periplasmic chaperone and the outer membrane usher proteins, respectively. These proteins mediate the translocation of the pilus structural components across the outer cell membrane during which time the pilus is assembled (Klemm and Christiansen 1990; Jones et al. 1993). The function of the *fimI* gene is as yet unknown; however, it has been implicated in pilus biogenesis (Valenski et al. 2003).

The final two genes of the *fim* gene cluster *fimB* and *fimE* encode regulatory proteins that control the phase variation of type 1 pili (Klemm 1986; Gally et al. 1996). Phase variation refers to the reversible on/off switch controlling the expression of type 1 pili in *E. coli*. Expression of type 1 pili varies as a result of the reversible inversion by site-specific recombination of a 314-bp regulatory element containing the promoter of the *fimA* gene (Abraham et al. 1985). The FimB and FimE proteins that function as site-specific recombinases mediate this DNA inversion (Klemm 1986; Gally et al. 1996). The FimB protein can turn the switch to both the on and off positions; however, FimE is specific for the switch from on to off (McClain et al. 1991; Gally et al. 1996). The importance of phase variation of type 1 pili in initiating infection of the bladder by UPEC has been well characterized using a murine model of ascending UTI (Hultgren et al. 1985; Schaeffer et al. 1987; Connell et al. 1996; Struve and Krogfelt 1999; Gunther et al. 2001; Bahrani-Mougeot et al. 2002; Gunther et al. 2002; Snyder et al. 2006). Type 1 expression was shown to be critical for infection of the lower urinary tract and strict control of phase variation was required for successful UPEC infection.

4.1.2   The FimH Adhesin

The FimH adhesin confers mannose-specific binding activity to type 1 pili (Abraham et al. 1987; Maurer and Orndorff 1987). FimH recognizes the terminal mannose moieties on many cell types and secreted glycoproteins, which include superficial bladder umbrella cells (Eden and Hansson 1978; Zhou et al. 2001; Duncan et al. 2004) and CD48 on mast cells and macrophages (Baorto et al. 1997; Malaviya et al. 1999; Shin et al. 2000). Colonization of the murine bladder requires the presence of FimH, and immunization with FimH was shown to protect against colonization and infection by UPEC in both murine and primate models of UTI (Langermann et al. 1997, 2000). Scanning electron microscopy (SEM) and high-resolution transmission electron microscopy (TEM) revealed that type 1 pili are in intimate contact with the uroplakin-coated superficial bladder epithelium via the tips of type 1 pili directly contacting the uroplakin-coated membrane, termed the asymmetric unit membrane (Mulvey et al. 1998). Uroplakins are proteins that cover the apical surface of superficial umbrella cells and are organized in hexagonal plaques that give strength to the bladder epithelium to help create a permeability barrier (Sun et al. 1996). Although earlier studies implicated the uroplakins UP1a and UP1b as targets for FimH binding

(Wu et al. 1996), recent in vitro studies using mouse uroepithelial plaques and purified recombinant FimH in a FimC-FimH complex showed unambiguously that uroplakin UP1a is the unique bacterial receptor for FimH adhesin binding to uroepithelial cells (Zhou et al. 2001).

Most commensal and pathogenic strains of *E. coli* with type 1 pili bind trimannose receptors via the FimH adhesin (Sokurenko et al. 1995). However, in about 70% of UPEC strains, higher tropism for the uroepithelium is conferred by the ability of FimH variants in these strains to bind monomannose as well as trimannose receptors (Sokurenko et al. 1995). Monomannose receptors are abundant on the uroplakin UP1a that coats the apical surface of superficial umbrella cells (Zhou et al. 2001). All UPEC strains with type 1 pili bind trimannose receptors with high affinity; however, affinities for monomannose binding by the FimH adhesin can vary greatly in these strains. FimH with low affinity for monomannose binding are found among fecal *E. coli* isolates, but naturally occurring variants with high affinity are most often found in UPEC strains (Sokurenko et al. 1995, 1997, 1998; Zhou et al. 2001). This ability to bind monomannose with high affinity is thought to provide a selective advantage during pathogenesis by increasing binding affinity specifically for the uroepithelium. Interestingly, an epidemiological study of a large panel of commensal and pathogenic *E. coli* strains indicated that UPEC isolates also have the highest frequency of mutator strains as compared to commensal strains and other pathogenic strain types (Denamur et al. 2002). No evidence was found to suggest that growth advantage in urine or increased antibiotic resistance provided a selective advantage that resulted in higher numbers of mutators among the UPEC strains. Although no difference was observed for the ability of a UPEC strain and its isogenic mutS⁻ mutant to individually infect a mouse in an experimental murine model of UTI, in competition assays, the mutS⁻ strain outcompeted the UPEC strain (Labat et al. 2005). Therefore, it was hypothesized that the increased mutation rate in mutator strains provides a mechanism by which mutations that alter binding specificity can accumulate and may confer a selective advantage to UPEC during pathogenesis.

4.1.3 UPEC Invasion of Host Cells

Not only are type 1 pili used by UPEC for adherence, they are also used for invasion of bladder epithelial cells. High-resolution TEM and SEM were used to show that bladder epithelial cells internalize UPEC in vivo via interactions between FimH and UP1a (Mulvey et al. 1998). In the same study, host bladder cells appeared to be zippering around and swallowing the attached bacteria, indicating that perhaps interaction between FimH and UP1a was the first step in internalization of UPEC strains. Gentamicin protection assays were subsequently used to show that UPEC and *E. coli* K12 strains expressing FimH but not isogenic FimH⁻ mutant strains are internalized by a human bladder epithelial cell line (Martinez et al. 2000). Adherence alone was not sufficient to cause internalization, as binding of P-piliated strains was not able to mediate invasion. FimC-FimH complexes conjugated to latex beads were then used to show that the presence of FimH was sufficient to cause uptake into a human bladder epithelial cell line. Beads coated with BSA or with FimC alone were not internalized. Therefore it was concluded that type 1-piliated strains mediate invasion of bladder epithelial cells via a mechanism dependent on the FimH adhesin. Moreover, internalization of type 1-piliated strains was dependent on localized host actin rearrangement as the actin polymerization inhibitor cytochalasin D could inhibit invasion of a bladder cell line in vitro (Martinez et al. 2000). Furthermore, FimC-FimH-coated latex bead uptake into bladder cells could also be blocked by the addition of cytochalasin D. In addition, inhibitors of tyrosine kinases and phosphoinositide 3-kinase (PI 3-kinase) blocked internalization of type 1-piliated bacteria or FimC-FimH-coated latex beads, indicating the involvement of signaling pathways in actin movement and bacterial engulfment. Inhibitors of bacterial invasion also inhibited the formation of complexes between focal adhesin kinase and PI 3-kinase, implicating this complex in the signaling cascade that results in actin cytoskeleton rearrangements during invasion.

More recent studies have indicated that type 1-piliated bacteria associate with plasma membrane microdomains known as lipid rafts (Duncan et al. 2004). Caveolae, a subtype of lipid raft with a cave-like appearance found in the plasma membrane was shown to be associated with intracellular bacteria during UPEC invasion and disruptors of the lipid rafts inhibited bacterial invasion. The involvement of the caveolin-1 protein indicative of caveolae was specifically determined via the use of RNA interference of caveolin-1 that led to reduced bacterial invasion of bladder epithelial cells in vitro. Finally, it was shown that UP1a was associated with lipid rafts in mouse bladder epithelial cells, thus demonstrating that lipid rafts mediate UPEC invasion of bladder cells. Ultimately, these data demonstrated that the FimH adhesin acts as an invasin to mediate bacterial invasion of host cells.

Type 1 piliated *E. coli* can also bind to and invade both macrophage (Baorto et al. 1997) and bone marrow-derived mast cells (Malaviya et al. 1999; Shin et al. 2000), which may be one source of bacteria causing chronic UTIs. FimH-mediated binding to macrophage cells via interaction with CD48 allows the bacteria to be internalized and survive within the macrophages (Baorto et al. 1997). Opsonized bacteria are generally phagocytosed and killed. The vesicles in which the UPEC strains are found after uptake differ between the opsonized cells and type 1 pili-expressing cells taken up via FimH-mediated invasion. Although the exact reason is unclear, this implies a fundamental difference between the two modes of entry, which allows for survival of the bacteria taken up during FimH-mediated entry into the macrophages. CD48 also mediates entry into mast cells (Malaviya et al. 1999; Shin et al. 2000). Binding of FimH to CD48 on mast cells triggers innate host immunity and the release of tumor necrosis factor alpha (TNFα), which can lead to clearance of the infection in a mouse model (Malaviya et al. 1999; Shin et al. 2000). However, in some cases the recruitment of caveolae has been documented for UPEC entry into both macrophage and mast cells, and caveolae disruptors inhibit FimH-mediated bacterial entry into both cell types (Baorto et al. 1997; Shin et al. 2000). Since intracellular compartments composed of caveolar material do not fuse with endosomes, the bacteria contained in these caveolar vesicles can survive within both macrophage and mast cells and may function as a source of persistent bacteria.

4.1.4   Avoidance of Host Defenses by UPEC and Persistence

It was originally proposed that internalization of UPEC into bladder cells was an innate defense mechanism of the host cell. More recent studies, however, suggested that the process of internalization is used by UPEC to avoid host defenses (Mulvey et al. 1998, 2001). These defenses include urine flow, the secretion of adhesin-binding competitors such as the Tamm-Horsfall protein and secretory IgA, the secretion of chemokines resulting in recruitment of neutrophils to apical site of the bladder mucosa, and exfoliation of superficial bladder cells (McTaggart et al. 1990; Wold et al. 1990; Sobel 1997; Haraoka et al. 1999; Pak et al. 2001; Svanborg et al. 2001). Invasion of host cells provides a haven from most of these defenses. Moreover, UPEC sequestered in superficial bladder cells are protected from antibiotic treatments that sterilize urine (Mulvey et al. 1998; Hvidberg et al. 2000) and are provided a rich environment in which the bacteria can replicate (Mulvey et al. 2001). Therefore, all of the defenses except exfoliation of the superficial bladder cells can be avoided by UPEC by invasion of host cells.

Turnover of the cells lining the lumen of the bladder generally occurs very slowly (Jost 1989). However, the superficial bladder cells undergo a massive exfoliation in response to adherence of type 1-piliated strains of E. coli and large numbers of bacteria can be found on exfoliated bladder epithelium in the urine of humans and mice (Elliott et al. 1985; McTaggart et al. 1990; Mulvey et al. 1998). In a murine model of UTI, within 2 h after transurethral infection by UPEC, superficial bladder cells are infected by individual UPEC cells and begin to exfoliate (Mulvey et al. 1998, 2001). By 6 h after infection, large masses of intracellular bacteria are apparent within many of the superficial bladder cells and massive exfoliation was also apparent. Induction of exfoliation is dependent on FimH-mediated attachment because both UPEC and type 1-piliated E. coli K-12 strains caused exfoliation but P-piliated or Fim⁻ mutant strains did not (Mulvey et al. 1998). Bladder epithelial cell exfoliation observed in these studies occurred through an apoptotic mechanism involving both DNA fragmentation and the activation of proteolytic enzymes, termed caspases (cysteine-containing aspartate-specific proteases), and inhibitors of caspases prevented exfoliation, but also decreased the level of UPEC clearance from the bladder.

Although it initially appeared that exfoliation of infected bladder cells was an efficient method for removing UPEC from the bladder, this process is actually used by UPEC to form a persistent reservoir. By 24-48 h after invasion and subsequent exfoliation of superficial bladder cells, UPEC is found to resist gentamicin treatment ex vivo, indicating that it exists intracellularly within the immature basal cells of the bladder (Mulvey et al. 1998). Moreover, this ability to persist intracellularly was specific to UPEC isolates as UPEC titers remained constant throughout the course of a 48 h infection in vitro despite gentamicin treatment, but titers of type 1-piliated *E. coli* K-12 strains decreased continuously (Mulvey et al. 2001). Furthermore, these studies showed that intracellular replication and the ability to escape from host cells was required for persistence of UPEC. Finally, TEM examination of a UPEC-infected bladder cell line maintained in gentamicin containing medium in vitro showed the presence of large intracellular inclusions in the tissue culture cells similar to foci of intracellular UPEC observed in mouse bladders in vivo (Mulvey et al. 2001). Type 1 piliated K-12 strains did not form these same inclusions, indicating that they do not multiply efficiently in bladder epithelial cells. These inclusions have been termed intracellular bacterial communities and form as a result of type 1 pili-mediated adherence and invasion of bladder epithelial cells. A detailed description of these intracellular communities can be seen below in Sect. 7.1.

4.2   Type 1 Pili Involvement in Abiotic Biofilm Formation

Many motile laboratory strains of *E. coli* are able to form biofilms on abiotic surfaces such as polyvinylchloride (PVC), polypropylene, polycarbonate, and borosilicate glass when grown statically in rich medium at room temperature (Pratt and Kolter 1998). Therefore, type 1-mediated biofilm formation may contribute to the ability of UPEC to withstand antibiotic treatments and host antimicrobial defenses in the urinary tract. To gain an understanding of the factors involved in formation of *E. coli* biofilms, Pratt and Kolter used transposon mutagenesis to generate mutants defective in biofilm formation on abiotic surfaces (Pratt and Kolter 1998). Mutants capable of motility but still severely defective in biofilm formation were isolated and determined to fall within the *fim* gene cluster. Independent insertions were found within *fimB*, *fimA*, *fimC*, *fimD*, and *fimH*. Microscopic analysis of PVC surfaces on which the *fim* mutants were grown statically in rich medium revealed that *fim* mutants were so severely defective in initial attachment that most of the surface contained no attached *E. coli*. This indicated that type 1 pili are essential in initial attachment to abiotic surfaces. Furthermore, FimH was shown to mediate this attachment as a nonmetabolizable mannose analog α-methyl-D-mannoside-inhibited biofilm formation by the type 1-piliated wild type *E. coli* strain 2K1056. FimH to surface interactions were concluded to be direct and involve nonspecific binding to abiotic surfaces. The role of FimH in mediating attachment to abiotic surfaces was supported by the isolation of both natural and engineered FimH variants that allow *E. coli* to form biofilms under hydrodynamic flow (HDF) conditions (Schembri and Klemm 2001a). *E. coli* with the wild type FimH adhesin and commensal *E. coli* strains could not form biofilms under the same HDF conditions. HDF shear force conditions are thought to better mimic the natural environment during UTI, and interestingly, a G73E variant of FimH that was identified as a functional alteration of FimH involved in biofilm formation was previously identified as a natural FimH variant pathoadaptive for UTI (Sokurenko et al. 1994).

The same FimH mutant library also yielded FimH variants that were capable of mediating autoaggregation of *E. coli* (Schembri et al. 2001). Phase contrast microscopy of cells expressing the FimH variants showed that they form large tight clusters of cells that might aid in forming microcolonies during biofilm formation. These results were again supported by the isolation of natural FimH variants from UPEC strains that exhibited the ability to autoaggregate, arguing that this phenotype may be relevant to UTI pathogenesis (Schembri et al. 2001). Furthermore, autoaggregation in general appears to be important for biofilm formation as other self-aggregating cell surface structures such as Antigen 43 (Ag43) and curli are also associated with biofilm formation in *E. coli* (Vidal et al. 1998; Danese et al. 2000; Kjaergaard et al. 2000; Prigent-Combaret et al. 2000). In support of this idea, DNA microarrays showed that both type 1 pili and Ag43 are more highly expressed in biofilm populations than in planktonic populations (Schembri et al. 2003b). Therefore, the presence of type 1 pili may facilitate the colonization of urinary tract catheters and other implants by mediating biofilm formation and autoaggregation. Finally, *E. coli* strains carrying transfer constitutive IncF plasmids were shown to form mature mushroom-shaped structures similar to those of *P. aeruginosa* biofilms in continuous-flow cell cultures in glucose minimal medium at 30°C (Reisner et al. 2003). The presence of type 1 pili was found to be dispensable for the biofilm maturation observed as neither fimbriated nor afimbriated variants had any effect on biofilm maturation by *E. coli* strains that were plasmid-free or carrying F plasmid.

The regulation of the FimB and FimE recombinases may also function as one regulatory checkpoint for biofilm formation in *E. coli*. Phase variation of type 1 pili is subject to multiple global regulators including LrpA (Blomfield et al. 1993; Gally et al. 1994), IHF (Blomfield et al. 1997), H-NS (Olsen and Klemm 1994; Donato et al. 1997; O'Gara and Dorman 2000) and LrhA (Olsen and Klemm 1994; Donato et al. 1997; O'Gara and Dorman 2000; Lehnen et al. 2002; Blumer et al. 2005). Both LrpA and IHF are required for the efficient switching by both the FimB and FimE recombinases, whereas H-NS affects only the FimB-mediated recombination event. LrhA, however, affects only the activation of the FimE recombinase that is required to turn off *fimA* production and subsequently the production of type 1 pili (Blumer et al. 2005). Mutation of LrhA therefore leads to decreased *fimE* expression and subsequent increased transcription of *fimA* that ultimately results in increased biofilm formation. Overexpression of LrhA abolishes biofilm formation.

4.3   Flagella and Motility

Many studies have implicated motility, flagella and/or chemotaxis in biofilm formation in bacteria (Deflaun et al. 1994; Korber et al. 1994; O'Toole and Kolter 1998a, 1998b; Watnick et al. 2001). To test their importance for biofilm formation by *E. coli*, Pratt and Kolter introduced defined mutations that affect flagellar function into the K-12 strain ZK1056 (Pratt and Kolter 1998). A microtiter dish assay was used with flagellar mutations (*fliC::kan*, *flhD::kan*), motility mutations (Δ*motA*, Δ*motB* and Δ*motAB*) and chemotaxis mutations (Δ*cheA-Z::kan*) to assess their ability to form biofilms in LB medium grown at 30°C. Nonchemotactic strains were shown to be no different from wild type in formation of biofilms, indicating that chemotaxis was dispensable for biofilm formation by *E. coli*. In contrast, nonmotile strains containing motility and flagellar mutations were severely defective in initial steps of biofilm formation. Microscopic analysis indicated that this was due to the lack of adherence to PVC and that those cells that do attach form small clusters. Furthermore, biofilms did eventually form over time, indicating that motility appears to be most critical for initial attachment by *E. coli* on abiotic surfaces under static growth conditions in rich medium. Therefore, it was proposed that motility initially promotes cell-to-surface contact by overcoming repulsive forces and at a later stage is used to colonize a surface (Pratt and Kolter 1998). The role of motility during UTI pathogenesis, however, is still unclear. Flagella are not entirely required for colonization of the urinary tract during UTI by UPEC; however, motility does appear to contribute to UPEC fitness during pathogenesis in a mouse cystitis model (Lane et al. 2005; Wright et al. 2005).

4.4   Antigen 43

Many UPEC strains (~60%) express the cell surface adhesin Antigen 43 (Ag43) (Owen et al. 1996). Ag43 is an autotransporter protein encoded by the *flu* gene (also known as *agn43*), which is a self-recognizing adhesin (Diderichsen 1980; Hasman et al. 1999). Ag43 confers the ability to autoaggregate on *E. coli* in static liquid medium, resulting in settling of the cells (Diderichsen 1980). Because of this self-aggregative phenotype, Ag43 was tested for its ability to form biofilms (Danese et al. 2000). Ag43 presence was found to enhance biofilm formation of *E. coli* on PVC in glucose-minimal medium at 30°C by inducing microcolony formation while the *agn43* null allele was shown to have a limited ability to form biofilms compared to its wild type parental strain. Further studies indicated that Ag43 expression could mediate interactions between different bacterial species on glass under continuous flow growth conditions (Kjaergaard et al. 2000). These data therefore indicated that Ag43 is involved in providing both cell-to-surface as well as cell-to-cell contacts. Recent studies have shown that Ag43 is expressed specifically during the biofilm mode of growth (Schembri et al. 2003b). Interestingly, Ag43 was also shown to be expressed by UPEC within the intracellular pods that appear during IBC formation in a murine cystitis model (Anderson et al. 2003), suggesting that like type 1 pili, Ag43 may be involved in both abiotic biofilm development and biofilm formation in living tissue. Furthermore, Ag43-mediated aggregation was shown to be protective against oxidizing agents such as hydrogen peroxide in a manner characteristic of the protective nature of the biofilm mode of growth (Schembri et al. 2003a).

The expression of Ag43 is phase-variable and is controlled by the concerted action of OxyR (negative regulation) and Dam methylation (positive regulation) (Henderson and Owen 1999; Haagmans and van der Woude 2000). OxyR binds to a site found upstream of the *flu* gene to repress its transcription. Dam methylation of three GATC sequences overlapping the OxyR binding site blocks OxyR binding and thereby activates transcription of *flu* (Haagmans and van der Woude 2000; Waldron et al. 2002). OxyR is a sensor of cellular oxidative stress and responds to the redox status of the cell (Zheng et al. 1998). OxyR exists in one of two forms either oxidized or reduced, and only the reduced form can efficiently repress *flu* gene expression (Henderson and Owen 1999; Haagmans and van der Woude 2000). Moreover, fimbriation itself appears to influence expression of Ag43 as DNA microarray studies comparing *fim⁺* and *fim⁻* strains indicated that *flu* gene expression was increased by approximately 20-fold when type 1 pili expression was absent (Schembri et al. 2002). Indeed, it has been proposed that expression of thiol-disulfide containing fimbria such as type 1 pili results in a net cellular oxidation that is countered by the reduction of OxyR into the form that efficiently represses *flu* expression (Schembri and Klemm 2001b). Because the expression of Ag43 is mediated by the redox status of OxyR, mutant OxyR proteins locked in either a reduced or an oxidized state were used to study Ag43 expression on biofilm formation (Schembri et al. 2003a). The reduced OxyR protein repressed Ag43 expression and formed poor biofilms on polystyrene microtiter plates, whereas the oxidized form activated Ag43 expression and formed significantly better biofilms. A second gene *rfaH*, a transcriptional antiterminator, was also shown to regulate *flu* expression (Beloin et al. 2006). RfaH negatively regulates *flu* expression and therefore, inactivation of *rfaH* resulted in increased *flu* expression and better biofilm formation by both *E. coli* K-12 and UPEC strains in flow cultures in minimal medium. Although unclear exactly how RfaH influences *flu* transcription, it was shown by RT-PCR that a *rfaH* mutation has no effect on *oxyR* or *dam* transcript levels. Therefore, the mode of RfaH action on *flu* transcription does not occur via OxyR or Dam methylase action, and has yet to be elucidated. Additionally, both fimbriation and capsule formation by either *E. coli* K-12 and/or UPEC strains have been shown to sterically inhibit autoaggregation mediated by the shorter Ag43 adhesin (Hasman et al. 1999; Schembri et al. 2004). Capsule can protect bacteria from host defenses but then they cannot adhere or invade without adhesins, which suggests that regulation of both adhesin and capsule production may be required by UPEC during pathogenesis. Ag43 autoaggregation was also found to inhibit flagellar-based motility (Ulett et al. 2006) and this may also have implications for UPEC pathogenesis and biofilm formation, although the role of flagella in these processes remain somewhat unclear.

Ag43 is made up of two domains, an α-module that consists of a passenger domain and the β-module that is the autotransporter domain. Structure-function studies of Ag43 using domain swapping and linker scanning mutagenesis identified residues responsible for autoaggregation in the amino-terminal third of the passenger domain (α-module) (Klemm et al. 2004). Analysis of the primary sequence of the protein showed that it is likely that ionic interactions in interacting α-modules are responsible for the autoaggregation phenotype. Ag43 is found in many *E. coli*

⌐ains and often in multiple copies in the genome (Roche et al. 2001). Although some natural variants do not mediate autoaggregation, all variants studied were found to promote biofilm formation with different efficiencies (Klemm et al. 2004). Interestingly, the two variants from the UPEC strain CFT073 did not mediate cell-cell aggregation; however, they formed more robust biofilms than those variants that did mediate autoaggregation. Therefore, it was proposed than cell aggregation and biofilm formation may be distinct features of the Ag43 protein but the primary function of Ag43 may be the formation of biofilms. In support of this idea, it was noted in this same study that in the CFT073 strain, both of the *flu* genes are associated with pathogenicity islands rather than the chromosomal locus of the *flu* gene found in the *E. coli* K-12 strain MG1655.

4.5 Curli

Curli are proteinaceous cell surface filaments composed of two proteins: a major subunit CsgA and a minor subunit CsgB (Olsen et al. 1989; Bian and Normark 1997). The insoluble curli filament consisting of curlin produced by the CsgA protein is formed at the bacterial cell surface by the CsgB protein that acts as a nucleator of filament formation (Bian and Normark 1997). The involvement of curli in biofilm formation was discovered as a result of the isolation of an *E. coli* K-12 strain carrying the mutant *ompR234* allele (Vidal et al. 1998). The presence of the mutant *ompR234* allele resulted in a biofilm-forming phenotype, and a knockout of either *csgA* or *ompR* caused the loss of adherence. The mutant OmpR protein activates expression of CsgD, a transcriptional activator of the *csgA* gene resulting in an increase in CsgA protein levels and promotion of biofilm formation. Furthermore, the overexpression of CsgA resulted in mature highly developed biofilm structures on polystyrene in glucose minimal medium at 30°C relative to biofilms formed by the isogenic wild type parent (Vidal et al. 1998). Similar results were observed when the same lesions were introduced into clinical isolates taken from patients suffering from catheter-associated bacteremia, indicating that curli involvement in biofilm formation may be relevant to UPEC pathogenesis and catheter-associated UTIs. In an attempt to understand what factors are involved in biofilm formation by curli, the role of flagellar motility and colanic acid production was examined (Prigent-Combaret et al. 2000). Biofilms formed by the *ompR234* strain with and without the flagellin gene *fliC* were examined using SEM. Both the FliC⁺ and FliC⁻ strains carrying the *ompR234* allele formed similar thick biofilms on plastic thermanox coverslips and polystyrene in glucose minimal medium. These data indicated that in curli-overproducing strains, flagella are dispensable for initial adhesion and biofilm formation. Examination of *ompR234* K-12 strains that either do or do not produce colanic acid (CA) indicated that while CA was necessary for a thick well-developed biofilm, strains that overexpress curli without CA production are able to support biofilm formation. Finally, SEM and TEM with negative staining indicated that the curli form a dense meshwork of intertwined fimbria between cells and with the abiotic surface, suggesting that curli mediate cell-cell and cell-surface interactions. Examination of clinical isolates by SEM and TEM once again supported these results, indicating that curli involvement in biofilm formation may be relevant to UPEC pathogenesis. In addition, it was suggested that curli may be better adapted to form biofilms on abiotic surfaces such as catheters given that curli expression is favored by conditions that are found outside the host (i.e., poor nutrient availability and 30°C) (Prigent-Combaret et al. 2000). There is, however, some indication that although curli is not expressed on solid or liquid medium at 37°C, it is expressed in biofilms at 37°C and is able to support limited biofilm development at that temperature (Kikuchi et al. 2005). Furthermore, a comparison of the curli-expressing strain YMel and the curli-deficient strain YMel-1 showed that only the YMel strain was capable of forming mature mushroom-shaped biofilms, implying that curli are required for biofilm maturation (Kikuchi et al. 2005). The YMel curli-expressing strain was also shown to better adhere to human uroepithelial cells in vitro more than the curli-deficient strain YMel-1, once again implicating curli expression in UPEC pathogenesis.

5 *P. mirabilis*: **Role of Mannose-Resistant Fimbriae**

The MR/P fimbriae are surface appendages that confer hemagglutination in a mannose-resistant manner and are important in colonization and pathogenicity in mouse models of ascending urinary tract infections (Li et al. 2001). The Mrp locus is comprised of the *mrpA-J* genes (Bahrani and Mobley 1994). The *mrpA-H* genes encode Mrp structural components and other proteins necessary for assembly. The MrpJ product acts as a repressor of flagellin expression and acts to downregulate flagellin when Mrp fimbriae are being expressed (Li et al. 2001). The *mrpI* gene is divergently transcribed and encodes a recombinase that directs the inversion of a small promoter containing region upstream from *mrpA* (Li et al. 2002). In the ON position, the promoter drives expression of the *mrp* operon and in the OFF position, the promoter is in a divergent orientation (Li et al. 2002). In strains with MR/P locked in the ON orientation, *P. mirabilis* growing in sterile urine formed biofilms that were significantly more developed up to 48 h of incubation than either wild type or strains locked in the MR/P OFF state (Jansen et al. 2004). However, in 7-day-old biofilms, the wild type biofilm was significantly thicker than either the MR/P ON or OFF strains. Therefore, the expression of MR/P fimbriae promoted the early stages of biofilm development, possibly by facilitating the cell-cell interactions important for microcolony formation. In fact, MR/P ON strains are more autoaggregative in liquid than wild type (Jansen et al. 2004). However, expression of MR/P was detrimental to the later stages of biofilm development. This may be due to MR/P interactions between cells that serve to inhibit cell movement within the biofilm. Immunization of mice with the MR/P protein, protected animals against *P. mirabilis* in experimental urinary tract infections (Li et al. 2004). This emphasizes the importance of the MR/P fimbriae in the ability of *P. mirabilis* to cause urinary tract infections.

6   *K. pneumoniae*: Role of Type-3 Fimbriae

Approximately 8% of urinary tract infections are caused by *K. pneumoniae* (Tambyah et al. 2002). Two types of pili, type-1 and type-3, are produced in *K. pneumoniae* (Clegg and Gerlach 1987). The type-3 pilus is composed of a MrkA protein that comprises the fimbral shaft and an adhesion MrkD. In a search for mutations that affected biofilm formation, Langstraat et al. demonstrated that MrpA, but not MrpD was required for biofilm formation on abiotic surfaces (Langstraat et al. 2001). In a separate study, Di Martino et al. noted a strong correlation with expression of type-3 pili and the ability to form biofilms on abiotic surfaces (Di Martino et al. 2003). However, since indwelling devices become coated with host-derived substances (Donlan 2001), the above studies do not directly address the clinical relevance of biofilm formation. More recent studies examined the role of *Klebsiella*-specific factors on the ability to form biofilms on surfaces coated with extracellular matrix or collagen (Jagnow and Clegg 2003). The MrkD adhesin, but not the MrkA shaft protein was found to mediate biofilm formation on collagen and matrix-coated surfaces. Therefore, MrkA and MrkD may have distinct roles in biofilm formation on different surfaces. In a screen using signature-tagged mutagenesis to identify transposon insertions that caused a defect in biofilm formation on extracellular matrix material, insertions were identified in genes required for: (1) capsular biosynthesis, (2) transcriptional regulation (LuxR, LysR, and Crp-like genes) and (3) the sugar phosphotransferase (PTS) system. (Boddicker et al. 2006). The identification of a CRP-like protein is consistent with a previous study indicating that biofilm development in *Klebsiella* is under catabolite repression (Jackson et al. 2002).

7   Biofilm Formation on and in Urinary Tissues

7.1   *Intracellular Bacterial Communities*

7.1.1   Intracellular Pods

UPEC strains are not only capable of efficient adherence to and invasion of bladder epithelial cells, but they are also highly successful at forming persistent intracellular reservoirs that escape host defenses and antibiotic treatments (Mulvey et al. 2001). Recent work by Anderson et al. (2003) has indicated that UPEC strains accomplish these feats by the formation of biofilm-like pods or intracellular bacterial communities (IBCs) within the host bladder tissue (Fig. 1). Investigations of the replication of UPEC within the superficial bladder epithelial cells led to the discovery of these intracellular pods on the bladders of mice infected with UPEC. Bacterial replication within the superficial bladder epithelial cells resulted in tightly packed biofilm-like pods jutting into the

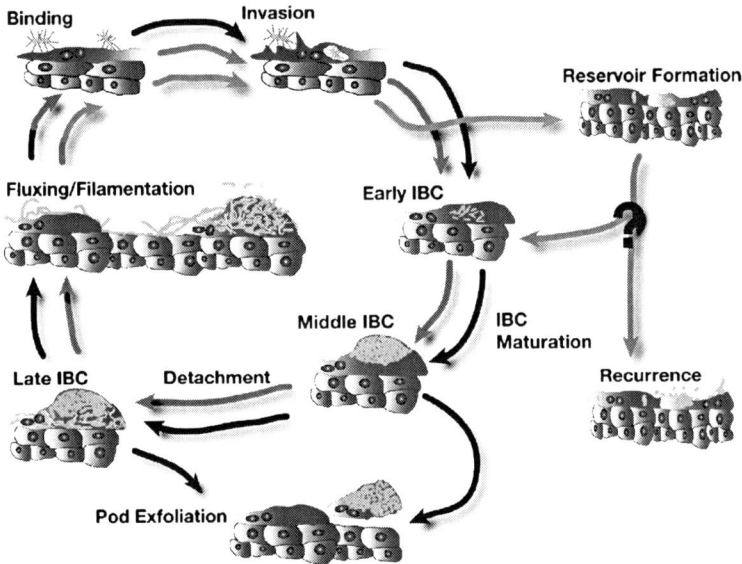

Fig. 1 Stages in intracellular pod formation in uropathogenic *E. coli*. (Reprinted from Justice et al. 2004, with permission form the National Academy of Sciences)

bladder lumen. Neither type 1-piliated *E. coli* K-12 strains nor *fimH* mutant strains produced pods, indicating that while type-1 pili are necessary for host-cell invasion, additional factors are required for UPEC pathogenesis (Anderson et al. 2003). Using time-lapse fluorescence videomicroscopy of mouse bladder explants infected with green fluorescent protein (GFP) producing UPEC, Justice et al. (2004) demonstrated that the bacteria within these intracellular pods undergo a continuous developmental program leading to maturation of the intracellular bacterial communities. This program can be divided into four distinct phases that result in dispersal to new sites of infection and establishment of a persistent reservoir, which closely parallels biofilm formation on abiotic surfaces (Fig. 1).

7.1.2   Early IBCs

The first phase of IBC formation begins at 1-3 h after infection with binding of type 1-piliated UPEC and invasion of superficial bladder epithelial cells (Fig. 1) (Justice et al. 2004). The bacteria in this phase are nonmotile rod-shaped cells that proliferate rapidly with a doubling time of approximately 30-35 min. Bacterial growth continues for up to 8 h postinfection to form loosely organized colonies free within the cytoplasm that resemble microcolonies of abiotic biofilms.

7.1.3 Middle IBCs

Middle IBCs form at 6-8 h postinfection producing characteristic pods on the mouse bladder lumenal surface (Fig. 1) (Justice et al. 2004). This stage is characterized by a reduction of cell growth and more strikingly, a reduction in cell size, resulting in a coccoid morphology for all of the UPEC within the pod. Each pod corresponds to a single superficial epithelial cell tightly packed with bacteria forming an intracellular biofilm. These pods share many of the definitive characteristics of abiotic biofilms. First, within the pods, a fibrous polysaccharide matrix reminiscent of a glycocalyx surrounds the bacteria (Anderson et al. 2003). Moreover, each bacterium is individually compartmentalized within this matrix and interacts with it via multiple fibers expressed on its surface. Next, immunofluorescent staining was used to show that both type 1 pili and Ag43 are expressed within the intracellular pods in a heterogeneous fashion reminiscent of gene expression within abiotic biofilms (Anderson et al. 2003). Finally, encasement of UPEC in these biofilm-like pods provides protection from antibiotic treatment and from host defenses similar to protection of bacteria found within abiotic biofilms. Polymorphonuclear leukocytes (PMNs) can discriminate with high accuracy cells infected by UPEC from cells uninfected by UPEC, but they cannot penetrate into the intracellular pod (Justice et al. 2004). Moreover, even when they gain access to the pod interior, the rapid proliferation during the early IBC formation provides sufficient numbers of bacteria to overwhelm the ability of the PMNs to engulf the entire bacterial population.

7.1.4 Late IBCs

During late IBC formation at approximately 12 h postinfection, the UPEC are found to flux out of the cells by regaining their rod-shaped morphology, becoming motile, and bursting out of the pods (Fig. 1) (Justice et al. 2004). The fluxing motility parallels the detachment of abiotic biofilms. The morphological change during this phase was not due to exposure to a rich medium environment because the same phenotype was also observed when the mouse bladder explants were exposed to saline buffer. Fluxing appeared to be necessary for UPEC to infect either adjacent superficial bladder cells or the underlying naïve bladder cells. Although the fluxing observed during this stage appeared characteristic of flagellar-based motility, subsequent studies showed that fluxing did not involve flagella because a UPEC mutant deficient in flagellin expression ($\Delta fliC$) was able to form as many pods as a wild type UPEC strain in a murine cystitis model (Wright et al. 2005). The expression of flagella, however, did confer a subtle advantage in co-challenge infections with the wild type UPEC and $\Delta fliC$ mutant, although it was unclear exactly why. These results are supported by similar results in infection studies of flagellar and motility mutants of UPEC (*fliC*, *fliA*, and *motAB*) in a murine cystitis model (Lane et al. 2005).

7.1.5   UPEC Filament Formation and Reinfection

The final phase of IBC formation occurs between 24 and 48 h postinfection and results in the filamentation of UPEC (Fig. 1) (Justice et al. 2004). Although unclear if filamentation occurs within the IBC or on the bladder surface, filamentation was determined to be critical for escape from innate host defenses. Filamentous bacteria were observed to be resistant to engulfment by the PMNs recruited to infected bladder cells, while rod-shaped cells were easily engulfed and eliminated. Filamentous bacteria were also seen to septate and form rod-shaped daughter cells that were susceptible to elimination by the PMNs; however, those that escaped could also reinfect new cells. Previous studies have shown that the filamentous bacteria may interact with adjacent or underlying cells to cause new infections (Mulvey et al. 2001). The appearance of filaments also coincided with the appearance of small groups (usually pairs) of UPEC in newly infected healthy superficial bladder cells (Justice et al. 2004). These cells were indicative of a second round of infection that progressed through all of the stages, as indicated above, although with much longer kinetics of infection. Moreover, after exfoliation of the bladder cells at 36-48 h postinfection, most of the superficial bladder cells were observed to be smaller than normal and once again, many had small groups of bacteria intracellularly. These clusters appeared to be quiescent bacteria that continued to produce GFP for at least 12 days after infection. These quiescent bacteria are proposed to be the cells responsible for persistent and recurrent infections that characterize UTIs caused by UPEC.

Recently, it has been suggested that host actin cytoskeletal rearrangements during bladder epithelial cell maturation modulate growth and resurgence of quiescent UPEC found in naïve bladder epithelial cells (Eto et al. 2006). In terminally differentiated superficial bladder cells, actin is found mostly associated with the basolateral surface of the cell, whereas in immature bladder epithelial cells, the actin is found throughout the cell and along the cell periphery (Romih et al. 1999). In a murine cystitis model using immunofluorescent microscopy, IBCs in the cell cytosol were found to be only weakly associated with actin, whereas UPEC within naïve bladder cells were found within actin-lined vacuoles (Eto et al. 2006). For these reasons, it was thought that perhaps actin was involved in suppressing growth of UPEC in naïve bladder cells. Therefore, a bladder epithelial cell line 5637 resembling the immature basal or intermediate bladder epithelial cells with respect to their actin cytoskeleton was used to study the involvement of actin in UPEC growth and resurgence. This cell line rarely formed IBCs as large as those seen during infection of terminally differentiated bladder cells in a murine cystitis model (Mulvey et al. 2001; Anderson et al. 2003). Disruptors of actin polymerization and/or the actin cytoskeleton stimulated both intracellular growth and efflux of UPEC in the 5637 cell line in vitro as determined by gentamicin protection assays (Eto et al. 2006). Moreover, UPEC in the 5637 cell line was found sequestered within late endosome-like vacuoles. However, robust intracellular growth and IBC formation by UPEC usually occurs in the cytoplasm (Mulvey et al. 2001; Anderson et al. 2003; Justice et al. 2004). Disruption of host actin resulted in increased clusters of bacteria and release from vacuoles into the host cytoplasm (Eto et al. 2006). However, these clusters of UPEC were smaller than the IBCs previously observed. To achieve more efficient release of UPEC from intracellular vacuoles, host membrane was treated with the membrane permeabilizing glycoside saponin. Saponin treatment stimulated IBC formation by UPEC but not by type 1-piliated *E. coli*, indicating that UPEC-specific factors are essential for pod formation. A model explaining how actin could affect the fate of UPEC during pathogenesis within the bladder suggested that infection of immature basal or intermediate epithelial cells results in UPEC trafficking into late endosome and lysosome-like acidic vacuoles enmeshed within actin. This process results in limited growth of UPEC until terminal differentiation of the bladder cells leads to host actin rearrangements that trigger the release of quiescent UPEC from the membrane-bound intracellular vacuoles. The alternate pathway in which UPEC infects superficial bladder cells directly results in the IBC formation described above (Mulvey et al. 2001; Anderson et al. 2003). This model suggests that both the interaction with actin and the membrane barrier of the intracellular vacuole results in quiescence of UPEC, and ultimately release into the host cytoplasm is required for growth and subsequent IBC formation (Eto et al. 2006).

7.2 Chronic Prostatitis

An additional infection of the urinary tract that is associated with biofilm formation is chronic bacterial prostatitis in men. The most commonly encountered bacteriological agent in prostatitis is *E. coli*, followed by other members of the *Enterobacteriaceae* (*Proteus* and *Klebsiella*) and coagulase-negative *Staphylococci* (Domingue and Hellstrom 1998). These infections are notoriously difficult to treat with antibiotic therapy. Studies by Nickel and Costerton demonstrated that prostate biopsy samples from chronically infected patients contained exopolysaccharide-encased microcolonies that were attached to the walls of the prostate ducts (Nickel and Costerton 1993). In chronic staphylococcal prostatitis, biofilm-like microcolonies were attached to the prostate in patients that were refractory to antibiotic therapy (Nickel and Costerton 1992). Finally, in a recent study, a total of 377 *E. coli* isolates obtained from a variety of urinary tract infections (cystitis, pyelonephritis, and prostatitis) were examined for biofilm-forming abilities by standard crystal violet staining of cells grown in microtiter wells. The isolates from prostatitis cases exhibited significantly greater biofilm formation than other isolates (Kanamaru et al. 2006).

8 Summary

Despite great strides in our understanding of the pathogenesis of UTI-causing organisms, urinary tract infections will continue to represent a major human health problem. Described in this chapter are some ways in which different UTI-causing bacteria use the biofilm mode of growth to gain a foothold in and cause infection of the urinary tract. The classical antimicrobial therapies available that can effectively kill UTI-causing bacteria are designed for killing during the planktonic mode of growth and therefore are relatively ineffective against UTI biofilms. Awareness that biofilms are not only formed on abiotic surfaces, but can also be present in and on living tissues allows for a rational and concerted effort to devise strategies to counter the refractory nature of these infections. By understanding the molecular characteristics of device-related biofilms and chronic infections caused by biofilms associated with tissues, new preventative and therapeutic strategies can be targeted to specifically treat or promote the dissolution of biofilms. It seems clear that the biofilm phenotype must be considered in any future development of treatments to prevent the emergence and recurrence of UTIs.

Acknowledgements PNR is supported by grants MCB0406047 from the National Science Foundation, a Merit Review from the Department of Veterans Affairs, and a Research Career Scientist Award from the Department of Veterans Affairs

References

Abraham JM, Freitag CS, Clements JR, Eisenstein BI (1985) An invertible element of DNA controls phase variation of type 1 fimbriae of *Escherichia coli*. Proc Natl Acad Sci U S A 82:5724-5727

Abraham SN, Goguen JD, Sun D, Klemm P, Beachey EH (1987) Identification of two ancillary subunits of *Escherichia coli* type 1 fimbriae by using antibodies against synthetic oligopeptides of *fim* gene products. J Bacteriol 169:5530-5536

Abrahams HM, Stoller ML (2003) Infection and urinary stones. Curr Opin Urol 13:63-67

Allison C, Lai HC, Hughes C (1992) Co-ordinate expression of virulence genes during swarm-cell differentiation and population migration of *Proteus mirabilis*. Mol Microbiol 6:1583-1591

Anderson GG, Palermo JJ, Schilling JD, Roth R, Heuser J, Hultgren SJ (2003) Intracellular bacterial biofilm-like pods in urinary tract infections. Science 301:105-107

Bagshaw SM, Laupland KB (2006) Epidemiology of intensive care unit acquired urinary tract infections. Curr Opin Infect Dis 19:67-71

Bahrani FK, Mobley HL (1994) *Proteus mirabilis* MR/P fimbrial operon: genetic organization, nucleotide sequence, and conditions for expression. J Bacteriol 176:3412-3419

Bahrani-Mougeot FK, Buckles EL, Lockatell CV, Hebel JR, Johnson DE, Tang CM, Donnenberg MS (2002) Type 1 fimbriae and extracellular polysaccharides are preeminent uropathogenic *Escherichia coli* virulence determinants in the murine urinary tract. Mol Microbiol 45:1079-1093

Baorto DM, Gao Z, Malaviya R, Dustin ML, van der Merwe A, Lublin DM, Abraham SN (1997) Survival of FimH-expressing enterobacteria in macrophages relies on glycolipid traffic. Nature 389:636-639

Beloin C, Michaelis K, Lindner K, Landini P, Hacker J, Ghigo JM, Dobrindt U (2006) The transcriptional antiterminator RfaH represses biofilm formation in *Escherichia coli*. J Bacteriol 188:1316-1331

Bian Z, Normark S (1997) Nucleator function of CsgB for the assembly of adhesive surface organelles in *Escherichia coli*. EMBO J 16:5827-5836

Bichler KH, Eipper E, Naber K Braun V Zimmmermann R, Lahme S (2002) Urinary infection stones. Int J Antimicrob Agents 19:488-498

Blomfield IC, Calie PJ, Eberhardt KJ, McClain MS, Eisenstein BI (1993) Lrp stimulates phase variation of type 1 fimbriation in *Escherichia coli* K-12. J Bacteriol 175:27-36

Blomfield IC, Kulasekara DH, Eisenstein BI (1997) Integration host factor stimulates both FimB- and FimE-mediated site-specific DNA inversion that controls phase variation of type 1 fimbriae expression in *Escherichia coli*. Mol Microbiol 23:705-717

Blumer C, Kleefeld A, Lehnen D, Heintz M, Dobrindt U, Nagy G, Michaelis K, Emody L, Polen T, Rachel R, Wendisch VF, Unden G (2005) Regulation of type 1 fimbriae synthesis and biofilm formation by the transcriptional regulator LrhA of *Escherichia coli*. Microbiology 151:3287-3298

Boddicker JD, Anderson RA, Jagnow J, Clegg S (2006) Signature-tagged mutagenesis of *Klebsiella pneumoniae* to identify genes that influence biofilm formation on extracellular matrix material. Infect Immun 74:4590-4597

Brinton CC Jr (1965) The structure, function, synthesis and genetic control of bacterial pili and a molecular model for DNA, RNA transport in Gram-negative bacteria. Trans N Y Acad Sci 27:1003-1054

Burton E, Gawande PV, Yakandawala N, LoVetri K, Zhanel GG, Romeo T, Friesen AD, Madhyastha S (2006) Antibiofilm activity of GlmU enzyme inhibitors against catheter-associated uropathogens. Antmicrob Agents Chemother 50:1835-1840

Clapham L, McLean RJC, Nickel JC, Downey J, Costerton JW (1990) The influence of bacteria on struvite crystal habit and its importance in urinary stone formation. J Crystal Growth 104:475-484

Choog S, Whitfield H (2000) Biofilms and their role in infections in urology. BJU Int 86:935-941

Clegg S, Gerlach GF (1987) Enterobacterial fimbriae. J Bacteriol 169:934-938

Connell I, Agace W, Klemm P, Schembri M, Marild S, Svanborg C (1996) Type 1 fimbrial expression enhances *Escherichia coli* virulence for the urinary tract. Proc Natl Acad Sci U S A 93:9827-9832

Danese PN, Pratt LA, Dove SL, Kolter R (2000) The outer membrane protein, antigen 43, mediates cell-to-cell interactions within *Escherichia coli* biofilms. Mol Microbiol 37:424-432

Deflaun MF, Marshall BM, Kulle EP, Levy SB (1994) Tn5 insertion mutants of *Pseudomonas fluorescens* defective in adhesion to soil and seeds. Appl Environ Microbiol 60:2637-2642

Denamur E, Bonacorsi S, Giraud A, Duriez P, Hilali F, Amorin C, Bingen E, Andremont A, Picard B, Taddei F, Matic I (2002) High frequency of mutator strains among human uropathogenic *Escherichia coli* isolates. J Bacteriol 184:605-609

Diderichsen B (1980) *flu*, a metastable gene controlling surface properties of *Escherichia coli*. J Bacteriol 141:858-867

Di Martino P, Cafferini N, Joly B, Darfeuillle-Michaud A (2003) *Klebsiella pneumoniae* type 3 pili facilitate adherence and biofilm formation on abiotic surfaces. Res Microbiol 154:9-16

Domingue GJ, Hellstrom WG (1998) Prostatitis. Clin Microbiol Rev 11:604-613

Donato GM, Lelivelt MJ, Kawula TH (1997) Promoter-specific repression of *fimB* expression by the *Escherichia coli* nucleoid-associated protein H-NS. J Bacteriol 179:6618-6625

Donlan RW (2001) Biofilm formation: a clinically relevant microbiological process. Clin Infect Dis 33:1387-1392

Donlan RM, Costerton JW (2002) Biofilms: survival mechanisms of clinically relevant microorganisms. Clin Microbiol Rev 15:167-193

Dumanski AJ, Hedelin H, Edin-Liljegren A, Beauchemin D, McLean RJC (1994) Unique ability of the *Proteus mirabilis* capsule to enhance mineral growth in infectious urinary calculi. Infect Immun 62:2998-3003

Duncan MJ, Li GJ, Shin JS, Carson JL, Abraham SN (2004) Bacterial penetration of bladder epithelium through lipid rafts. J Biol Chem 279:18944-18951

Eden CS, Hansson HA (1978) *Escherichia coli* pili as possible mediators of attachment to human urinary tract epithelial cells. Infect Immun 21:229-237

Elliott TS, Reed L, Slack RC, Bishop MC (1985) Bacteriology and ultrastructure of the bladder in patients with urinary tract infections. J Infect 11:191-199

Emori TG, Gaynes RP (1993) An overview of nosocomial infections, including the role of the microbiology laboratory. Clin Microbiol Rev 6:428-442

Eto DS, Sundsbak JL, Mulvey MA (2006) Actin-gated intracellular growth and resurgence of uropathogenic *Escherichia coli*. Cell Microbiol 8:704-717

Foxman B, Zhang L, Palin K, Tallman P, Marrs CF (1995) Bacterial virulence characteristics of *Escherichia coli* isolates from first-time urinary tract infection. J Infect Dis 171:1514-1521

Foxman B (2003) Epidemiology of urinary tract infections: incidence, morbidity and economic costs. Dis Mon 49:53-70

Gally DL, Leathart J, Blomfield IC (1996) Interaction of FimB, FimE with the *fim* switch that controls the phase variation of type 1 fimbriae in *Escherichia coli* K-12. Mol Microbiol 21:725-738

Gally DL, Rucker TJ, Blomfield IC (1994) The leucine-responsive regulatory protein binds to the *fim* switch to control phase variation of type 1 fimbrial expression in *Escherichia coli* K-12. J Bacteriol 176:5665-5672

Griffith DP, Musher DM, Itin C (1976) Urease. The primary cause of infection-induced urinary stones. Invest Urol 13:346-350

Gunther IN, Snyder JA, Lockatell V, Blomfield I, Johnson DE, Mobley HL (2002) Assessment of virulence of uropathogenic *Escherichia coli* type 1 fimbrial mutants in which the invertible element is phase-locked on or off. Infect Immun 70:3344-3354

Gunther NW, Lockatell V, Johnson DE, Mobley HL (2001) In vivo dynamics of type 1 fimbria regulation in uropathogenic *Escherichia coli* during experimental urinary tract infection. Infect Immun 69:2838-2846

Haagmans W, van der Woude M (2000) Phase variation of Ag43 in *Escherichia coli*: Dam-dependent methylation abrogates OxyR binding and OxyR-mediated repression of transcription. Mol Microbiol 35:877-887

Hagberg L, Jodal U, Korhonen TK, Lidin-Janson G, Lindberg U, Svanborg Eden C (1981) Adhesion, hemagglutination, and virulence of *Escherichia coli* causing urinary tract infections. Infect Immun 31:564-570

Haraoka M, Hang L, Frendeus B, Godaly G, Burdick M, Strieter R, Svanborg C (1999) Neutrophil recruitment and resistance to urinary tract infection. J Infect Dis 180:1220-1229

Hasman H, Chakraborty T, Klemm P (1999) Antigen-43-mediated autoaggregation of *Escherichia coli* is blocked by fimbriation. J Bacteriol 181:4834-4841

Henderson IR, Owen P (1999) The major phase-variable outer membrane protein of *Escherichia coli* structurally resembles the immunoglobulin A1 protease class of exported protein and is regulated by a novel mechanism involving Dam and oxyR. J Bacteriol 181:2132-2141

Hooton TM, Stamm WE (1997) Diagnosis and treatment of uncomplicated urinary tract infection. Infect Dis Clin North Am 11:551-581

Hultgren SJ, Porter TN, Schaeffer AJ, Duncan JL (1985) Role of type 1 pili and effects of phase variation on lower urinary tract infections produced by *Escherichia coli*. Infect Immun 50:370-377

Hvidberg H, Struve C, Krogfelt KA, Christensen N, Rasmussen SN, Frimodt-Moller N (2000) Development of a long-term ascending urinary tract infection mouse model for antibiotic treatment studies. Antimicrob Agents Chemother 44:156-163

Jackson DW, Simecka JW, Romeo T (2002) Catabolite repression of *Escherichia coli* biofilm formation. J Bacteriol 184:3406-3410

Jagnow J, Clegg S (2003) *Klebsiella pneumoniae* MrkD-mediated biofilm formation on extracellular matrix- and collagen coated surfaces. Microbiology 149:2397-2405

Jansen AM, Lockatell V, Johnson DE, Mobley HLT (2004) Mannose resistant *Proteus* like fimbriae are produced by most *Proteus mirabilis* strains infecting the urinary tract, dictate the in-vivo localization of bacteria and contribute to biofilm formation. Infect Immun 72:7294-7305

Johnson DE, Lockatell CV, Russell RG, Hebel JR, Island MD, Stapleton A, Stamm WE, Warren JW (1998) Comparison of *Escherichia coli* strains recovered from human cystitis and pyelonephritis infections in transurethrally challenged mice. Infect Immun 66:3059-3065

Johnson JR (1991) Virulence factors in *Escherichia coli* urinary tract infection. Clin Microbiol Rev 4:80-128

Jones BV, Mahenthiralingam E, Sabbuba NA, Stickler DJ (2005) Role of swarming in the formation of crystalline *Proteus mirabilis* biofilms on urinary catheters. J Med Microbiol 54:807-813

Jones CH, Pinkner JS, Nicholes AV, Slonim LN, Abraham SN, Hultgren SJ (1993) FimC is a periplasmic PapD-like chaperone that directs assembly of type 1 pili in bacteria. Proc Natl Acad Sci U S A 90:8397-8401

Jones CH, Pinkner JS, Roth R, Heuser J, Nicholes AV, Abraham SN, Hultgren SJ (1995) FimH adhesin of type 1 pili is assembled into a fibrillar tip structure in the Enterobacteriaceae. Proc Natl Acad Sci. U S A 92:2081-2085

Jost SP (1989) Cell cycle of normal bladder urothelium in developing and adult mice. Virchows Archiv 57:27-36

Justice SS, Hung C, Theriot JA, Fletcher DA, Anderson GG, Footer MJ, Hultgren SJ (2004) Differentiation and developmental pathways of uropathogenic *Escherichia coli* in urinary tract pathogenesis. Proc Natl Acad Sci U S A 101:1333-1338

Kajander EO, Ciftcioglu N (1998) Nanobacteria: an alternative mechanism for pathogenic intra- and extracellular calcification and stone formation. Proc Natl Acad Sci U S A 95:8274-8279

Kalsi J, Arya M, Wilson P, Mundy A (2003) Hospital acquired urinary tract infection. Int J Clin Prac 57:388-391

Kanamaru S, Kurazono H, Terai A, Monden K, Kumon H, Mizunoe Y, Ogawa O, Yamamoto S (2006) Increased biofilm formation in *Escherichia coli* isolated from acute prostatitis. Int J Antimicrob Agents S28:S21-S25

Kikuchi T, Mizunoe Y, Takade A, Naito S, Yoshida S (2005) Curli fibers are required for development of biofilm architecture in *Escherichia coli* K-12 and enhance bacterial adherence to human uroepithelial cells. Microbiol Immunol 49:875-884

Kjaergaard K, Schembri MA, Ramos C, Molin S, Klemm P (2000) Antigen 43 facilitates formation of multispecies biofilms. Environ Microbiol 2:695-702

Klemm P (1986) Two regulatory *fim* genes, *fimB* and *fimE*, control the phase variation of type 1 fimbriae in *Escherichia coli*. EMBO J 5:1389-1393

Klemm P, Christiansen G (1987) Three *fim* genes required for the regulation of length and mediation of adhesion of *Escherichia coli* type 1 fimbriae. Mol Gen Genet 208:439-445

Klemm P, Christiansen G (1990) The *fimD* gene required for cell surface localization of *Escherichia coli* type 1 fimbriae. Mol Gen Genet 220:334-338

Klemm P, Hjerrild L, Gjermansen M, Schembri MA (2004) Structure-function analysis of the self-recognizing Antigen 43 autotransporter protein from *Escherichia coli*. Mol Microbiol 51:283-296

Knirel YA, Vinogradov EV, Shashkov AS, Sidorczyk Z, Rozalski A, Korber DR, Lawrence JR, Caldwell DE (1994) Effect of motility on surface colonization and reproductive success of *Pseudomonas fluorescens* in dual-dilution continuous culture and batch culture systems. Appl Environ Microbiol 60:1421-1429

Krogfelt KA, Bergmans H, Klemm P (1990) Direct evidence that the FimH protein is the mannose-specific adhesin of *Escherichia coli* type 1 fimbriae. Infect Immun 58:1995-1998

Labat F, Pradillon O, Garry L, Peuchmaur M, Fantin B, Denamur E (2005) Mutator phenotype confers advantage in *Escherichia coli* chronic urinary tract infection pathogenesis. FEMS Immunol Med Microbiol 44:317-321

Lane MC, Lockatell V, Monterosso G, Lamphier D, Weinert J, Hebel JR, Johnson DE, Mobley HL (2005) Role of motility in the colonization of uropathogenic *Escherichia coli* in the urinary tract. Infect Immun 73:7644-7656

Langermann S, Mollby R, Burlein JE, Palaszynski SR, Auguste CG, DeFusco A, Strouse R, Schenerman MA, Hultgren SJ, Pinkner JS, Winberg J, Guldevall L, Soderhall M, Ishikawa K, Normark S, Koenig S (2000) Vaccination with FimH adhesin protects cynomolgus monkeys from colonization and infection by uropathogenic *Escherichia coli*. J Infect Dis 181:774-778

Langermann S, Palaszynski S, Barnhart M, Auguste G, Pinkner JS, Burlein J, Barren P, Koenig S, Leath S, Jones CH, Hultgren SJ (1997) Prevention of mucosal *Escherichia coli* infection by FimH-adhesin-based systemic vaccination. Science 276:607-611

Langstraat J, Bohse M, Clegg S (2001) Type 3 fimbrial shaft (MrkA) of *Klebsiella pneumoniae*, but not the fimbral adhesion (MrkD), facilitates biofilm formation. Infect Immun 69:5805-5812

Lehnen D, Blumer C, Polen T, Wackwitz B, Wendisch VF, Unden G (2002) LrhA as a new transcriptional key regulator of flagella, motility and chemotaxis genes in *Escherichia coli*. Mol. Microbiol 45:521-532

Li X, Rasko DA, Lockatell CV, Johnson DE, Mobley HL (2001) Repression of bacterial motility by a novel fimbrial gene product. EMBO J 20:4854-4862

Li X, Lockatell CV, Johnson DE, Mobley HL (2002) Identification of MrpI as the sole recombinase that regulates the phase variation of MR/P fimbria, a bladder colonization factor of uropathogenic *Proteus mirabilis*. Mol Microbiol 45:865-874

Li X, Zhao HJ, Lockatell CV, Drachenberg CB, Johnson DE, Mobley HLT (2002) Visualization of *Proteus mirabilis* within the matrix of urease-induced bladder stones during experimental urinary tract infection. Infect Immun 70:389-394

Li X, Lockatell CV, Johnson DE, Lane MC, Warren JW, Mobley HL (2004) Development of an intranasal vaccine to prevent urinary tract infection by *Proteus mirabilis*. Infect Immun 72:66-75

Maki DG, Tambyah PA (2001) Engineering out the risk of infection with urinary catheters. Emerg Infect Dis 7:342-347

Malaviya R, Gao Z, Thankavel K, van der Merwe PA, Abraham SN (1999) The mast cell tumor necrosis factor alpha response to FimH-expressing *Escherichia coli* is mediated by the glycosylphosphatidylinositol-anchored molecule CD48. Proc Natl Acad Sci U S A 96:8110-8115

Marrs CF, Zhang L, Foxman B (2005) *Escherichia coli* mediated urinary tract infections: are there distinct uropathogenic *E. coli* (UPEC) pathotypes? FEMS Microbiol Lett 252:183-190

Martinez JJ, Mulvey MA, Schilling JD, Pinkner JS, Hultgren SJ (2000) Type 1 pilus-mediated bacterial invasion of bladder epithelial cells. EMBO J 19:2803-2812

Maurer L, Orndorff PE (1987) Identification and characterization of genes determining receptor binding and pilus length of *Escherichia coli* type 1 pili. J Bacteriol 169:640-645

McClain MS, Blomfield IC, Eisenstein BI (1991) Roles of *fimB* and *fimE* in site-specific DNA inversion associated with phase variation of type 1 fimbriae in *Escherichia coli*. J Bacteriol 173:5308-5314

McLean RJ, Nickel JC, Beveridge TJ, Costerton JW (1989) Observations of the ultrastructure of infected kidney stones. J Med Microbiol 29:1-7

McTaggart LA, Rigby RC, Elliott TS (1990) The pathogenesis of urinary tract infections associated with *Escherichia coli*, *Staphylococcus saprophyticus* and *S epidermidis*. J Med Microbiol 32:135-141

Mireles JR, Toguchi A, Harshey RM (2001) *Salmonella enterica* serovar *typhimurium* swarming mutants with altered biofilm forming properties: surfactin inhibits biofilm formation. J Bacteriol 183:5848-5854

Morris NS, Stickler DJ, Winters C (1997) Which indwelling catheters resist encrustation by *Proteus mirabilis* biofilms? Br J Urol 80:58-63

Morris N, Stickler DJ (1998) The effect of urease inhibitors on the encrustation of urethral catheters. Urol Res 26:275-279

Morris NS, Stickler DJ, McLean RJC (1999) The development of bacterial biofilms on indwelling urethral catheters. World J Urol 17:345-350

Morris NS, Stickler DJ (2001) Does drinking cranberry juice produce urine inhibitory to the development of crystalline, catheter blocking *Proteus mirabilis* biofilms. BJU Int 88:192-197

Mulvey MA, Lopez-Boado YS, Wilson CL, Roth R, Parks WC, Heuser J, Hultgren SJ (1998) Induction and evasion of host defenses by type 1-piliated uropathogenic *Escherichia coli*. Science 282:1494-1497

Mulvey MA, Schilling JD, Hultgren SJ (2001) Establishment of a persistent *Escherichia coli* reservoir during the acute phase of a bladder infection. Infect Immun 69:4572-4579

Nickel JC, Gristina AG, Costerton JW (1985) Electron microscopic study of an infected Foley catheter. Can J Surg 28:50-51

Nickel JC, Olson M, McLean RJ, Grant SW, Costerton JW (1987) An ecological study of infected urinary stone genesis in an animal model. Br J Urol 59:21-30

Nickel JC, Downey JA, Costerton JW (1989) Ultrastructural study of microbiologic colonization of urinary catheters. 134:284-291

Nickel JC, Costerton JW (1992) Coagulase-negative Staphylococcus in chronic prostatitis. J Urol 147:398-400

Nickel JC, Costerton JW (1993) Bacterial localization in antibiotic-refractory chronic bacterial prostatitis. Prostate 23:107-114

Nickel C, Costerton W, McClean RJC, Olson M (1994) Bacterial biofilms: influence on the pathogenesis, diagnosis and treatment of urinary tract infections. Antimicrob Chemother 33:31-41

Nicolle LE (2005) Catheter-related urinary tract infection. Drugs Aging 22:627-639

O'Gara J, Dorman CJ (2000) Effects of local transcription and H-NS on inversion of the *fim* switch of *Escherichia coli*. Mol Microbiol 36:457-466

O'Toole GA, Kolter R (1998a) Flagellar and twitching motility are necessary for *Pseudomonas aeruginosa* biofilm development. Mol Microbiol 30:295-304

O'Toole GA, Kolter R (1998b) Initiation of biofilm formation in *Pseudomonas fluorescens* WCS365 proceeds via multiple, convergent signalling pathways: a genetic analysis. Mol Microbiol 28:449-461

Oelschlaeger TA, Dobrindt U, Hacker J (2002) Virulence factors of uropathogens. Curr Opin Urol 12:33-38

Ohkawa M, Sugata T, Sawaki M, Nakashima T, Fuse H, Hisazumi H (1990) Bacterial and crystal adherence to the surfaces of indwelling urethral catheters. J Urol 143:717-721

Olsen A, Jonsson A, Normark S (1989) Fibronectin binding mediated by a novel class of surface organelles on *Escherichia coli*. Nature 338:652-655

Olsen PB, Klemm P (1994) Localization of promoters in the *fim* gene cluster and the effect of H-NS on the transcription of *fimB* and *fimE*. FEMS Microbiol Lett 116:95-100

Owen P, Meehan M, de Loughry-Doherty H, Henderson I (1996) Phase-variable outer membrane proteins in *Escherichia coli*. FEMS Immunol Med Microbiol 16:63-76

Parsek MR, Singh PK (2003) Bacterial biofilms: an emerging link to disease pathogenesis. Annu Rev Microbiol 57:677-701

Pak J, Pu Y, Zhang ZT, Hasty DL, Wu XR (2001) Tamm-Horsfall protein binds to type 1 fimbriated *Escherichia coli* and prevents *E coli* from binding to uroplakin Ia and Ib receptors. J Biol Chem 276:9924-9930

Pratt LA, Kolter R (1998) Genetic analysis of *Escherichia coli* biofilm formation: roles of flagella, motility, chemotaxis and type I pili. Mol Microbiol 30:285-293

Prigent-Combaret C, Prensier G, Le Thi TT, Vidal O, Lejeune P, Dorel C (2000) Developmental pathway for biofilm formation in curli-producing *Escherichia coli* strains: role of flagella, curli and colanic acid. Environ Microbiol 2:450-464

Radziejewska J, Kaca W (1993) Structural study of O-specific polysaccharides of Proteus. J Carbohydrate Chem 12:379-414

Rather PN (2005) Swarmer cell differentiation in *Proteus mirabilis*. Environ Microbiol 7:1065-1073

Reid G, Denstedt JD, Yang YS, Lam D, Nause C (1992) Microbial adhesion and biofilm formation on urethral stents in-vitro and in-vivo. J Urol 148:1592-1594

Reid G (1999) Biofilms in infectious disease and on medical devices. Int J Antimicrob Agents 11:223-239

Reisner A, Haagensen JAJ, Schembri MA, Zechner EL, Molin S (2003) Development and maturation of *Escherichia coli* K-12 biofilms. Mol Microbiol 48:933-946

Roche A, McFadden J, Owen P (2001) Antigen 43, the major phase-variable protein of the *Escherichia coli* outer membrane, can exist as a family of proteins encoded by multiple alleles. Microbiology 147:161-169

Romih R, Veranic P, Jezernik K (1999) Actin filaments during terminal differentiation of urothelial cells in the rat urinary bladder. Histochem Cell Biol 112:375-380

Rosch W, Lugauer S (1999) Catheter associated infections in urology: possible use of silver impregnated catheters and the Erlanger silver catheter 27:S74-S77

Russell PW, Orndorff PE (1992) Lesions in two *Escherichia coli* type 1 pilus genes alter pilus number and length without affecting receptor binding. J Bacteriol 174:5923-5935

Sabbuba N, Hughes G, Stickler DJ (2002) The migration of *Proteus mirabilis* and other urinary tact pathogens over Foley catheters. BJU Int 89:55-60

Schaeffer AJ, Schwan WR, Hultgren SJ, Duncan JL (1987) Relationship of type 1 pilus expression in *Escherichia coli* to ascending urinary tract infections in mice. Infect Immun 55:373-380

Schembri MA, Christiansen G, Klemm P (2001) FimH-mediated autoaggregation of *Escherichia coli*. Mol Microbiol 41:1419-1430

Schembri MA, Dalsgaard D, Klemm P (2004) Capsule shields the function of short bacterial adhesins. J Bacteriol 186:1249-1257

Schembri MA, Hjerrild L, Gjermansen M, Klemm P (2003a) Differential expression of the *Escherichia coli* autoaggregation factor antigen 43. J Bacteriol 185:2236-2242

Schembri MA, Kjaergaard K, Klemm P (2003b) Global gene expression in *Escherichia coli* biofilms. Mol Microbiol 48:253-267

Schembri MA, Klemm P (2001a) Biofilm formation in a hydrodynamic environment by novel FimH variants and ramifications for virulence. Infect Immun 69:1322-1328

Schembri MA, Klemm P (2001b) Coordinate gene regulation by fimbriae-induced signal transduction. EMBO J 20:3074-3081

Schembri MA, Ussery DW, Workman C, Hasman H, Klemm P (2002) DNA microarray analysis of *fim* mutations in *Escherichia coli*. Mol Genet Genomics 267:721-729

Shin JS, Gao Z, Abraham SN (2000) Involvement of cellular caveolae in bacterial entry into mast cells. Science 289:785-788

Silverstein A, Donatucci CF (2003) Bacterial biofilms and implantable prosthetic devices. Int J Impot Res 5:S150-S154

Snyder JA, Lloyd AL, Lockatell CV, Johnson DE, Mobley HL (2006) Role of phase variation of type 1 fimbriae in a uropathogenic *Escherichia coli* cystitis isolate during urinary tract infection. Infect Immun 74:1387-1393

Sobel JD (1997) Pathogenesis of urinary tract infection. Role of host defenses. Infect Dis Clin North Am 11:531-549

Sokurenko EV, Chesnokova V, Doyle RJ, Hasty DL (1997) Diversity of the *Escherichia coli* type 1 fimbrial lectin. Differential binding to mannosides and uroepithelial cells. J Biol Chem 272:17880-17886

Sokurenko EV, Chesnokova V, Dykhuizen DE, Ofek I, Wu XR, Krogfelt KA, Struve C, Schembri MA, Hasty DL (1998) Pathogenic adaptation of *Escherichia coli* by natural variation of the FimH adhesin. Proc Natl Acad Sci U S A 95:8922-8926

Sokurenko EV, Courtney HS, Maslow J, Siitonen A, Hasty DL (1995) Quantitative differences in adhesiveness of type 1 fimbriated *Escherichia coli* due to structural differences in *fimH* genes. J Bacteriol 177:3680-3686

Sokurenko EV, Courtney HS, Ohman DE, Klemm P, Hasty DL (1994) FimH family of type 1 fimbrial adhesins: functional heterogeneity due to minor sequence variations among *fimH* genes. J Bacteriol 176:748-755

Stickler D, Ganderton L, King J, Nettleton J, Winters C (1993) *Proteus mirabilis* biofilms and the encrustation of urethral catheters. Urol Res 21:407-411

Stickler D, Morris N, Moreno MC, Sabbuba N (1998) Studies on the formation of crystalline bacterial biofilms on urethral catheters. Eur J Clin Microbiol Infect Dis 17:649-652

Stickler D, Hughes G (1999) Ability of *Proteus mirabilis* to swarm over urethral catheters. Eur J Clin Microbiol Infect Dis 18:206-208

Stickler DJ (2002) Susceptibility of antibiotic resistant Gram-negative bacteria to biocides: a perspective from the study of catheter biofilms. Symp Ser Appl Microbiol 31:163S-170S

Stickler DJ, Jones GL, Russell AD (2003a) Control of encrustation and blockage of Foley catheters. Lancet 361:1435-1437

Stickler DJ, Young R, Jones G, Sabbuba N, Morris N (2003b) Why are Foley catheters so vulnerable to encrustation and blockage by crystalline bacterial biofilm? Urol Res 31:306-311

Struve C, Krogfelt KA (1999) In vivo detection of *Escherichia coli* type 1 fimbrial expression and phase variation during experimental urinary tract infection. Microbiology 145:2683-2690

Sun TT, Zhao H, Provet J, Aebi U, Wu XR (1996) Formation of asymmetric unit membrane during urothelial differentiation. Mol Biol Rep 23:3-11

Svanborg C, Bergsten G, Fischer H, Frendeus B, Godaly G, Gustafsson E, Hang L, Hedlund M, Karpman D, Lundstedt AC, Samuelsson M, Samuelsson P, Svensson M, Wullt B (2001) The 'innate' host response protects and damages the infected urinary tract. Ann Med 33:563-570

Takeuchi H, Takayama H, Konishi T, Tomoyoshi T (1984) Scanning electron microscopy detects bacteria within infection stones. J Urol 132:67-69

Tenke P, Kovacs B, Jackel M, Nagy E (2006) The role of biofilm infection in urology. World J Urol 24:13-20

Torzewska A, Staczek P, Rozalski A (2003) Crystallization of urine mineral components may depend on the chemical nature of *Proteus* endotoxin polysaccharides. J Med Microbiol 52:471-477

Trautner BW, Darouiche RO (2004) Role of biofilm in catheter-associated urinary tract infection. Am J Infect Control 32:177-183

Ulett GC, Webb RI, Schembri MA (2006) Antigen-43-mediated autoaggregation impairs motility in *Escherichia coli*. Microbiology 152:2101-2110

Valenski ML, Harris SL, Spears PA, Horton JR, Orndorff PE (2003) The product of the *fimI* gene is necessary for *Escherichia coli* type 1 pilus biosynthesis. J Bacteriol 185:5007-5011

Vidal O, Longin R, Prigent-Combaret C, Dorel C, Hooreman M, Lejeune P (1998) Isolation of an *Escherichia coli* K-12 mutant strain able to form biofilms on inert surfaces: involvement of a new *ompR* allele that increases curli expression. J Bacteriol 180:2442-2449

Wagenlehner FME, Naber KG (2006) The treatment of bacterial urinary tract infections: presence and future. Eur Urol 49:235-244

Waldron DE, Owen P, Dorman CJ (2002) Competitive interaction of the OxyR DNA-binding protein and the Dam methylase at the antigen 43 gene regulatory region in *Escherichia coli*. Mol Microbiol 44:509-520

Watnick PI, Lauriano CM, Klose KE, Croal L, Kolter R (2001) The absence of a flagellum leads to altered colony morphology, biofilm development and virulence in *Vibrio cholerae* O139. Mol Microbiol 39:223-235

Wold AE, Mestecky J, Tomana M, Kobata A, Ohbayashi H, Endo T, Eden CS (1990) Secretory immunoglobulin A carries oligosaccharide receptors for *Escherichia coli* type 1 fimbrial lectin. Infect Immun 58:3073-3077

Wright KJ, Seed PC, Hultgren SJ (2005) Uropathogenic *Escherichia coli* flagella aid in efficient urinary tract colonization. Infect Immun 73:7657-7668

Wu XR, Sun TT, Medina JJ (1996) In vitro binding of type 1-fimbriated *Escherichia coli* to uroplakins Ia and Ib: relation to urinary tract infections. Proc Natl Acad Sci U S A 93:9630-9635

Yamamoto S, Tsukamoto T, Terai A, Kurazono H, Takeda Y, Yoshida O (1995) Distribution of virulence factors in *Escherichia coli* isolated from urine of cystitis patients. Microbiol Immunol 39:401-404

Zheng M, Aslund F, Storz G (1998) Activation of the OxyR transcription factor by reversible disulfide bond formation. Science 279:1718-1721

Zhou G, Mo WJ, Sebbel P, Min G, Neubert TA, Glockshuber R, Wu XR, Sun TT, Kong XP (2001) Uroplakin Ia is the urothelial receptor for uropathogenic *Escherichia coli:* evidence from in vitro FimH binding. J Cell Sci 114:4095-4103

Shifting Paradigms in *Pseudomonas aeruginosa* Biofilm Research

A. H. Tart and D. J. Wozniak(✉)

Abstract Biofilms formed by *Pseudomonas aeruginosa* have long been recognized as a challenge in clinical settings. Cystic fibrosis, endocarditis, device-related infections, and ventilator-associated pneumonia are some of the diseases that are considerably complicated by the formation of bacterial biofilms, which are resistant to most current antimicrobial therapies. Due to intense research efforts, our understanding of the molecular events involved in *P. aeruginosa* biofilm formation, maintenance, and antimicrobial resistance has advanced significantly. Over the years, several dogmas regarding these multicellular structures have emerged. However, more recent data reveal a remarkable complexity of *P. aeruginosa* biofilms and force investigators to continually re-evaluate previous findings. This chapter provides examples in which paradigms regarding *P. aeruginosa* biofilms have been challenged, reflecting the need to critically re-assess what is emerging in this rapidly growing field. In this process, several avenues of research have been opened that will ultimately provide the foundation for the development of preventative measures and therapeutic strategies to successfully treat *P. aeruginosa* biofilm infections.

D. J. Wozniak
Department of Microbiology and Immunology, Wake Forest University School Medicine,
Medical Center Blvd., Winston Salem, NC 27157, USA
dwozniak@wfubmc.edu

T. Romeo (ed.), *Bacterial Biofilms.*
Current Topics in Microbiology and Immunology 322.
© Springer-Verlag Berlin Heidelberg 2008

1 Introduction

The rod-shaped motile Gram-negative bacterium *Pseudomonas aeruginosa* is an environmental organism that rarely causes disease in individuals with a healthy immune system. However, serious infection frequently develops in immunocompromised hosts and cystic fibrosis (CF) patients (Deretic et al. 1995; Govan and Deretic 1996; Ramsey and Wozniak 2005). *P. aeruginosa* is a predominant cause of opportunistic nosocomial infections, accounting for 10% of hospital-acquired infections, with case fatality due to bacteremia as high as 50% (Richards et al. 1999; Hancock and Speert 2000). In addition, its high intrinsic resistance to antibiotics, its remarkable ability to develop resistance during antimicrobial treatment, and its preferred biofilm-mode of growth considerably complicates therapies aimed at eradicating both acute and chronic *P. aeruginosa* infections (Hancock and Speert 2000; Mah and O'Toole 2001; Donlan and Costerton 2002; Hoffman et al. 2005).

Over the past decade, the study of biofilms has gained attention when it was appreciated that biofilms contribute substantially to infectious disease (Potera 1999). Cystic fibrosis, ventilator-associated pneumonia, device-related infections, and endocarditis are examples of diseases that are considerably complicated by *P. aeruginosa* biofilms. Colonization of a CF patient with *P. aeruginosa*, along with the emergence of alginate-producing mucoid variants, is considered a poor prognostic indicator, as these infections are impossible to eradicate with current therapeutic strategies (Govan and Deretic 1996; Ramsey and Wozniak 2005). Matrix-encased bacteria indicative of biofilms are found in the bronchioles of CF patients during later stages of the disease (Lam et al. 1980; Baltimore et al. 1989; Hoiby et al. 2001). In addition, quorum-sensing signals can be detected in the CF lung, which is yet another piece of evidence for the presence of *P. aeruginosa* biofilms (Singh et al. 2000). In the case of *P. aeruginosa*-induced endocarditis, symptoms are initially transient (Donlan and Costerton 2002). When biofilms are eventually identified as the cause of disease, treatment becomes tremendously difficult. Usually, immediate valve replacement and long-term administration of high-dose antibiotics are required. However, even these drastic measures are often futile, leading to significant morbidity and mortality (Gavin et al. 2003).

Due to their clinical importance, *P. aeruginosa* biofilms are one of the best-studied single-species biofilms. Over the years, certain themes regarding *P. aeruginosa* biofilms have emerged and been considered dogmatic. However, new advances in the field challenge many of the original findings. This has been the case with regard to the role of motility in *P. aeruginosa* biofilm formation, antimicrobial resistance strategies in biofilms, and the composition of the biofilm matrix. The studies reviewed in this chapter emphasize the importance of using caution when interpreting and evaluating data that is emerging in this rapidly expanding field of research.

2   Role of *P. aeruginosa* Motility in the Formation of Biofilms

The first evidence for a role of bacterial motility in the formation of *P. aeruginosa* biofilms was provided in 1998 by O'Toole and Kolter (1998). Using a static attachment model, it was shown that both flagellar and twitching motility are involved in the development of biofilms. In this study, wild type cells attached to the PVC surface and formed a monolayer of cells followed by the appearance of dispersed microcolonies. An isogenic flagellar mutant, on the other hand, did not attach to the substrate. This finding supported the hypothesis that flagella and/or motility are required for initial cell-surface interactions. In the same report, an isogenic mutant lacking type IV pili attached to the surface and formed a monolayer, but did not develop microcolonies, which indicated that type IV pili are required in the biofilm maturation process. Thus, the seminal study by O'Toole and Kolter (1998) provided compelling evidence that the *P. aeruginosa* motility status affects the initiation and maturation of biofilms.

More recent studies reveal that the findings described above may be somewhat generalized. For example, in contrast to O'Toole's (O'Toole and Kolter 1998) observation, Klausen et al. (2003a) discovered that flagellar mutants of *P. aeruginosa* can, in fact, form a biofilm. In this study, however, bacteria were grown as biofilms in flow chambers irrigated with a minimal medium containing citrate as carbon source rather than in a static model with glucose minimal medium (O'Toole and Kolter 1998). Confocal scanning laser microscopy (CSLM) revealed that under the conditions used by Klausen et al. (2003a), both the parental strain and the isogenic flagellar mutant formed biofilms, albeit with distinct structural differences. While the wild-type strain formed a flat "carpet-like" biofilm structure, the flagellar mutant biofilm was thicker and "hilly" (Klausen et al. 2003a). With regard to type IV pili, Klausen et al. (2003a) observed that both the wild type and the isogenic pilus mutant are able to form biofilms in the dynamic flow chamber system but display dramatic structural differences. Under the conditions used, wild type cells formed a uniform flat biofilm without microcolonies, whereas the pilus mutant formed a biofilm with distinct microcolonies. These observations appeared opposite those made by O'Toole (O'Toole and Kolter 1998) where the wild type cells formed microcolonies over time, which were absent in the type IV pili mutant. The discrepancies are likely due to the media used as other reports reveal that biofilms grown in the presence of glucose (Stewart et al. 1993; Davies et al. 1998) are different from those grown in the presence of citrate (Heydorn et al. 2000; Heydorn et al. 2002).

Another important question addresses the contribution of *P. aeruginosa* motility in the formation of the distinctive mushroom-like structures separated by water-filled channels that are frequently observed in biofilms (Lawrence et al. 1991; deBeer et al. 1994). Over the years, several plausible hypotheses have been proposed to explain how these structures are assembled. It has been suggested, for example, that bacterial cells are forced to differentiate as a result of nutrient-limiting conditions depending on their location within the biofilm (Wimpenny and Colasanti 1997;

Picioreanu et al. 1998). A second hypothesis implies that the differentiation of *P. aeruginosa* in the complex multicellular structures is mediated through the expression of specific genes, which directly control the spatial organization of cells (Stoodley et al. 2002). A third hypothesis proposes that cell-to-cell signaling is required (Davies et al. 1998). More recent data reveal that bacterial motility plays an important role in the development of the multicellular structures. For instance, a model by Shrout et al. (2006) suggests that nutritional conditions dictate the contributions of quorum sensing and swarming motility on biofilm formation and can result in structurally distinct biofilms. Moreover, there is compelling evidence that type IV pili-mediated motility is crucial for the proper development of the mushroom-like structures. In an inventive series of experiments, Klausen and colleagues (2003b) used differentially fluorescently labeled wild type and type IV pilin *P. aeruginosa* isolates to show that these multicellular structures are formed in a sequential process involving motile and nonmotile subpopulations of *P. aeruginosa*. In this model, nonmotile bacteria form the mushroom stalks, whereas the caps are assembled by migrating cells that ascend the stalks and aggregate on the tops via a type IV pili-driven mechanism. A subsequent study discovered that the motile and nonmotile subpopulations exhibit differential sensitivity to the membrane-active antimicrobial agents colistin and sodium dodecyl sulfate (SDS) (Haagensen et al. 2007).

Collectively, the data discussed above demonstrate that *P. aeruginosa* motility plays a central role in biofilm formation and development. However, the studies also clearly show that the experimental conditions employed (i.e., biofilm reactor, growth conditions) strongly dictate how and to what extent motility affects the biofilm. This implies that certain environmental parameters promote the expression of specific subsets of genes, which allow the bacterium to form a biofilm most suitable for the conditions encountered. This scenario provides a plausible explanation for the remarkable ability of *P. aeruginosa* to successfully form biofilms in an unusually wide range of ecological niches.

3 Resistance of *P. aeruginosa* Biofilms to Antimicrobial Treatment

One of the most perplexing aspects of biofilms is their remarkable resistance to antimicrobial agents. This phenomenon has been attributed to a variety of factors, including (1) impaired access of antibiotics due to the protective matrix encasing the biofilm (Costerton et al. 1999; Gilbert et al. 2002), (2) reduced growth rates of bacteria within hypoxic/anoxic zones of the biofilm (Gilbert et al. 1990), perhaps in conjunction with the inhibition of microbial activity within these zones (Worlitzsch et al. 2002; Yoon et al. 2002), and (3) induced and/or increased expression of particular resistance genes (Costerton et al. 1999; Gilbert et al. 2002). However, more recent studies challenge and/or expand these suppositions.

The idea of the exopolysaccharide as a protective shield from antimicrobials is appealing and was commonly accepted as a plausible explanation for the unusual resistance of *P. aeruginosa* biofilms to antimicrobials. However, newer evidence does not support this hypothesis. Several studies have shown that the diffusion of many antibiotics into the biofilm is not impeded (Gordon et al. 1988; Shigeta et al. 1997; Ishida et al. 1998). Quinolones, for instance, readily enter *P. aeruginosa* biofilms (Shigeta et al. 1997; Vrany et al. 1997; Ishida et al. 1998) and should thus be able to kill biofilm-growing bacteria. Interestingly, however, a successful breach of the exopolysaccharide barrier by a given antibiotic does not automatically result in eradication of *P. aeruginosa* in the biofilm. For example, Brooun et al. (2000) observed that the quinolone ofloxacin effectively penetrates biofilm-grown *P. aeruginosa* at clinically feasible concentrations (5 µg/ml). Upon administration, the antibiotic resulted in a drop in viable cells. However, a small percentage of cells were unaffected. Higher concentrations of the antibiotic did not have an effect, which was evidenced by the constant number of super-resistant cells at ofloxacin concentrations as high as 100 µg/ml. In addition, other groups reported significant differences in the bactericidal activity of various quinolones despite their ability to successfully penetrate the matrix (Vrany et al. 1997; Ishida et al. 1998). Interestingly, Ishida et al. (1998) found levofloxacin to be significantly more bactericidal than ciprofloxacin, whereas Vrany et al (1997) observed the opposite effect, which may be attributed to the experimental conditions used in the studies.

Another factor that needs to be considered with regard to the ability of antibiotics to cross the polysaccharide barrier is the rate of penetration. Jefferson (2005) proposed that as an antibiotic diffuses through the matrix, cells are exposed to different concentrations of the drug. Thus, the bacteria may have time to mount a protective response to the antibiotic. In support of this idea, Jefferson et al. (2005) showed that vancomycin binds quickly to cells on the biofilm surface but requires more than 1 h to bind to bacteria located within the deepest layer of the biofilm. While this particular study examined *Staphylococcus aureus* biofilms, it is quite possible that this mechanism translates to *P. aeruginosa*. In fact, there is evidence that subinhibitory (sub-MIC) levels of antibiotics affect *P. aeruginosa* biofilm formation. For example, Fonseca et al. (2004) observed that treatment of *P. aeruginosa* biofilms with sub-MIC concentrations of a piperacillin/tazobactam mixture affects biofilm morphology and results in a decrease of biofilm formation, an increase in sensitivity to oxidative stress, and a decrease in type IV pili-mediated twitching motility. Wozniak and Keyser (2004) observed that subinhibitory levels of macrolides also have an effect on biofilm architecture and result in reduced type IV pili-mediated twitching motility. Other studies have shown that aminoglycosides used at sub-MIC concentrations can actually induce *P. aeruginosa* biofilm formation, presumably as a consequence of stress (Hoffman et al. 2005). More recently, Linares et al. (2006) proposed a model that designates antibiotics as signaling molecules. The study demonstrated that exposure of *P. aeruginosa* biofilms to subinhibitory levels of particular antibiotics modulates the expression of virulence determinants including motility and type III secretion systems.

A second hypothesis attempting to explain the exceptionally high resistance of *P. aeruginosa* biofilms to antimicrobial killing involves slow growth of bacteria within the biofilm. However, this explanation may only be plausible with regard to antibiotics that kill rapidly dividing cells such as carbenicillin and other β-lactams. It does not account for resistance to antibiotics such as quinolones, which are able to successfully eradicate nongrowing cells. The observation that stationary-phase planktonic cells (which are in a state of slow growth) remain susceptible to antibiotics also argues against this hypothesis (Brooun et al. 2000). However, the latter finding has been challenged by Spoering and Lewis (2001), who reported that both biofilm-grown and stationary-phase planktonic cells are significantly more resistant to antibiotic killing than logarithmic phase planktonic cells. In addition, this study identified a super-resistant subpopulation of *P. aeruginosa*, which was impervious to the antibiotics used. Spoering and Lewis (2001) also provided evidence that the formation and/or maintenance of the super-resistant subpopulation is dependent on the density of the stationary-phase planktonic culture. The latter finding would help explain the discrepancy found in the literature regarding the antibiotic resistance status of stationary phase planktonic *P. aeruginosa* vs biofilm-grown cells: previous studies seeking to compare antibiotic resistance of biofilm-grown vs planktonic bacteria tended to dilute cells in order to analyze similar cell numbers from both populations, which would obviously have a detrimental effect on any density-dependent components involved in antibiotic resistance.

A very interesting report was recently published by Kaneko et al. (2007), who showed that slow-growing *P. aeruginosa* within a biofilm may actually be targeted by gallium, a transition metal. The study was based on the observation that many biological systems are unable to distinguish between gallium and iron (Chitambar and Narasiham 1991). Thus, gallium may be used in a Trojan horse approach to interfere with and limit iron metabolism. In theory, this would have a detrimental effect on bacteria, and there is evidence to support this hypothesis (Bullen et al. 2005). Kaneko et al. (2007) demonstrated that gallium is able to kill established *P. aeruginosa* in biofilms in a dose-dependent manner. Surprisingly, the slow-growing bacteria located in the center of the biofilm, which are usually unaffected by antibiotics, were particularly susceptible to the bactericidal effects of gallium. Another study found that in addition to gallium, cells located in the middle of the biofilm are also sensitive to killing by colistin and SDS (Haagensen et al. 2007).

Several recent studies have suggested that oxygen limitation may play an important role in *P. aeruginosa* biofilm development both in vitro and in vivo (Worlitzsch et al. 2002; Yoon et al. 2002; Walters et al. 2003). Based on these findings, others have reported that low levels of oxygen result in slow growth of cells within the biofilm and subsequent recalcitrance to antibiotic treatment (Walters et al. 2003; Borriello et al. 2004). However, a more recent study showed that growth of *P. aeruginosa* under anaerobic conditions did not reduce the ability of ceftazidime, meropenem, aztreonam, piperacillin, or piperacillin/tazobactam to inhibit planktonic growth (Field et al. 2005).

A popular explanation for antimicrobial resistance of *P. aeruginosa* is the induced and/or increased expression of particular resistance genes, which may confer a super-resistant phenotype. Interestingly, the phenomenon of super-resistant bacterial subpopulations is nothing new. In 1944, Bigger (Bigger 1944) reported that penicillin treatment of a population of Staphylococci did not sterilize the culture, but rather left a small subpopulation of bacteria that was impervious to the antibiotic. The cells within this subpopulation were defined as persisters. Regrowth of these persisters resulted in a population like the original one with respect to growth inhibition by penicillin and the generation of persisters. While this was a noteworthy discovery, its clinical significance was questioned. Thus, there is relatively little information regarding the molecular basis of persister cells. However, with the expanding study of biofilms, persisters have been met with a rekindled interest as they may offer a logical explanation for the remarkable antibiotic resistance of biofilms. In this context, Lewis (2007) has proposed a persister-based model for the resistance phenomenon in *P. aeruginosa* biofilms: upon administration of a bactericidal antibiotic to treat a biofilm-based infection, the entire population, with the exception of the persisters, is killed. If the antibiotic is withdrawn, the biofilm reforms and the cycle begins anew. This model is attractive as it accounts for the extraordinary resistance of biofilms to antibiotics as well as for the relapsing nature of biofilm infections. However, while the presence of persisters within biofilms has been established, much work is needed to decipher the unusual phenotype of these super-resistant cells. In this regard, several reports and computer modeling studies have provided mechanistic insights into the basis for antibiotic tolerance of biofilm-grown *P. aeruginosa* (Drenkard and Ausubel 2002; Mah et al. 2003; Hoffman et al. 2005; Szmolay et al. 2005; Chambless et al. 2006).

Together, the work discussed in this section reflects the intriguing complexity of the antimicrobial resistance phenomenon of *P. aeruginosa* biofilms. It appears that this opportunistic pathogen has evolved a diverse array of strategies that allow it to counteract the detrimental effects of various classes of antimicrobial agents. Moreover, it becomes clear that the development of therapeutic strategies that can successfully overcome the remarkable antimicrobial resistance of *P. aeruginosa* biofilms is certainly not trivial.

4 Composition of the *P. aeruginosa* Biofilm Matrix

It is now widely recognized that the structural integrity of biofilms depends on an extracellular matrix, which is produced by the bacterial cells constituting a given biofilm (Branda et al. 2005). Although the production of an extracellular substance is a commonality among biofilms, there is remarkable diversity in their individual composition. Table 1 summarizes existing data and features regarding components of the *P. aeruginosa* matrix. It is important to note that the relative ratios and abundance of these elements depends on a variety of factors, including strain background, the growth medium used, and the biofilm reactor system.

Table 1 Components of the *P. aeruginosa* biofilm matrix

Biofilm matrix component	Features	References
Alginate	Mannuronate and guluronate polymer	Linker and Jones 1966
	High–level synthesis associated with isolates from CF airway	Govan and Deretic 1996
	Widely conserved among *P. aeruginosa* strains	Goldberg et al. 1993; Wolfgang et al. 2003
	Expressed in *mucA* mutant *P. aeruginosa*	Martin et al. 1993; Govan and Deretic 1996
	Not essential for biofilm formation in non–mucoid strains but affects biofilm architecture and resistance phenotype	Hentzer et al. 2001; Nivens et al. 2001; Wozniak et al. 2003; Stapper et al. 2004
	Tightly regulated	Govan and Deretic 1996; Ramsey and Wozniak 2005
Psl	Mannose–rich	Friedman and Kolter 2004b; Matsukawa and Greenberg 2004
	Required for surface interactions	Friedman and Kolter 2004b; Jackson et al. 2004; Matsukawa and Greenberg 2004; Overhage et al. 2005
	Overproduced in variants from aged biofilms	Kirisits et al. 2005
	Maintains biofilm structure post–attachment	Ma et al. 2006
	Regulated by c–di–GMP	D'Argenio et al. 2002; Goodman et al. 2004; Hickman et al. 2005; Ventre et al. 2006
Pel	Glucose–rich	Friedman and Kolter 2004b
	Required for pellicle formation in some *P. aeruginosa* strains	Friedman and Kolter 2004a
	Overproduced in variants from aged biofilms	Kirisits et al. 2005
	Regulated by c–di–GMP	D'Argenio et al. 2002; Goodman et al. 2004; Hickman et al. 2005; Ventre et al. 2006
DNA	Major component of in vitro biofilm matrix	Sutherland 2001; Matsukawa and Greenberg 2004; Steinberger and Holden 2005
	Connects cells within the biofilm matrix	Allesen–Holm et al. 2006
	Localized release by a subpopulation of bacteria within the biofilm	Allesen–Holm et al. 2006
	DNase treatment prevents biofilm formation	Whitchurch et al. 2002
	Iron levels can mediate DNA release	Yang et al. 2007
CupA fimbriae	Chaperone–usher family adhesin	Vallet et al. 2001
Membrane vesicles	Derived from outer membrane. Package virulence factors and bind antibiotics.	Schooling and Beveridge 2006

Alginate is perhaps the most extensively studied *P. aeruginosa* exopolysaccharide. It forms a linear polymer consisting of β-D-mannuronic and α-l-guluronic acid residues (Linker and Jones 1966; Ramsey and Wozniak 2005) and is associated with mucoid *P. aeruginosa* isolates recovered from patients suffering from cystic fibrosis (Govan and Deretic 1996; Ramsey and Wozniak 2005; Hoiby 2006). Historically, alginate has been considered the sole component of the *P. aeruginosa* biofilm matrix. This is not surprising since early biofilm studies utilized mucoid *P. aeruginosa* strains. Additionally, many reports show that the presence of alginate as well as its modification strongly influences the physical properties of mucoid biofilms (Hentzer et al. 2001; Nivens et al. 2001; Stapper et al. 2004). However, the hypothesis that the matrix of *P. aeruginosa* biofilms consist of only alginate was challenged by the discovery that most mucoid strains of this opportunistic pathogen harbored one or more mutations that resulted in the constitutive production of the exopolysaccharide (Govan and Deretic 1996). Thus, extracellular matrices produced by non-mucoid *P. aeruginosa* were analyzed with regard to their chemical composition. Interestingly, the exopolysaccharide embedding biofilms formed by common laboratory strains such as PAO1 and PA14 contained little detectable alginate (Wozniak et al. 2003). Moreover, inactivation of genes required for alginate biosynthesis in these strains neither affected the formation nor the structure of the biofilms in vitro (Wozniak et al. 2003; Stapper et al. 2004). This led to the hypothesis that in *P. aeruginosa*-based lung infections in CF patients, biofilm formation may precede the appearance of mucoid isolates. In this context, it was also speculated that the transition of *P. aeruginosa* from a non-mucoid to a mucoid phenotype might involve a switch from a yet unidentified exopolysaccharide to alginate.

Subsequently, several laboratories concomitantly identified two genetic loci, *pel* and *psl*, which encoded putative biochemically distinct polysaccharides (Friedman and Kolter 2004a, 2004b; Jackson et al. 2004; Matsukawa and Greenberg 2004; Overhage et al. 2005). There is evidence that these exopolysaccharides are involved in both early- and late-stage biofilm development (Friedman and Kolter 2004b; Jackson et al. 2004; Matsukawa and Greenberg 2004; Vasseur et al. 2005; Ma et al. 2006). Overproduction of Psl and Pel results in distinct alterations in colony morphology, biofilm architecture, and auto aggregation properties (Friedman and Kolter 2004a; Kirisits et al. 2005; Ma et al. 2006). An exciting recent discovery is that both polysaccharides seem to be regulated by bis-(3′, 5′)-cyclic dimeric guanosine monophosphate (c-di-GMP), a ubiquitous second messenger molecule found in bacteria (D'Argenio et al. 2002; Goodman et al. 2004; Hickman et al. 2005; Ventre et al. 2006). Levels of c-di-GMP are regulated by the opposing activities of GGDEF diguanylate cyclase and EAL phosphodiesterase proteins. Interestingly, c-di-GMP signaling seems to play an important role in the control of biofilm development in other Gram-negative bacteria, including *Yersinia*, *Salmonella*, *Vibrio*, and *Escherichia coli* (Jenal and Malone 2006).

Aside from polysaccharides, the *P. aeruginosa* biofilm matrix contains a considerable amount of DNA (Sutherland 2001; Whitchurch et al. 2002; Matsukawa and Greenberg 2004; Steinberger and Holden 2005). While early biofilms (≤60 h) can be dissociated by DNase treatment, older biofilms (≥ 84 h) seem to be recalcitrant to such treatment (Whitchurch et al. 2002). Allesen-Holm et al. (2006) discovered that the DNA found in the biofilm matrix is comprised of random chromosomal DNA, which is released by a subpopulation of cells within the biofilm, either through cell lysis or the generation of DNA-containing membrane vesicles (Schooling and Beveridge 2006). Presumably, this process is regulated via quorum sensing (Allesen-Holm et al. 2006) and can also be modulated by iron (Yang et al. 2007). Moreover, the extracellular DNA seems to localize in a time-dependent manner in the stalks of the mushroom-shaped microcolonies. Particularly high concentrations of DNA are found in the outer parts of the stalk, thus forming a border between the stalk- and the cap-forming *P. aeruginosa* subpopulations (Allesen-Holm et al. 2006). It is important to note, however, that in the context of CF, the source of DNA in the biofilm matrix is probably not solely of bacterial origin. In this context, Walker et al. (2005) suggest that DNA and actin derived from necrotic neutrophils become part of the biofilm matrix, which may ultimately enhance *P. aeruginosa* biofilm development in vivo. In any case, the presence of considerable amounts of DNA in the biofilm matrix provides a partial explanation for the therapeutic benefits of inhaled nebulized recombinant DNase I in the treatment of chronic lung infections in CF patients (Bollert et al. 1999; Ratjen et al. 2005).

5 Concluding Remarks

In the last decade, the study of bacterial biofilms and surface-associated communities has been met with rekindled interest. It is now recognized that many of the early findings were rather generalized and that biofilms are much more complex and dynamic than originally anticipated. This chapter reviews past and current *P. aeruginosa* biofilm research and provides insight into how older paradigms are challenged by newer and sometimes conflicting observations. The reports show how strain background as well as the choice of biofilm reactors and/or growth medium can substantially influence the outcome of a given experiment and reflect the ability of *P. aeruginosa* to successfully adapt to various environmental conditions. However, the broad spectrum of results obtained in these studies also reminds us that our understanding of *P. aeruginosa* biofilm formation, architecture, and resistance phenotype is rudimentary and that we have merely scratched its surface.

Acknowledgements D.J.W. is supported by Public Health Service grants AI061396 and HL58334 and A.H.T. by American Heart Association grant 0515325U.

References

Allesen-Holm M, Barken KB, Yang L, Klausen M, Webb JS, Kjelleberg S, Molin S, Givskov M, Tolker-Nielsen T (2006) A characterization of DNA release in *Pseudomonas aeruginosa* cultures and biofilms. Mol Microbiol 59:1114–1128

Baltimore R, Christie C, Smith G (1989) Immunohistological localization of *Pseudomonas aeruginosa* in lungs from patients with cystic fibrosis. Implications for the pathogenesis of progressive lung deterioration. Am Rev Respir Dis 140:1650–1661

Bigger JW (1944) Treatment of staphylococcal infections with penicillin. Lancet 2:497–500

Bollert FG, Paton JY, Marshall TG, Calvert J, Greening AP, Innes JA (1999) Recombinant DNase in cystic fibrosis: a protocol for targeted introduction through n-of-1 trials. Scottish Cystic Fibrosis Group. Eur Respir J 13:107–113

Borriello G, Werner E, Roe F, Kim AM, Ehrlich GD, Stewart PS (2004) Oxygen limitation contributes to antibiotic tolerance of *Pseudomonas aeruginosa* in biofilms. Antimicrob Agents Chemother 48:2659–2664

Branda SS, Vik A, Friedman L, Kolter R (2005) Biofilms: the matrix revisited. Trends Microbiol 13:20–26

Brooun A, Liu S, Lewis K (2000) A dose-response study of antibiotic resistance in *Pseudomonas aeruginosa* biofilms. Antimicrob Agents Chemother 44:640–646

Bullen JJ, Rogers HJ, Spalding PB, Ward CG (2005) Iron and infection: the heart of the matter. FEMS Immunol Med Microbiol 43:325–330

Chambless JD, Hunt SM, Stewart PS (2006) A three-dimensional computer model of four hypothetical mechanisms protecting biofilms from antimicrobials. Appl Environ Microbiol 72:2005–2013

Chitambar CR, Narasiham J (1991) Targeting iron-dependent DNA synthesis with gallium and transferrin-gallium. Pathobiology 59:3–10

Costerton JW, Stewart PS, Greenberg EP (1999) Bacterial biofilms: a common cause of persistent infection. Science 284:1318–1322

D'Argenio DA, Calfee MW, Rainey PB, Pesci EC (2002) Autolysis and autoaggregation in *Pseudomonas aeruginosa* colony morphology mutants. J Bacteriol 184:6481–6489

Davies DG, Parsek MR, Pearson JP, Iglewski BH (1998) The involvement of cell-to-cell signals in the development of bacterial biofilm. Science 280:295–298

deBeer DS, Stoodley P, Roe F, Lewandowski Z (1994) Effects of biofilms structures on oxygen distribution and mass transport. Biotechn Bioeng 43:1131–1138

Deretic V, Schurr MJ, Yu H (1995) *Pseudomonas aeruginosa*, mucoidy and the chronic infection phenotype in cystic fibrosis. Trends Microbiol 3:351–356

Donlan RM, Costerton JW (2002) Biofilms: survival mechanisms of clinically relevant microorganisms. Clin Microbiol Rev 15:176–193

Drenkard E, Ausubel FM (2002) *Pseudomonas* biofilm formation and antibiotic resistance are linked to phenotypic variation. Nature 416:740–743

Field TR, White A, Elborn JS, Tunney MM (2005) Effect of oxygen limitation on the in vitro antimicrobial susceptibility of clinical isolates of *Pseudomonas aeruginosa* grown planktonically and as biofilms. Eur J Clin Microbiol Infect Dis 24:677–687

Fonseca AP, Extremina C, Fonseca AF, Sousa JC (2004) Effect of subinhibitory concentration of piperacillin/tazobactam on *Pseudomonas aeruginosa*. J Med Microbiol 53:903–910

Friedman L, Kolter R (2004a) Genes involved in matrix formation in *Pseudomonas aeruginosa* PA14 biofilms. Mol Microbiol 51:675–690

Friedman L, Kolter R (2004b) Two genetic loci produce distinct carbohydrate-rich structural components of the *Pseudomonas aeruginosa* biofilm matrix. J Bacteriol 186:4457–4465

Gavin PJ, Suseno MT, Cook FV, Peterson LR, Thomson RB Jr (2003) Left-sided endocarditis caused by *Pseudomonas aeruginosa*: successful treatment with meropenem and tobramycin. Microbiol Infect Dis 47:427–430

Gilbert P, Collier PJ, Brown MR (1990) Influence of growth rate on susceptibility to antimicrobial agents: biofilms, cell cycle, dormancy, and stringent response. Antimicrob Agents Chemother 34:1865–1868

Gilbert P, Maira-Litran T, McBain AJ, Rickard AH, Whyte FW (2002) The physiology and collective recalcitrance of microbial biofilm communities. Adv Microb Physiol 46:202–256

Goodman AL, Kulasekara B, Rietsch A, Boyd D, Smith RS, Lory S (2004) The signaling network reciprocally regulates genes associated with acute infection and chronic persistence in *Pseudomonas aeruginosa*. Dev Cell 7:745–754

Gordon CA, Hodges NA, Marriott C (1988) Antibiotic interaction and diffusion through alginate and exopolysaccharide of cystic fibrosis-derived *Pseudomonas aeruginosa*. J Antimicrob Chemother 22:667–674

Govan RJW, Deretic V (1996) Microbial pathogenesis in cystic fibrosis: mucoid *Pseudomonas aeruginosa* and *Burkholderia cepacia*. Microbiol Rev 60:539–574

Haagensen JAJ, Klausen M, Ernst RK, Miller SI, Folkesson A, Tolker-Nielsen T, Molin S (2007) Differentiation and distribution of colistin- and sodium dodecyl sulfate-tolerant cells in *Pseudomonas aeruginosa* biofilms. J Bacteriol 189:28–37

Hancock REW, Speert DP (2000) Antibiotic resistance in *Pseudomonas aeruginosa*: mechanisms and impact on treatments. Drug Resist Update 3:247–255

Hentzer M, Teitzel GM, Balzer GJ, Heydorn A, Molin S, Givskov M, Parsek MR (2001) Alginate overproduction affects *Pseudomonas aeruginosa* biofilm structure and function. J Bacteriol 183:5395–5401

Heydorn A, Nielsen AT, Hentzer M, Sternberg C, Givskov M, Ersboll BKM, Molin S (2000) Quantification of biofilm structures by the novel computer program by the novel computer program COMSTAT. Microbiology 146:2395–2407

Heydorn A, Ersboll BK, Kato J, Hentzer M, Parsek MR, Tolker-Nielsen T, Givskov M, Molin S (2002) Statistical analysis of *Pseudomonas aeruginosa* biofilm development: impact of mutations in genes involved in twitching motility, cell-to-cell signaling, and stationary-phase sigma factor expression. Appl Environ Microbiol 68:2008–2017

Hickman JW, Tifrea DF, Harwood CS (2005) A chemosensory system that regulates biofilm formation through modulation of cyclic diguanylate levels. Proc Natl Acad Sci U S A 102:14422–14427

Hoffman LR, D'Argenio DA, MacCoss MJ, Zhang Z, Jones RA, Miller SI (2005) Aminoglycoside antibiotics induce bacterial biofilm formation. Nature 436:1171–1175

Hoiby N (2006) *P. aeruginosa* in cystic fibrosis patients resist host defenses, antibiotics. Microbe 1:571–577

Hoiby N, Johansen HK, Moser C, Song Z, Ciofu O, Kharazmi A (2001) *Pseudomonas aeruginosa* and the in vitro and in vivo biofilm mode of growth. Microbes Infect 3:23–35

Ishida H, Ishida Y, Kurosaka Y, Otani T, Sato K, Kobayashi H (1998) In vitro and in vivo activities of levofloxacin against biofilm-producing *Pseudomonas aeruginosa*. Antimicrob Agents Chemother 42:1641–1645

Jackson KD, Starkey M, Kremer S, Parsek MR, Wozniak DJ (2004) Identification of *psl*, a locus encoding a potential exopolysaccharide that is essential for *Pseudomonas aeruginosa* PAO1 biofilm formation. J Bacteriol 186:4466–4475

Jefferson KK, Goldmann DA, Pier GB (2005) Use of confocal microscopy to analyze the rate of vancomycin penetration through *Staphylococcus aureus* biofilms. Antimicrob Agents Chemother 49:2467–2473

Jenal U, Malone J (2006) Mechanisms of cyclic-di-GMP signaling in bacteria. Annu Rev Genet 40:385–407

Kaneko Y, Thoendel M, Olakami O, Britigan BE, Singh PK (2007) The transition metal gallium disrupts *Pseudomonas aeruginosa* iron metabolism and has antimicrobial and antibiofilm activity. J Clin Invest 117:877–888

Kirisits MJ, Prost L, Starkey M, Parsek MR (2005) Characterization of colony morphology variants isolated from *Pseudomonas aeruginosa* biofilms. Appl Environ Microbiol 71:4809–4821

Klausen M, Aaes-Jorgensen A, Molin S, Tolker-Nielsen T (2003a) Biofilm formation by *Pseudomonas aeruginosa* wild type, flagella, and type IV pili mutants. Mol Microbiol 48:1511–1524

Klausen M, Aaes-Jorgensen A, Molin S, Tolker-Nielsen T (2003b) Involvement of bacterial migration in the development of complex multicellular structures in *Pseudomonas aeruginosa* biofilms. Mol Microbiol 50:61–68

Lam J, Chan R, Lam K, Costerton JW (1980) Production of mucoid microcolonies by *Pseudomonas aeruginosa* within infected lungs in cystic fibrosis. Infect Immun 28:546–556

Lawrence JR, Korber DR, Hoyle BD, Costerton JW, Caldwell DE (1991) Optional sectioning of microbial biofilms. J Bacteriol 173:6558–6567

Lewis K (2007) Persister cells, dormancy and infectious disease. Nat Rev Microbiol 5:48–56

Linares JF, Gustafsson I, Baquero F, Martinez JL (2006) Antibiotics as intermicrobial signaling agents instead of weapons. Proc Natl Acad Sci U S A 103:19484–19489

Linker A, Jones RS (1966) A new polysaccharide resembling alginic acid isolated from pseudomonads. J Biol Chem 241:3845–3851

Ma L, Jackson KD, Landry RM, Parsek MR, Wozniak DJ (2006) Analysis of *Pseudomonas aeruginosa* conditional Psl variants reveals roles for the Psl polysaccharide in adhesion and maintaining biofilm structure post attachment. J Bacteriol 188:8213–8221

Mah TF, O'Toole GA (2001) Mechanisms of biofilm resistance to antimicrobial agents. Trends Microbiol 9:34–39

Mah TF, Pitts B, Pellok B, Walker GC, Stewart PS, O'Toole GA (2003) A genetic basis for *Pseudomonas aeruginosa* biofilm antibiotic resistance. Nature 426:306–310

Matsukawa M, Greenberg EP (2004) Putative exopolysaccharide synthesis genes influence *Pseudomonas aeruginosa* biofilm development. J Bacteriol 186:4449–4456

Nivens DE, Ohman DE, Williams J, Franklin MJ (2001) Role of alginate and its O-acetylation in the formation of *Pseudomonas aeruginosa* microcolonies and biofilms. J Bacteriol 183:1047–1057

O'Toole GA, Kolter R (1998) Flagellar and twitching motility are necessary for *Pseudomonas aeruginosa* biofilm development. Mol Microbiol 30:295–304

Overhage J, Schemionek M, Webb JS, Rehm BHA (2005) Expression of the *psl* operon in *Pseudomonas aeruginosa* PAO1 biofilms: PslA performs an essential function in biofilm formation. Appl Environ Microbiol 71:4407–4413

Picioreanu C, van Loosdrecht MCM, Heijnen JJ (1998) Mathematical modeling of biofilm structure with a hybrid differential-discrete cellular automaton approach. Biotechn Bioeng 58:101–116

Potera C (1999) Forging a link between biofilms and disease. Science 283:1837–1839

Ramsey DM, Wozniak DJ (2005) Understanding the control of *Pseudomonas aeruginosa* alginate synthesis and the prospects for management of chronic infections in cystic fibrosis. Mol Microbiol 56:309–322

Ratjen F, Paul K, van Koningsbruggen S, Breitenstein S, Rietschel E, Nikolaizik W (2005) DNA concentrations in BAL fluid of cystic fibrosis patients with early lung disease: influence of treatment with streptodornase alpha. Pediatr Pulmonol 39:1–4

Richards MJ, Edwards JR, Culver DH, Gaynes RP (1999) Nosocomial infections in medical intensive care units in the United States. National nosocomial infections surveillance system. Crit Care Med 27:887–892

Schooling SR, Beveridge TJ (2006) Membrane vesicles: an overlooked component of the matrices of biofilms. J Bacteriol 188:5945–5957

Shigeta M, Tanaka G, Komatsuzawa H, Sugai M, Suginaka H, Usui T (1997) Permeation of antimicrobial agents through *Pseudomonas aeruginosa* biofilms: a simple method. Chemotherapy 43:340–345

Shrout J, Chopp DL, Just CL, Hentzer M, Givskov M, Parsek MR (2006) The impact of quorum-sensing and swarming motility on *Pseudomonas aeruginosa* biofilm formation is nutritionally conditional. Mol Microbiol 62:1264–1277

Singh PK, Schaefer AL, Parsek MR, Moninger TO, Welsh MJ, Greenberg EP (2000) Quorum-sensing signals indicate that cystic fibrosis lungs are infected with bacterial biofilms. Nature 407:762–764

Spoering AL, Lewis K (2001) Biofilms and planktonic cells of *Pseudomonas aeruginosa* have similar resistance to killing by antimicrobials. J Bacteriol 183:6746–6751

Stapper AP, Narasimhan G, Ohman DE, Barakat JH, Hentzer M, Molin S, Kharazmi A, Hoiby N, Mathee K (2004) Alginate production affects *Pseudomonas aeruginosa* biofilm development and architecture, but is not essential for biofilm formation. J Med Microbiol 53:679–690

Steinberger RE, Holden PA (2005) Extracellular DNA in single- and multiple-species unsaturated biofilms. Appl Environ Microbiol 71:5404–5410

Stewart PS, Peyton BM, Drury WJ, Murga R (1993) Quantitative observations of the heterogeneities in *Pseudomonas aeruginosa* biofilms. Appl Environ Microbiol 59:327–329

Stoodley P, Sauer K, Davies DG, Costerton JW (2002) Biofilms as complex differentiated communities. Annu Rev Microbiol 56:187–209

Sutherland IW (2001) The biofilm matrix – an immobilized by dynamic microbial environment. Trends Microbiol 9:222–227

Szmolay B, Klapper I, Dockery J, Stewart PS (2005) Adaptive responses to antimicrobial agents in biofilms. Environ Microbiol 7:1186–1191

Vasseur P, Vallet-Gely I, Soscia C, Genin S, Filloux A (2005) The *pel* genes of the *Pseudomonas aeruginosa* PAK strain are involved at early and late stages of biofilm formation. Microbiology 151:985–997

Ventre I, Goodman AL, Vallet-Gely I, Vasseur P, Soscia C, Molin S, Bleves SL, Lazdunski A, Lory S, Filloux A (2006) Multiple sensors control reciprocal expression of *Pseudomonas aeruginosa* regulatory RNA and virulence genes. Proc Natl Acad Sci U S A 103:171–176

Vrany J, Stewart PS, Suci P (1997) Comparison of recalcitrance to ciprofloxacin and levofloxacin exhibited by *Pseudomonas aeruginosa* biofilms displaying rapid-transport characteristics. Antimicrob Agents Chemotherap 41:1352–1358

Walker TS, Tomlin KL, Worthen GS, Poch KR, Lieber JG, Saavedra MT, Fessler MB, Malcolm KC, Vasil ML, Nick JA (2005) Enhanced *Pseudomonas aeruginosa* biofilm development mediated by human neutrophils. Infect Immun 73:3693–3701

Walters MC III, Roe F, Bugnicourt A, Franklin MJ, Stewart PS (2003) Contributions of antibiotic penetration, oxygen limitation, and low metabolic activity to tolerance of *Pseudomonas aeruginosa* biofilms to ciprofloxacin and tobramycin. Antimicrob Agents Chemotherap 47:317–323

Whitchurch CB, Tolker-Nielsen T, Ragas PC, Mattick JS (2002) Extracellular DNA required for bacterial biofilm formation. Science 295:1487

Wimpenny JWT, Colasanti R (1997) A unifying hypothesis for the structure of microbial biofilms based on cellular automaton model. FEMS Microbiol Ecol 22:1–16

Worlitzsch D, Tarran R, Ulrich M, Schwab U, Cekici A, Meyer K, Birrer P, Bellon G, Berger J, Weiss T, Botzenhart K, Yankaskas J, Randell S, Boucher R, Doring G (2002) Effects of reduced mucus oxygen concentration in airway *Pseudomonas* infections of cystic fibrosis patients. J Clin Invest 109:317–325

Wozniak DJ, Keyser RA (2004) Effects of subinhibitory concentrations of macrolide antibiotics on *Pseudomonas aeruginosa*. Chest 125:62S–69S

Wozniak DJ, Wyckoff TJO, Starkey M, Keyser RA, Azadi P, O'Toole GA, Parsek MR (2003) Alginate is not a significant component of the exopolysaccharide matrix of PA14 and PAO1 *Pseudomonas aeruginosa* biofilms. Proc Natl Acad Sci U S A 100:7907–7912

Yang L, Barken KB, Skindersoe ME, Christensen AB, Givskov M, Tolker-Nielsen T (2007) Effects of iron on DNA release and biofilm development by *Pseudomonas aeruginosa*. Microbiology 153:1318–1328

Yoon SS, Hennigan RF, Hilliard GM, Ochsner UA, Parvatiyar K, Kamani MC, Allen HL, KeKievit TR, Gardener PR, Schwab U, Rowe JJ, Iglewski BH, McDermott TR, Mason RP, Wozniak DJ, Hancock REW, Parsek MR, Noah TL, Boucher RC, Hassett DJ (2002) *Pseudomonas aeruginosa* anaerobic respiration in biofilms: relationships to cystic fibrosis pathogenesis. Dev Cell 3:593–603

Staphylococcal Biofilms

M. Otto

Abstract *Staphylococcus epidermidis* and *Staphylococcus aureus* are the most frequent causes of nosocomial infections and infections on indwelling medical devices, which characteristically involve biofilms. Recent advances in staphylococcal molecular biology have provided more detailed insight into the basis of biofilm

M. Otto

Laboratory of Human Bacterial Pathogenesis, National Institute of Allergy and Infectious Diseases, The National Institutes of Health, Rocky Mountain Laboratories, Hamilton, MT, USA

motto@niaid.nih.gov

T. Romeo (ed.), *Bacterial Biofilms.*
Current Topics in Microbiology and Immunology 322.
© Springer-Verlag Berlin Heidelberg 2008

formation in these opportunistic pathogens. A series of surface proteins mediate initial attachment to host matrix proteins, which is followed by the expression of a cationic glucosamine-based exopolysaccharide that aggregates the bacterial cells. In some cases, proteins may function as alternative aggregating substances. Furthermore, surfactant peptides have now been recognized as key factors involved in generating the three-dimensional structure of a staphylococcal biofilm by cell-cell disruptive forces, which eventually may lead to the detachment of entire cell clusters. Transcriptional profiling experiments have defined the specific physiology of staphylococcal biofilms and demonstrated that biofilm resistance to antimicrobials is due to gene-regulated processes. Finally, novel animal models of staphylococcal biofilm-associated infection have given us important information on which factors define biofilm formation in vivo. These recent advances constitute an important basis for the development of anti-staphylococcal drugs and vaccines.

1   Introduction

Staphylococci are recognized as the most frequent causes of biofilm-associated infections. This exceptional status among biofilm-associated pathogens is due to the fact that staphylococci are frequent commensal bacteria on the human skin and mucous surfaces (and those of many other mammals). Thus, staphylococci are among the most likely germs to infect any medical device that penetrates those surfaces, such as when being inserted during surgery (Vuong and Otto 2002).

For a long time, research on the molecular basis of biofilm formation was focused on Gram-negative pathogens, predominantly *Pseudomonas aeruginosa*, which is more easily accessible to molecular genetic investigation. More recently, advances in staphylococcal molecular biology have allowed researchers to determine the molecular basis of biofilm formation in staphylococci. In addition, animal models of staphylococcal biofilm-associated infection have been established. Therefore, we now find staphylococci, and particularly *S. epidermidis*, among the best studied clinically relevant biofilm-forming organisms.

This review will give an overview of the role of staphylococci in biofilm-associated human diseases and focus on the mechanism of biofilm development and the molecular basis of virulence in biofilm-forming *S. epidermidis* and *S. aureus*.

2   Biofilms and Staphylococcal Infections

The Nosocomial Infections Surveillance System (http://www.cdc.gov/ncidod/hip/NNIS/2004NNISreport.pdf) recognizes *S. aureus* and CoNS (coagulase-negative staphylococci, i.e., *S. epidermidis* and most other staphylococci other than *S. aureus*) as the most frequently isolated nosocomial pathogens from intensive care unit patients. An extremely high percentage of these isolates are resistant to methicillin (89% CoNS compared to 59.5% for *S. epidermidis*). In addition to specific antibiotic resistance, which is based on the acquisition of genetic resistance factors and may be chromosomally, or more often plasmid-encoded, staphylococci have nonspecific mechanisms of resistance, of which biofilm formation is undoubtedly the most important.

2.1   S. epidermidis *Infections on Indwelling Medical Devices*

S. epidermidis is known as an opportunistic pathogen because it predominantly causes infection in immunocompromised individuals such as intravenous drug abusers, AIDS patients, patients receiving immunosuppressive therapy and premature newborns (Vadyvaloo and Otto 2005). In otherwise healthy patients, *S. epidermidis* causes infection only after penetration of the skin or mucous membranes, which can occur by trauma, inoculation, or implantation of medical devices. These patients may develop septicemia or endocarditis (Arber et al. 1994). As *S. epidermidis* makes up a significant part of the normal bacterial flora of the human skin and mucous membranes, it is probably easily introduced as a contaminant during the surgical implantation of the polymeric device. Notably, a device-related infection of *S. epidermidis* characteristically involves biofilm formation, which generally is considered the most important factor involved in the pathogenesis of *S. epidermidis*.

2.2   S. aureus *Biofilm-Associated Infection*

S. aureus only colonizes a certain percentage of healthy adults permanently or transiently (van Belkum 2006). The reasons for these differences are not understood, but may involve yet undiscovered host factors that predispose for *S. aureus* colonization. Thus, whether indwelling medical devices are contaminated with *S. aureus* during insertion depends significantly on the carrier, be it the patient or health care personnel. To some extent, biofilm-associated infections with *S. aureus* are similar to those with *S. epidermidis*. However, the involvement of *S. aureus* usually requires more intensive care. Often, *S. aureus* biofilm-associated infections are difficult to treat with antibiotics and devices need to be replaced more frequently than those infected with *S. epidermidis* (Jones et al. 2001). In addition, they represent a reservoir of dissemination of *S. aureus* infection to other sites in the human body. In this regard, it is critical from a perspective of molecular pathogenesis, whether biofilm-forming *S. aureus* strains are genetically different from those involved in more serious infections, or - alternatively - whether they are in a different physiological status and might thus develop a more aggressive behavior when spreading within the body.

2.3   Other Staphylococci

Similar to *S. epidermidis*, most other staphylococci have a benign relationship with their host and develop from commensals to pathogens only after damage of a natural barrier such as the skin. In comparison with *S. epidermidis* and *S. aureus*, biofilm-associated infections with other staphylococci are far less frequent. It is not known if this is due to a difference in virulence or abundance on the human skin, or - which appears most likely - a combination of both factors. CoNS found in humans colonize different parts of the human skin and mucous membranes, with each species having a certain predominance on specific parts of the body (Kloos and Schleifer 1986). Notably, every species of CoNS that has been characterized as a resident of the human body (*S. epidermidis*, *S. capitis*, *S. hominis*, *S. haemolyticus*, *S. saccharolyticus*, *S. warneri*, *S. lugdunensis*, *S. saprophyticus*, *S. cohnii*) has also at least once been connected to an infection. The specific sites and frequency of infection seem to be related to those of normal colonization. *S. saprophyticus*, for example, is often found in the inguinal and perineal areas and is a common cause of urinary tract infections (Kloos and Schleifer 1986). In these infections, biofilm formation is probably a crucial determinant of disease, although this remains to be investigated. In general, the specific molecular determinants of biofilm formation in CoNS may be different from *S. epidermidis* and *S. aureus*, but appear to use the same basic mechanisms.

2.4   Interaction of Staphylococci with Other Pathogens in Mixed-Species Medical Biofilms

In contrast to many other medical biofilms, such as multispecies dental plaque formation, biofilm-associated infections with staphylococci are usually not mixed with other species (Arciola et al. 2005). In addition, it is rare to find more than one strain in an infection. A possible explanation for this phenomenon is interspecies communication by quorum-sensing signals, which in staphylococci leads to interspecies inhibition of virulence factor expression (Ji et al. 1997). Similarly, bacterial interference by quorum-sensing signals may explain why *P. aeruginosa* outgrows *S. aureus* and other bacterial pathogens in progressed lung infections (Renders et al. 2001; Qazi et al. 2006). However, these phenomena are poorly understood and there may be a simpler explanation based on the evolutionary adaptation of the bacteria to a specific environment, such as of *S. epidermidis* on the skin.

3   The Molecular Basis of Biofilm Formation in Staphylococci

Research conducted in many biofilm-forming organisms has revealed that the development of a biofilm is a two-step process involving an initial attachment and a subsequent maturation phase, which are physiologically different from each other

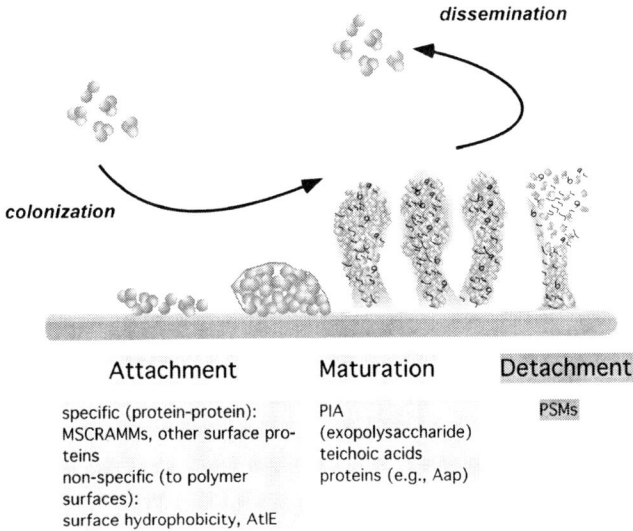

dissemination

colonization

Attachment	Maturation	Detachment
specific (protein-protein): MSCRAMMs, other surface proteins	PIA (exopolysaccharide)	PSMs
non-specific (to polymer surfaces): surface hydrophobicity, AtlE	teichoic acids proteins (e.g., Aap)	

Fig. 1 Phases of biofilm development in staphylococci. Biofilms form by initial attachment to a surface, which can occur on tissues or after covering of an abiotic surface by host matrix proteins in the human body (specific, protein-protein interaction) or directly to an abiotic surface (nonspecific). Subsequently, biofilms grow and mature. The molecules that connect the cells in a staphylococcal biofilm are predominantly the exopolysaccharide PIA, teichoic acids, and some proteins such as the accumulation-associated protein Aap. Finally, cell clusters detach. Detachment is facilitated by expression of the surfactant-like PSM peptides, which are also important in producing the three-dimensional structure of the biofilm. During infection, attachment is a crucial part of the colonization on host tissues or on indwelling medical devices, whereas detachment is a prerequisite for the dissemination of an infection

and require phase-specific factors. A final detachment (or dispersal) phase involves the detachment of single cells or cell clusters by various mechanisms and is believed to be crucial for the dissemination of the bacteria, in the case of pathogens to new infection sites in the human body (Fig. 1).

3.1 Attachment

In the human body, the attachment to human matrix proteins represents the first step of biofilm formation. *S. epidermidis* and *S. aureus* express dozens of so-called MSCRAMMs (microbial surface components recognizing adhesive matrix molecules) that have the capacity to bind to human matrix proteins such as fibrinogen or fibronectin, and often combine binding capacity for several different matrix proteins (Patti et al. 1994). MSCRAMMs have a common structure that includes an exposed binding domain, a cell-wall spanning domain, which often has a repeat structure, and a domain that is responsible for the covalent or noncovalent attachment to the bacterial surface. Covalent attachment is catalyzed by a family of enzymes called sortases that link a conserved motif of the MSCRAMMS to peptidoglycan (Marraffini et al. 2006). The most important one is sortase A, which recognizes an LPXTG motif at the C-terminus of the surface protein sequence (Mazmanian et al. 1999). *S. aureus* strains have a wider variety of LPXTG-type MSCRAMMs (~20), compared to approximately 12 in *S. epidermidis* (Gill et al. 2005). The only functional equivalents between the two species appear to be several members of the serine-aspartate-repeat family (Sdr proteins). This family comprises several surface proteins that have a characteristic serine-aspartate repeat cell-wall-spanning domain (McCrea et al. 2000). In addition, both species have the accumulation-associated protein Aap and several noncovalently bound surface proteins, such as the autolysin Atl, in common.

The forces that govern the attachment of noncovalently bound MSCRAMMs to the surface of staphylococci are not well understood (Navarre and Schneewind 1999). The most important examples are autolysins, which often represent some of the most abundant proteins on the staphylococcal cell surface. There is some evidence to suggest that autolysins are noncovalently attached to teichoic acids (Peschel et al. 2000). These enzymes, in addition to their primary role in cell wall turnover, also facilitate attachment to plastic surfaces and harbor binding sites for human matrix proteins (Heilmann et al. 1997, 2003). Thus, they have a crucial bifunctional importance for bacterial attachment. Similar to the autolysins, the lipase GehD has a primary catalytic role, but there is evidence to suggest that it has an additional adhesive function (Bowden et al. 2002).

Staphylococci are known for their extraordinary ability to stick to plastic surfaces. While this ability has been the basis for most of the in vitro biofilm research conducted in staphylococci (and in other biofilm-forming pathogens), it is not clear if direct attachment to plastic plays a significant role in the pathogenesis of medical device-associated infection. Host matrix proteins cover the devices soon after insertion and therefore the specific interaction between these proteins and MSCRAMMs most likely is of much greater importance for colonization. The classic microtiter plate assay for biofilm formation on abiotic surfaces has been a valuable tool, especially in large screens for biofilm-related factors. However, it is far from representing the detailed characteristics of biofilm-associated infection in vivo and might have led to an overestimation of the importance of some molecules in biofilm formation. It should thus optimally be accompanied by more elaborate in vitro methods, such as flow cells and confocal laser scanning microscopy, and animal models of biofilm-associated infection. For example, subcutaneous infection models with catheter tubing (Rupp et al. 1999a) or tissue-cage models (Zimmerli et al. 1982) have been used successfully to monitor staphylococcal biofilm-associated infection.

3.2 Maturation

The maturation phase of biofilm formation is characterized by (1) intercellular aggregation that can be accomplished by a variety of molecules such as adhesive proteins or - usually polysaccharide-based - exopolymers and (2) biofilm structuring forces that lead to the typical three-dimensional appearance of mature biofilms with its mushroom-like cell towers surrounding fluid-filled channels.

3.3 Adhesive Forces: Aggregation

In staphylococci, the main molecule responsible for intercellular adhesion is the polysaccharide intercellular adhesin (PIA), which is also called poly-*N*-acetylglucosamine (PNAG) according to its chemical composition (Mack et al. 1996). It is a partially deacetylated polymer of beta-1-6-linked *N*-acetylglucosamine, which together with other polymers such as teichoic acids and proteins forms the major part of what has often been called slime, the extracellular matrix of biofilm-forming staphylococci (Fig. 2). More recently, PIA homologs

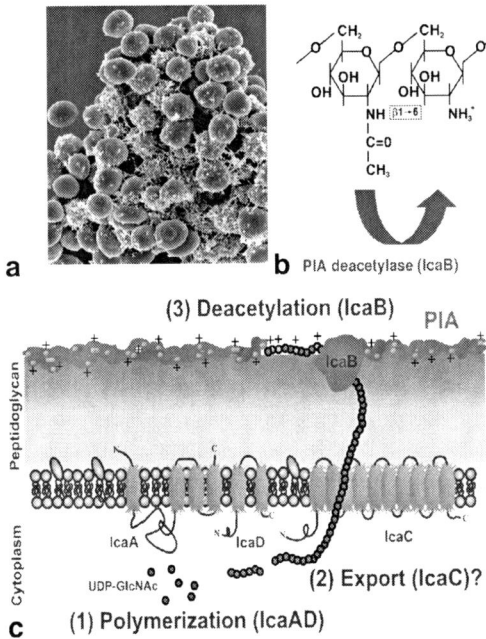

Fig. 2 The biofilm exopolysaccharide PIA. **a** PIA covers staphylococcal cells and sticks them together as the major component of the extracellular matrix (backscatter scanning electron microscopic picture of *S. epidermidis*). **b** PIA is a homopolymer of beta 1-6-linked N-acetylglucosamine, of which about 10%-20% of residues are deacetylated. **c** The biosynthesis of PIA in *S. epidermidis* occurs in three steps: (*1*) IcaA adds GlcNAc moieties from UDP-GlcNAc to the growing PIA chain. The IcaA transferase needs the presence of IcaD for full activity. (2) Presumably, the nascent PIA chain is then exported by IcaC. (3) After export, PIA is deacetylated by the surface-attached IcaB to introduce positive charges, which are crucial for its surface location and biological function

have been detected in a variety of biofilm-forming pathogens, suggesting that this polymer has a widespread function in biofilms and biofilm-associated infections (Darby et al. 2002; Kaplan et al. 2004b; Wang et al. 2004).

The deacetylation of *N*-acetylglucosamine residues in PIA is of major biological importance. It introduces a positively charged character in the otherwise neutral molecule by liberating free amino groups that become charged at neutral or acid pH, such as found in the natural habitat of staphylococci, the human skin (Vuong et al. 2004a). As the bacterial cell surface is negatively charged, PIA supposedly works like glue that sticks the cells together by electrostatic interaction. Teichoic acids may represent the negatively charged molecules that interact with PIA on the cell surface. Interestingly, the relative amounts of teichoic acids and PIA are subject to environmental control - the biological role of which is not yet understood (Sadovskaya et al. 2005).

PIA biosynthesis is accomplished by the products of the *ica* gene locus, which comprises an *N*-acetylglucosamine transferase (*icaA* and *icaD*), a PIA deacetylase (*icaB*), a putative PIA exporter (*icaC*), and a regulatory gene (*icaR*) (Heilmann et al. 1996; Gerke et al. 1998; Vuong et al. 2004a). Expression of the *ica* gene locus is regulated by a variety of environmental factors and regulatory proteins (see Sect. 4.3). The production of PIA and its deacetylation have been recognized as key virulence factors in *S. epidermidis* (Rupp et al. 1999a, 1999b, 2001; Vuong et al. 2004a; Fluckiger et al. 2005). Several animal models have confirmed this key role, although some conflicting results exist (Kristian et al. 2004). However, PIA production does not seem to be of universal importance for biofilm formation and biofilm-associated infection, as PIA-independent biofilm formation has been demonstrated (Rohde et al. 2005). Furthermore, some strains isolated from biofilm-associated infection do not have the *ica* genes (Arciola et al. 2006). Interestingly, invasiveness of noninvasive *S. epidermidis* that lack the *ica* operon can be restored by introduction of the *ica* genes (Li et al. 2005).

In cases of PIA-independent biofilm formation, adhesive proteins most likely substitute for PIA. The most important protein involved in PIA-independent biofilm formation appears to be Aap (Hussain et al. 1997). In a recent study, 27% of biofilm-forming strains isolated from the infection of prosthetic joint infections formed PIA-independent biofilms, in most of which biofilm formation appeared to be mediated by Aap (Rohde et al. 2007). In this study, *S. aureus* biofilms, in contrast, always seemed to be dependent on PIA. Furthermore, biofilm formation was less pronounced when exclusively dependent on proteins. Thus, although PIA does not have an absolutely universal importance for staphylococcal biofilms, this study confirms its key role in staphylococcal biofilm formation.

Aap is a 220-kD protein that needs to be proteolytically cleaved to a smaller 140-kD form to induce biofilm formation (Rohde et al. 2005) and has been suggested to interact with PIA (Hussain et al. 1997). Aap may be identical to the SSP-1 and SSP-2 proteins, which have been implicated in biofilm formation but whose identity was not investigated further (Veenstra et al. 1996). Interestingly, it was shown that SSP forms protein strands on the *S. epidermidis* surface, thus possibly contributing to cell-cell adhesion over greater distances. This capacity could explain how proteins contribute to the aggregation step of biofilm development. A very recent publication has in fact demonstrated that the formation of fibril-like structures on the *S. epidermidis* surface is dependent on Aap (Banner et al. 2007).

In *S. aureus* isolates from animals suffering from mastitis, a cell-wall-bound surface protein named biofilm-associated protein, Bap, is involved in adherence to a polystyrene surface, intercellular adhesion, and biofilm formation (Cucarella et al. 2001). There is evidence for the significance of Bap during infection of bovine mammary glands (Cucarella et al. 2004). A homolog of *bap* named *bhp* occurs in human strains of *S. epidermidis* (Zhang et al. 2003; Gill et al. 2005). Bap homologs are also found in other bacteria, suggesting that the Bap family of surface proteins may have widespread importance in biofilm formation (Latasa et al. 2005; Lasa and Penades 2006).

S. aureus and *S. epidermidis* contain teichoic acids (TA), which are commonly found in many Gram-positive bacteria (Hussain et al. 1992, 1993). TA can be linked to the cell wall, in which case they are referred to as cell-wall TA (WTA), or they can be linked to the cell membrane via a lipid anchor, known as lipoteichoic acid (LTA). TA consist of (1,3)-linked poly (glycerol/ribitol phosphate), substituted with glucose, *N*-acetylglucosamine, D-alanine, or 6-alanyl glucose at the position 2 of the glycerol residue. *S. epidermidis* TA significantly increase adhesion to fibronectin-coated surfaces, suggesting a probable role for TA in *S. epidermidis* virulence (Hussain et al. 2001). Furthermore, the importance of the D-alanylation of *S. aureus* TA in biofilm formation has been demonstrated (Gross et al. 2001).

3.4 Disruptive Forces: Biofilm Structuring

A mature biofilm has a specific three-dimensional structure, which has been described to consist of towers or mushrooms (Costerton et al. 1995). In between those towers, there are fluid-filled channels that are believed to have a vital function in delivering nutrients to cells in deeper biofilm layers. The mechanisms that lead to channel formation and biofilm structuring are far less well understood than those governing intercellular adhesion. Findings primarily achieved in *P. aeruginosa* indicate the involvement of cell-to-cell signaling, e.g., by quorum-sensing systems (Davies et al. 1998). The quorum-sensing controlled expression of the surfactant rhamnolipid appears to be the major mechanism for biofilm structuring in *P. aeruginosa* (Davey et al. 2003; Boles et al. 2005). In staphylococci, differential expression of the biofilm exopolysaccharide PIA might to some degree contribute to biofilm structuring. In contrast, enzymatic degradation of PIA, which appears to occur in other bacteria that express PIA homologs (Kaplan et al. 2003), is very likely not present in staphylococci.

Recent findings in my laboratory suggest that staphylococci use quorum-sensing-controlled surfactant peptides to structure biofilms in a mechanism similar to *P. aeruginosa*, but based on chemically different effector molecules. Phenol-soluble modulins (PSMs) are a class of peptides that have first been described as pro-inflammatory agents in *S. epidermidis* (Mehlin et al. 1999). All PSMs have a pronounced amphipathic alpha-helical character and thus strong surfactant-like

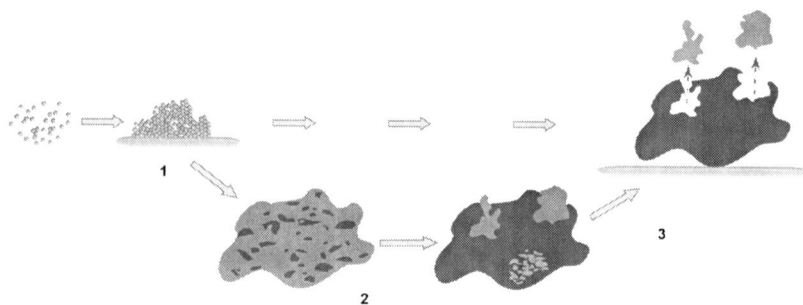

Fig. 3 Model of PSM function in biofilm structuring. (*1*) Cells actively expressing PSM beta peptides attach to a surface. (*2*) Later on, some cell clusters discontinue PSM beta expression for yet unknown reasons, possibly due to limited oxygen concentration. (*3*) Cell clusters with active PSM beta expression detach, leaving gaps in the biofilm, which ultimately leads to the typical structure of a biofilm with cell towers and fluid-filled channels

properties. They can be subdivided in two classes: those with a length of approximately 20 amino acids (alpha type) and those with a length of approximately 40-45 amino acids (beta type). Notably, under biofilm conditions, PSM expression is shifted to the beta type of PSM peptides, which are encoded in an operon (Yao et al. 2005). Recently, we found that expression of PSM beta peptides has a key role in biofilm development in *S. epidermidis*. During dynamic *S. epidermidis* biofilm formation in flow cells, expression of the PSM beta peptides leads to the detachment of cell clusters (unpublished results). This likely leads to the formation of holes in early biofilms and thereby to biofilm structuring (Fig. 3). Consistently, a PSM beta operon deletion strain forms a more compact biofilm than the isogenic wild type strain. PSM homologs also occur in *S. aureus* and other staphylococci (M. Otto, unpublished results). Whether they have the same role in biofilm development needs to be determined.

3.5 Detachment

Biofilm detachment is crucial for the dissemination of bacteria to other colonization sites. It may occur by the detachment of single cells or larger cell clusters. Several factors may contribute to detachment: (1) mechanical forces, such as flow in a blood vessel, (2) cessation of the production of biofilm building material, such as exopolysaccharide, and (3) detachment factors *sensu strictu*, such as enzymes that destroy the matrix, or surfactants. For all that we know, the latter factors are not different from those discussed as biofilm structuring agents. When produced at a high rate, these factors will cause detachment, especially at the biofilm surface area. In fact, controlled detachment maintains a certain biofilm thickness and governs a specific rate of biofilm dissemination. In staphylococci, this mechanism is controlled by the quorum-sensing system *agr* (see Sect. 4.4).

3.6   Cell Death and Extracellular DNA

Some more recent publications claim that controlled cell death in staphylococci contributes to biofilm development. While the phenomenon of controlled cell death in bacteria is still a controversial issue (Rice and Bayles 2003), an increased degree of cell lysis clearly appears to influence biofilm formation. Several regulators that control autolysis have been shown to affect biofilms, such as CidR (Yang et al. 2006). In the case of the CidA murein hydrolase regulator, the release of DNA, a process naturally involved in cell lysis, contributes to biofilm development (Rice et al. 2007). In fact, DNA has recently been frequently implicated in biofilm formation. As a polyanionic molecule, DNA has the capacity to link other molecules together in the biofilm matrix in a way similar to teichoic acids, notably including cationic polymers such as the genuine biofilm polymer PIA discussed above. Due to the conserved nature of the DNA molecule, it is to be expected that autolysis in general will have a similar impact on biofilm formation by that mechanism, which may also in part be responsible for observations made with the Atl type of autolysins (Heilmann et al. 1997; Heilmann and Gotz 1998).

4   Regulation of Biofilm Formation in Staphylococci

Biofilms are the common way of growth for a multitude of microorganisms. Thus, it is to be expected that biofilm growth is under the influence of a vast variety of regulatory mechanisms, just like planktonic growth. However, we lack knowledge on the specific metabolism of biofilms. Regulatory influences on biofilm factors *sensu strictu* will be discussed here, whereas our current knowledge on the physiology of staphylococcal biofilms as determined by transcriptional profiling will be discussed in Sect. 5. We will focus on regulators, for which a mechanism for the influence on biofilm formation has been described in more detail. There are several regulatory systems described in the literature, such as the *rbf* (regulator of biofilm formation) (Lim et al. 2004), for which this is still elusive. In addition, very recent reports suggest that the effect that has been described for the Trap regulator, allegedly affecting biofilm formation in response to a peptide called RIP (Balaban et al. 2003, 2007), is not genuine but due to a second site mutation, most likely in the *agr* system (Shaw et al. 2007; Tsang et al. 2007).

4.1   Environmental Influences

In the earlier literature, when the composition of the staphylococcal slime matrix was not yet known, one can find many reports on the influence of environmental changes on slime formation and biofilm formation as a whole. From a biological point of view, the influence of oxygen and iron limitation, and high osmolarity, appear to be the most crucial. More recently, knowledge of slime composition and the availability of reporter gene constructs have given a clearer picture of what controls specific biofilm factors.

4.2 Regulation of Attachment Factors

The classical notion of quorum-sensing regulation in *S. aureus* comprises the upregulation of adhesion factors such as MSCRAMMs when the cell density is low, a situation encountered during the beginning of a staphylococcal infection. After colonization has been accomplished, increasing activity of the *agr* quorum-sensing system is believed to abolish the expression of the no longer needed colonization factors (Novick 2003). Consistently, many MSCRAMMs are under negative regulation by *agr* in *S. aureus* (Patti et al. 1994). Real-time monitoring of *agr* activity during *S. aureus* infection using bioluminescence has provided a better understanding of quorum-sensing regulation in vivo (Wright et al. 2005), but results from biofilm-associated infection monitored in real-time are not yet available. In *S. epidermidis*, our knowledge of colonization factors and their regulation is more limited. Results obtained by transcriptional profiling (Yao et al. 2006) and measurement of MSCRAMM expression (Bowden et al. 2005) suggest that several MSCRAMMs do not follow the classical notion of *agr* downregulation.

Other attachment factors may be controlled by completely different regulation. For example, the autolysins are in general mainly expressed during times of high cell-wall turnover, as this is their primary function (Giesbrecht et al. 1976). It is not known how this influences biofilm development.

4.3 Regulation of Exopolysaccharide Synthesis

The regulation of PIA expression is probably the best studied among the regulatory influences on staphylococcal biofilm formation. Many previously found regulatory influences on biofilm formation as a whole could be attributed to a change of PIA expression, after tools to pinpoint the regulated targets had become available, such as PIA-specific antisera or *ica*-reporter gene fusion constructs. However, somewhat rashly, many researchers have equated the staphylococcal biofilm matrix with PIA, which as we now know is not completely valid. Thus, there is some confusion in the literature as to which factors have clearly been shown to influence specifically the production of PIA.

A clearly demonstrated influence on PIA expression has been shown for *N*-acetylglucosamine and glucosamine, the building blocks of PIA, probably as these molecules are readily available substrates for the biosynthesis of PIA (Gerke et al. 1998). Furthermore, anaerobiosis significantly increases PIA expression

(Cramton et al. 2001). This represents a very important finding for biofilm physiology, as oxygen concentration thus would limit biofilm formation in the oxygen-loaded arterial bloodstream. Also, it would lead to increased expression of PIA in an established biofilm, in which oxygen concentration decreases significantly with increasing depth. Finally, subinhibitory concentrations of specific antibiotics increase *ica* transcription in *S. epidermidis* (Rachid et al. 2000), a factor to be taken into account during therapy of staphylococcal biofilm-associated infection.

In *S. aureus* or in *S. epidermidis*, several global regulators have been shown to regulate *ica* transcription or PIA expression: the DNA-binding protein SarA and the alternative sigma factor SigB upregulate, whereas the quorum-sensing system *luxS* downregulates (Knobloch et al. 2001; Tormo et al. 2005; Xu et al. 2006). In contrast, *agr* does not regulate PIA expression (Vuong et al. 2000, 2003). The exact mechanism of *sarA* and *sigB* influence on *ica* transcription is complicated. Briefly, whereas the influence of *sigB* appears to occur via repression of *icaR* transcription (Knobloch et al. 2004), which in turn represses transcription of *icaADBC* (Conlon et al. 2002), *sarA* regulates the *icaA* promoter independently of *icaR* (Tormo et al. 2005).

The widespread insertion element IS256 can integrate in the *ica* genes, thus abolishing PIA production (Ziebuhr et al. 1999; Conlon et al. 2004). It has been speculated that IS256 thereby contributes to virulence by a mechanism of adaptation to changing environments during infection. The integration of IS256 in the *agr* operon might have a very similar function of environmental adaptation (Vuong et al. 2004b). Strains with IS256 integrated into *ica* and *agr* genes have been isolated from infection (Kozitskaya et al. 2004; Vuong et al. 2004b). In addition, the presence of IS256 appears to be correlated with the invasiveness of *S. epidermidis* strains (Gu et al. 2005; Kozitskaya et al. 2005). However, although IS256 might have a genuine role in the adaptation of the bacterial population to a different ecological niche, and thus to bacterial versatility, one can probably not call it a true regulator. The action of IS256 appears to be final: it has not been demonstrated to excise from a gene thus re-establishing its function in vivo.

4.4   Regulation of Phenol-Soluble Modulin Expression: agr

We have discussed how the quorum-sensing system *agr* represses surface protein expression after the initial attachment phase. The major *agr*-dependent control of biofilm development is, however, likely accomplished by the strict regulation of PSM expression. Expression of *agr* in a biofilm is limited to surface-exposed areas, where it is probably the key regulator that controls biofilm detachment by upregulating the expression of the PSM effector molecules (Vuong et al. 2004b). Yarwood et al. have used *gfp* expression to measure *agr* activity in *S. aureus* biofilms over time and have proposed a model for *agr*-dependent biofilm maintenance (Yarwood et al. 2004). We have speculated earlier that staphylococcal delta-toxin, one of the

PSMs, is a major effector of *agr*-controlled biofilm detachment in *S. aureus* (Vuong et al. 2000). However, recent research on *S. epidermidis* in our laboratory suggests that the PSMs of the beta type are more important in that regard (see Sect. 3.2.2), at least in this species.

As a consequence of the influence of *agr* on PSM expression, *agr* mutants from a thicker and more compact biofilm in vitro compared to isogenic wild type strains (Vuong et al. 2000, 2003). Furthermore, *agr* mutants occur naturally and can be isolated from biofilm-associated infections at a rate of approximately 25% (Vuong et al. 2004b). Most likely, the permanent disabling of *agr* regulation and the consequent excessive biofilm formation are of advantage to bacterial survival in specific stages or types of infection. Notably, mutations that produce *agr*-negative phenotypes are common and can also be seen in vitro where they occur at a high rate (Somerville et al. 2002).

5 Physiology of Staphylococcal Biofilms: Lessons from Transcriptional Profiling

After complete genome sequences of *S. aureus*, *S. epidermidis*, and other staphylococci had become available, transcriptional profiling of biofilm gene expression was soon initiated. Three transcriptional profiling-based manuscripts have been published, two on *S. aureus* (Beenken et al. 2004; Resch et al. 2005) and one on *S. epidermidis* (Yao et al. 2005), and in addition, proteomics were used to confirm results obtained by the microarray experiments (Resch et al. 2006). The general lessons learned from these studies are comparable, although differences exist that originate most likely from different experimental setups. In addition, it has to be taken into consideration that two studies (Resch et al. 2005, 2006) were conducted in the SA113 strain of *S. aureus*, which is a natural *agr* mutant.

First and foremost, staphylococcal biofilms have a physiological status that is characterized by a general downregulation of active cell processes, such as protein, DNA, and cell wall biosynthesis, which is typical of slow-growing cells. Other metabolic changes can be interpreted as a switch to fermentative processes such as acetoin metabolism, resulting from the low oxygen concentration in biofilms. Finally, the upregulation of urease and the arginine deiminase pathway, which ultimately produce ammonia compounds, has been explained as a switch to limit the deleterious effects of the reduced pH associated with anaerobic growth conditions (Beenken et al. 2004). In general, although similarities exist, a crucial finding of these experiments was that biofilms are physiologically different from planktonic cells in stationary growth phase. In addition, specific resistance mechanisms were found to be upregulated in staphylococcal biofilms (Yao et al. 2005). Thus, gene-regulatory effects add to the intrinsic structure-based resistance that biofilms have to antibiotics and other antibacterial agents (see Sect. 6).

6 The Molecular Basis of Biofilm Resistance to Host Defenses and Antibiotics

It has long been recognized that biofilms have dramatically increased resistance to antibiotics, and to key mechanisms of innate host defense, such as antimicrobial peptides (AMPs) and neutrophil phagocytosis (Costerton et al. 1999). However, the molecular basis of this phenomenon has only recently been further investigated. Two main mechanisms contribute to biofilm resistance: (1) prevention of the antibacterial substance from reaching its target, e.g., by limited diffusion or repulsion, and (2) the specific physiology of a biofilm, which limits the efficacy of antibiotics, mainly of those that target active cell processes and that may also include specific subpopulations of resistant cells (persisters) (Keren et al. 2004).

Limited diffusion of antibiotics through the extracellular biofilm matrix may be the mechanism of resistance to some antibiotics, such as ciprofloxacin in *P. aeruginosa* (Walters et al. 2003), whereas several others (e.g., rifampicin and vancomycin) have been shown to break through the exopolysaccharide layer of *S. epidermidis* (Dunne et al. 1993). A major role in preventing an antibacterial agent from reaching its target (often, the cytoplasm, the cytoplasmic membrane, or the peptidoglycan layer) is electrostatic repulsion or sequestration by surface polymers. Interestingly, PIA protects from cationic and anionic AMPs, and thus may use either mechanism for differently charged molecules (Vuong et al. 2004c). Similarly, the exopolymer poly-gamma-glutamic acid, which is present in *S. epidermidis* and a variety of other CoNS and is upregulated during biofilm formation, contributes to resistance to AMPs of either charge (Kocianova et al. 2005; S. Kocianova, unpublished results; for a review on biofilm resistance to AMPs see Otto 2006).

Phagocytosis, mainly by neutrophils, is a major mechanism by which the innate immune system eliminates invading microorganisms. It has been known for a long time that neutrophil phagocytosis is impaired against staphylococci in a biofilm. More recently, we could show that the exopolysaccharide PIA, and the exopolymer PGA, are specific molecules that shield cells from neutrophil phagocytosis, thus significantly contributing to biofilm resistance from elimination by innate host defense (Vuong et al. 2004c; Kocianova et al. 2005).

7 Possible Anti-biofilm Therapeutics

Biofilms are involved in a multitude of different infections and often contribute significantly to the difficulties encountered in treatment. Developing anti-biofilm drugs aims to combine these drugs with conventional antibiotics, thus restoring the efficacy that the latter show to bacteria in a non-biofilm status.

7.1   Interfering with Essential Staphylococcal Biofilm Factors

An ideal anti-biofilm drug in staphylococci would inactivate a factor that is indispensable for every case of staphylococcal biofilm-associated infection. However, such a factor very likely does not exist, because staphylococcal biofilm formation, as we now know, is multifactorial. Still, targeting the biosynthesis of a factor that appears to be involved in at least the majority of staphylococcal biofilm-associated infection, such as PIA, seems worth considering. Interestingly, some bacteria produce a PIA-degrading enzyme, which - although not present in staphylococci - can degrade staphylococcal PIA and destroy staphylococcal biofilms (Kaplan et al. 2004a). This PIAse, named dispersin B, has first been found in *Actinobacillus actinomycetemcomitans* and appears to have potential as an anti-biofilm drug (Kaplan et al. 2003). Similarly, although not biofilm-specific, the peptidoglycan-degrading enzyme lysostaphin is being evaluated for therapeutic use against biofilm-forming staphylococci (Wu et al. 2003).

7.2   Altering Adhesive Features of Indwelling Medical Devices

The surface of indwelling medical devices can be altered in attempts to decrease bacterial adhesion. However, staphylococci show great versatility and can still adhere to the modern polymers that are in use now. As an additional or complimentary approach, it has been proposed to coat indwelling medical devices with antibiotics or other antibacterial substances. These approaches had limited success, with one problem being plasmid-encoded resistance that is widespread in staphylococci. It is evident that because of the difficulties that staphylococci present to antibacterial therapy, considerable efforts need to be made in both the alteration of device surfaces and the molecular approaches to control staphylococcal biofilm formation.

7.3   Vaccination

Whether vaccination against staphylococcal infection is a promising means to control staphylococcal diseases is controversial. However, many antisera that have been raised for example to PIA (Kelly-Quintos et al. 2006) and several surface proteins, such as fibronectin-binding protein (Rennermalm et al. 2001), have proven effective in animal infection models. Nevertheless, many of these vaccines still need to be tested for their usefulness against biofilm-associated infections.

8 Conclusions and Outlook

Recent advances in our understanding of staphylococcal biofilm development have demonstrated that there are some key structural and regulatory factors that determine the form and physiology of *Staphylococcus* biofilms. Although not all staphylococcal biofilms depend on the expression of the exopolysaccharide PIA, it is by far the most crucial determinant that we know for biofilm-associated infection in staphylococci and possibly a variety of other pathogens. We are only beginning to comprehend the physiological role of the surfactant PSM peptides in biofilm structuring and it is to be expected that we will soon know better how these peptides contribute to the formation of biofilm structure. Additionally, the more recent development of real-time monitoring of biofilm-associated infection using animal models with bioluminescent bacteria will yield a better understanding of the detailed roles of biofilm factors during biofilm-associated infection in vivo. Finally, an even more intensive use of genome-wide approaches to understand biofilm physiology in greater detail will be a key step in our efforts to establish the molecular basis for the development of anti-staphylococcal drugs and vaccines.

References

Arber N, Pras E, Copperman Y et al (1994) Pacemaker endocarditis. Report of 44 cases and review of the literature. Medicine (Baltimore) 73:299-305

Arciola CR, An YH, Campoccia D, Donati ME, Montanaro L (2005) Etiology of implant orthopedic infections: a survey on 1027 clinical isolates. Int J Artif Organs 28:1091-1100

Arciola CR, Campoccia D, Baldassarri L et al (2006) Detection of biofilm formation in *Staphylococcus epidermidis* from implant infections. Comparison of a PCR-method that recognizes the presence of ica genes with two classic phenotypic methods. J Biomed Mater Res A 76:425-430

Balaban N, Giacometti A, Cirioni O et al (2003) Use of the quorum-sensing inhibitor RNAIII-inhibiting peptide to prevent biofilm formation in vivo by drug-resistant *Staphylococcus epidermidis*. J Infect Dis 187:625-630

Balaban N, Cirioni O, Giacometti A et al (2007) Treatment of *Staphylococcus aureus* biofilm infection by the quorum-sensing inhibitor RIP. Antimicrob Agents Chemother 51:2226-2229

Banner MA, Cunniffe JG, Macintosh RL et al (2007) Localized tufts of fibrils on *Staphylococcus epidermidis* NCTC 11047 are comprised of the accumulation-associated protein. J Bacteriol 189:2793-2804

Beenken KE, Dunman PM, McAleese F et al (2004) Global gene expression in *Staphylococcus aureus* biofilms. J Bacteriol 186:4665-4684

Boles BR, Thoendel M, Singh PK (2005) Rhamnolipids mediate detachment of *Pseudomonas aeruginosa* from biofilms. Mol Microbiol 57:1210-1223

Bowden MG, Visai L, Longshaw CM, Holland KT, Speziale P, Hook M (2002) Is the GehD lipase from *Staphylococcus epidermidis* a collagen binding adhesin? J Biol Chem 277:43017-43023

Bowden MG, Chen W, Singvall J et al (2005) Identification and preliminary characterization of cell-wall-anchored proteins of *Staphylococcus epidermidis*. Microbiology 151:1453-1464

Conlon KM, Humphreys H, O'Gara JP (2002) icaR encodes a transcriptional repressor involved in environmental regulation of *ica* operon expression and biofilm formation in *Staphylococcus epidermidis*. J Bacteriol 184:4400-4408

Conlon KM, Humphreys H, O'Gara JP (2004) Inactivations of *rsbU* and *sarA* by IS256 represent novel mechanisms of biofilm phenotypic variation in *Staphylococcus epidermidis*. J Bacteriol 186:6208-6219

Costerton JW, Lewandowski Z, Caldwell DE, Korber DR, Lappin-Scott HM (1995) Microbial biofilms. Annu Rev Microbiol 49:711-745

Costerton JW, Stewart PS, Greenberg EP (1999) Bacterial biofilms: a common cause of persistent infections. Science 284:1318-1322

Cramton SE, Ulrich M, Gotz F, Doring G (2001) Anaerobic conditions induce expression of polysaccharide intercellular adhesin in *Staphylococcus aureus* and *Staphylococcus epidermidis*. Infect Immun 69:4079-4085

Cucarella C, Solano C, Valle J, Amorena B, Lasa I, Penades JR (2001) Bap, a *Staphylococcus aureus* surface protein involved in biofilm formation. J Bacteriol 183:2888-2896

Cucarella C, Tormo MA, Ubeda C et al (2004) Role of biofilm-associated protein bap in the pathogenesis of bovine *Staphylococcus aureus*. Infect Immun 72:2177-2185

Darby C, Hsu JW, Ghori N, Falkow S (2002) *Caenorhabditis elegans*: plague bacteria biofilm blocks food intake. Nature 417:243-244

Davey ME, Caiazza NC, O'Toole GA (2003) Rhamnolipid surfactant production affects biofilm architecture in *Pseudomonas aeruginosa* PAO1. J Bacteriol 185:1027-1036

Davies DG, Parsek MR, Pearson JP, Iglewski BH, Costerton JW, Greenberg EP (1998) The involvement of cell-to-cell signals in the development of a bacterial biofilm. Science 280:295-298

Dunne WM Jr, Mason EO Jr, Kaplan SL (1993) Diffusion of rifampin and vancomycin through a *Staphylococcus epidermidis* biofilm. Antimicrob Agents Chemother 37:2522-2526

Fluckiger U, Ulrich M, Steinhuber A et al (2005) Biofilm formation, *icaADBC* transcription, and polysaccharide intercellular adhesin synthesis by staphylococci in a device-related infection model. Infect Immun 73:1811-1819

Gerke C, Kraft A, Sussmuth R, Schweitzer O, Gotz F (1998) Characterization of the N-acetylglucosaminyltransferase activity involved in the biosynthesis of the *Staphylococcus epidermidis* polysaccharide intercellular adhesin. J Biol Chem 273:18586-18593

Giesbrecht P, Wecke J, Reinicke B (1976) On the morphogenesis of the cell wall of staphylococci. Int Rev Cytol 44:225-318

Gill SR, Fouts DE, Archer GL et al (2005) Insights on evolution of virulence and resistance from the complete genome analysis of an early methicillin-resistant *Staphylococcus aureus* strain and a biofilm-producing methicillin-resistant *Staphylococcus epidermidis* strain. J Bacteriol 187:2426-2438

Gross M, Cramton SE, Gotz F, Peschel A (2001) Key role of teichoic acid net charge in *Staphylococcus aureus* colonization of artificial surfaces. Infect Immun 69:3423-3426

Gu J, Li H, Li M et al (2005) Bacterial insertion sequence IS256 as a potential molecular marker to discriminate invasive strains from commensal strains of *Staphylococcus epidermidis*. J Hosp Infect 61:342-348

Heilmann C, Gotz F (1998) Further characterization of *Staphylococcus epidermidis* transposon mutants deficient in primary attachment or intercellular adhesion. Zentralbl Bakteriol 287:69-83

Heilmann C, Schweitzer O, Gerke C, Vanittanakom N, Mack D, Gotz F (1996) Molecular basis of intercellular adhesion in the biofilm-forming *Staphylococcus epidermidis*. Mol Microbiol 20:1083-1091

Heilmann C, Hussain M, Peters G, Gotz F (1997) Evidence for autolysin-mediated primary attachment of *Staphylococcus epidermidis* to a polystyrene surface. Mol Microbiol 24:1013-1024

Heilmann C, Thumm G, Chhatwal GS, Hartleib J, Uekotter A, Peters G (2003) Identification and characterization of a novel autolysin (Aae) with adhesive properties from *Staphylococcus epidermidis*. Microbiology 149:2769-2778

Hussain M, Hastings JG, White PJ (1992) Comparison of cell-wall teichoic acid with high-molecular-weight extracellular slime material from *Staphylococcus epidermidis*. J Med Microbiol 37:368-375

Hussain M, Wilcox MH, White PJ (1993) The slime of coagulase-negative staphylococci: biochemistry and relation to adherence. FEMS Microbiol Rev 10:191-207

Hussain M, Herrmann M, von Eiff C, Perdreau-Remington F, Peters G (1997) A 140-kilodalton extracellular protein is essential for the accumulation of *Staphylococcus epidermidis* strains on surfaces. Infect Immun 65:519-524

Hussain M, Heilmann C, Peters G, Herrmann M (2001) Teichoic acid enhances adhesion of *Staphylococcus epidermidis* to immobilized fibronectin. Microb Pathog 31:261-270

Ji G, Beavis R, Novick RP (1997) Bacterial interference caused by autoinducing peptide variants. Science 276:2027-2030

Jones SM, Morgan M, Humphrey TJ, Lappin-Scott H (2001) Effect of vancomycin and rifampicin on methicillin-resistant *Staphylococcus aureus* biofilms. Lancet 357:40-41

Kaplan JB, Ragunath C, Ramasubbu N, Fine DH (2003) Detachment of *Actinobacillus actinomycetemcomitans* biofilm cells by an endogenous beta-hexosaminidase activity. J Bacteriol 185:4693-4698

Kaplan JB, Ragunath C, Velliyagounder K, Fine DH, Ramasubbu N (2004a) Enzymatic detachment of *Staphylococcus epidermidis* biofilms. Antimicrob Agents Chemother 48:2633-2636

Kaplan JB, Velliyagounder K, Ragunath C et al (2004b) Genes involved in the synthesis and degradation of matrix polysaccharide in *Actinobacillus actinomycetemcomitans* and *Actinobacillus pleuropneumoniae* biofilms. J Bacteriol 186:8213-8220

Kelly-Quintos C, Cavacini LA, Posner MR, Goldmann D, Pier GB (2006) Characterization of the opsonic and protective activity against *Staphylococcus aureus* of fully human monoclonal antibodies specific for the bacterial surface polysaccharide poly-N-acetylglucosamine. Infect Immun 74:2742-2750

Keren I, Kaldalu N, Spoering A, Wang Y, Lewis K (2004) Persister cells and tolerance to antimicrobials. FEMS Microbiol Lett 230:13-18

Kloos W, Schleifer KH (1986) Staphylococcus. In: Sneath PHA, Mair S, Sharpe ME, Holt JG (eds) Bergey's manual of systematic bacteriology. Williams and Wilkins, Baltimore

Knobloch JK, Bartscht K, Sabottke A, Rohde H, Feucht HH, Mack D (2001) Biofilm formation by *Staphylococcus epidermidis* depends on functional RsbU, an activator of the *sigB* operon: differential activation mechanisms due to ethanol and salt stress. J Bacteriol 183:2624-2633

Knobloch JK, Jager S, Horstkotte MA, Rohde H, Mack D (2004) RsbU-dependent regulation of *Staphylococcus epidermidis* biofilm formation is mediated via the alternative sigma factor sigmaB by repression of the negative regulator gene *icaR*. Infect Immun 72:3838-3848

Kocianova S, Vuong C, Yao Y et al (2005) Key role of poly-gamma-DL-glutamic acid in immune evasion and virulence of *Staphylococcus epidermidis*. J Clin Invest 115:688-694

Kozitskaya S, Cho SH, Dietrich K, Marre R, Naber K, Ziebuhr W (2004) The bacterial insertion sequence element IS256 occurs preferentially in nosocomial *Staphylococcus epidermidis* isolates: association with biofilm formation and resistance to aminoglycosides. Infect Immun 72:1210-1215

Kozitskaya S, Olson ME, Fey PD, Witte W, Ohlsen K, Ziebuhr W (2005) Clonal analysis of *Staphylococcus epidermidis* isolates carrying or lacking biofilm-mediating genes by multilocus sequence typing. J Clin Microbiol 43:4751-4757

Kristian SA, Golda T, Ferracin F et al (2004) The ability of biofilm formation does not influence virulence of *Staphylococcus aureus* and host response in a mouse tissue cage infection model. Microb Pathog 36:237-245

Lasa I, Penades JR (2006) Bap: a family of surface proteins involved in biofilm formation. Res Microbiol 157:99-107

Latasa C, Roux A, Toledo-Arana A et al (2005) BapA, a large secreted protein required for biofilm formation and host colonization of *Salmonella enterica* serovar Enteritidis. Mol Microbiol 58:1322-1339

Li H, Xu L, Wang J et al (2005) Conversion of *Staphylococcus epidermidis* strains from commensal to invasive by expression of the *ica* locus encoding production of biofilm exopolysaccharide. Infect Immun 73:3188-3191

Lim Y, Jana M, Luong TT, Lee CY (2004) Control of glucose- and NaCl-induced biofilm formation by *rbf* in *Staphylococcus aureus*. J Bacteriol 186:722-729

Mack D, Fischer W, Krokotsch A et al (1996) The intercellular adhesin involved in biofilm accumulation of *Staphylococcus epidermidis* is a linear beta-1,6-linked glucosaminoglycan: purification and structural analysis. J Bacteriol 178:175-183

Marraffini LA, Dedent AC, Schneewind O (2006) Sortases and the art of anchoring proteins to the envelopes of Gram-positive bacteria. Microbiol Mol Biol Rev 70:192-221

Mazmanian SK, Liu G, Ton-That H, Schneewind O (1999) *Staphylococcus aureus* sortase, an enzyme that anchors surface proteins to the cell wall. Science 285:760-763

McCrea KW, Hartford O, Davis S et al (2000) The serine-aspartate repeat (Sdr) protein family in *Staphylococcus epidermidis*. Microbiology 146:1535-1546

Mehlin C, Headley CM, Klebanoff SJ (1999) An inflammatory polypeptide complex from *Staphylococcus epidermidis*: isolation and characterization. J Exp Med 189:907-918

Navarre WW, Schneewind O (1999) Surface proteins of Gram-positive bacteria and mechanisms of their targeting to the cell wall envelope. Microbiol Mol Biol Rev 63:174-229

Novick RP (2003) Autoinduction and signal transduction in the regulation of staphylococcal virulence. Mol Microbiol 48:1429-1449

Otto M (2006) Bacterial evasion of antimicrobial peptides by biofilm formation. Curr Top Microbiol Immunol 306:251-258

Patti JM, Allen BL, McGavin MJ, Hook M (1994) MSCRAMM-mediated adherence of microorganisms to host tissues. Annu Rev Microbiol 48:585-617

Peschel A, Vuong C, Otto M, Gotz F (2000) The D-alanine residues of *Staphylococcus aureus* teichoic acids alter the susceptibility to vancomycin and the activity of autolytic enzymes. Antimicrob Agents Chemother 44:2845-2847

Qazi S, Middleton B, Muharram SH et al (2006) N-acylhomoserine lactones antagonize virulence gene expression and quorum sensing in *Staphylococcus aureus*. Infect Immun 74:910-919

Rachid S, Ohlsen K, Witte W, Hacker J, Ziebuhr W (2000) Effect of subinhibitory antibiotic concentrations on polysaccharide intercellular adhesin expression in biofilm-forming *Staphylococcus epidermidis*. Antimicrob Agents Chemother 44:3357-3363

Renders N, Verbrugh H, Van Belkum A (2001) Dynamics of bacterial colonisation in the respiratory tract of patients with cystic fibrosis. Infect Genet Evol 1:29-39

Rennermalm A, Li YH, Bohaufs L et al (2001) Antibodies against a truncated *Staphylococcus aureus* fibronectin-binding protein protect against dissemination of infection in the rat. Vaccine 19:3376-3383

Resch A, Rosenstein R, Nerz C, Gotz F (2005) Differential gene expression profiling of *Staphylococcus aureus* cultivated under biofilm and planktonic conditions. Appl Environ Microbiol 71:2663-2676

Resch A, Leicht S, Saric M et al (2006) Comparative proteome analysis of *Staphylococcus aureus* biofilm and planktonic cells and correlation with transcriptome profiling. Proteomics 6:1867-1877

Rice KC, Bayles KW (2003) Death's toolbox: examining the molecular components of bacterial programmed cell death. Mol Microbiol 50:729-738

Rice KC, Mann EE, Endres JL et al (2007) The *cidA* murein hydrolase regulator contributes to DNA release and biofilm development in *Staphylococcus aureus*. Proc Natl Acad Sci U S A 104:8113-8118

Rohde H, Burdelski C, Bartscht K et al (2005) Induction of *Staphylococcus epidermidis* biofilm formation via proteolytic processing of the accumulation-associated protein by staphylococcal and host proteases. Mol Microbiol 55:1883-1895

Rohde H, Burandt EC, Siemssen N et al (2007) Polysaccharide intercellular adhesin or protein factors in biofilm accumulation of *Staphylococcus epidermidis* and *Staphylococcus aureus* isolated from prosthetic hip and knee joint infections. Biomaterials 28:1711-1720

Rupp ME, Ulphani JS, Fey PD, Bartscht K, Mack D (1999a) Characterization of the importance of polysaccharide intercellular adhesin/hemagglutinin of *Staphylococcus epidermidis* in the pathogenesis of biomaterial-based infection in a mouse foreign body infection model. Infect Immun 67:2627-2632

Rupp ME, Ulphani JS, Fey PD, Mack D (1999b) Characterization of *Staphylococcus epidermidis* polysaccharide intercellular adhesin/hemagglutinin in the pathogenesis of intravascular catheter-associated infection in a rat model. Infect Immun 67:2656-2659

Rupp ME, Fey PD, Heilmann C, Gotz F (2001) Characterization of the importance of *Staphylococcus epidermidis* autolysin and polysaccharide intercellular adhesin in the pathogenesis of intravascular catheter-associated infection in a rat model. J Infect Dis 183:1038-1042

Sadovskaya I, Vinogradov E, Flahaut S, Kogan G, Jabbouri S (2005) Extracellular carbohydrate-containing polymers of a model biofilm-producing strain, *Staphylococcus epidermidis* RP62A. Infect Immun 73:3007-3017

Shaw LN, Jonnson IM, Singh VK, Tarkowski A, Stewart GC (2007) Inactivation of *traP* has no effect on the Agr quorum sensing system or virulence of *Staphylococcus aureus*. Infect Immun 75:4521-4527

Somerville GA, Beres SB, Fitzgerald JR et al (2002) In vitro serial passage of *Staphylococcus aureus*: changes in physiology, virulence factor production, and *agr* nucleotide sequence. J Bacteriol 184:1430-1437

Tormo MA, Marti M, Valle J et al (2005) SarA is an essential positive regulator of *Staphylococcus epidermidis* biofilm development. J Bacteriol 187:2348-2356

Tsang LH, Daily ST, Weiss EC, Smeltzer MS (2007) Mutation of *traP* in *Staphylococcus aureus* has no impact on expression of *agr* or biofilm formation. Infect Immun 75:4528-4533

Vadyvaloo V, Otto M (2005) Molecular genetics of *Staphylococcus epidermidis* biofilms on indwelling medical devices. Int J Artif Organs 28:1069-1078

Van Belkum A (2006) Staphylococcal colonization and infection: homeostasis versus disbalance of human (innate) immunity and bacterial virulence. Curr Opin Infect Dis 19:339-344

Veenstra GJ, Cremers FF, van Dijk H, Fleer A (1996) Ultrastructural organization and regulation of a biomaterial adhesin of *Staphylococcus epidermidis*. J Bacteriol 178:537-541

Vuong C, Otto M (2002) *Staphylococcus epidermidis* infections. Microbes Infect 4:481-489

Vuong C, Saenz HL, Gotz F, Otto M (2000) Impact of the *agr* quorum-sensing system on adherence to polystyrene in *Staphylococcus aureus*. J Infect Dis 182:1688-1693

Vuong C, Gerke C, Somerville GA, Fischer ER, Otto M (2003) Quorum-sensing control of biofilm factors in *Staphylococcus epidermidis*. J Infect Dis 188:706-718

Vuong C, Kocianova S, Voyich JM et al (2004a) A crucial role for exopolysaccharide modification in bacterial biofilm formation, immune evasion, and virulence. J Biol Chem 279:54881-54886

Vuong C, Kocianova S, Yao Y, Carmody AB, Otto M (2004b) Increased colonization of indwelling medical devices by quorum-sensing mutants of *Staphylococcus epidermidis* in vivo. J Infect Dis 190:1498-1505

Vuong C, Voyich JM, Fischer ER et al (2004c) Polysaccharide intercellular adhesin (PIA) protects *Staphylococcus epidermidis* against major components of the human innate immune system. Cell Microbiol 6:269-275

Walters MC 3rd, Roe F, Bugnicourt A, Franklin MJ, Stewart PS (2003) Contributions of antibiotic penetration, oxygen limitation, and low metabolic activity to tolerance of *Pseudomonas aeruginosa* biofilms to ciprofloxacin and tobramycin. Antimicrob Agents Chemother 47:317-323

Wang X, Preston JFI, Romeo T (2004) The *pgaABCD* locus of *Escherichia coli* promotes the synthesis of a polysaccharide adhesin required for biofilm formation. J Bacteriol 186:2724-2734

Wright JS 3rd, Jin R, Novick RP (2005) Transient interference with staphylococcal quorum sensing blocks abscess formation. Proc Natl Acad Sci U S A 102:1691-1696

Wu JA, Kusuma C, Mond JJ, Kokai-Kun JF (2003) Lysostaphin disrupts *Staphylococcus aureus* and *Staphylococcus epidermidis* biofilms on artificial surfaces. Antimicrob Agents Chemother 47:3407-3414

Xu L, Li H, Vuong C et al (2006) Role of the *luxS* quorum-sensing system in biofilm formation and virulence of *Staphylococcus epidermidis*. Infect Immun 74:488-496

Yang SJ, Dunman PM, Projan SJ, Bayles KW (2006) Characterization of the *Staphylococcus aureus* CidR regulon: elucidation of a novel role for acetoin metabolism in cell death and lysis. Mol Microbiol 60:458-468

Yao Y, Sturdevant DE, Otto M (2005) Genomewide analysis of gene expression in *Staphylococcus epidermidis* biofilms: insights into the pathophysiology of *S. epidermidis* biofilms and the role of phenol-soluble modulins in formation of biofilms. J Infect Dis 191:289-298

Yao Y, Vuong C, Kocianova S et al (2006) Characterization of the *Staphylococcus epidermidis* accessory-gene regulator response: quorum-sensing regulation of resistance to human innate host defense. J Infect Dis 193:841-848

Yarwood JM, Bartels DJ, Volper EM, Greenberg EP (2004) Quorum sensing in *Staphylococcus aureus* biofilms. J Bacteriol 186:1838-1850

Zhang YQ, Ren SX, Li HL et al (2003) Genome-based analysis of virulence genes in a non-biofilm-forming *Staphylococcus epidermidis* strain (ATCC 12228). Mol Microbiol 49:1577-1593

Ziebuhr W, Krimmer V, Rachid S, Lossner I, Gotz F, Hacker J (1999) A novel mechanism of phase variation of virulence in *Staphylococcus epidermidis*: evidence for control of the polysaccharide intercellular adhesin synthesis by alternating insertion and excision of the insertion sequence element IS256. Mol Microbiol 32:345-356

Zimmerli W, Waldvogel FA, Vaudaux P, Nydegger UE (1982) Pathogenesis of foreign body infection: description and characteristics of an animal model. J Infect Dis 146:487-497

Yersinia pestis Biofilm in the Flea Vector and Its Role in the Transmission of Plague

B. J. Hinnebusch(✉) and D. L. Erickson

Abstract Transmission by fleabite is a relatively recent evolutionary adaptation of *Yersinia pestis*, the bacterial agent of bubonic plague. To produce a transmissible infection, *Y. pestis* grows as an attached biofilm in the foregut of the flea vector. Biofilm formation both in the flea foregut and in vitro is dependent on an extracellular matrix (ECM) synthesized by the *Yersinia hms* gene products. The *hms* genes are similar to the *pga* and *ica* genes of *Escherichia coli* and *Staphylococcus epidermidis*, respectively, that act to synthesize a poly-β-1,6-*N*-acetyl-*d*-glucosamine ECM required for biofilm formation. As with extracellular polysaccharide production in many other bacteria, synthesis of the Hms-dependent ECM is controlled by

B. J. Hinnebusch
Laboratory of Zoonotic Pathogens, Rocky Mountain Laboratories, NIH, NIAID,
Hamilton, MT, 59840 USA
jhinnebusch@niaid.nih.gov

T. Romeo (ed.), *Bacterial Biofilms.*
Current Topics in Microbiology and Immunology 322.
© Springer-Verlag Berlin Heidelberg 2008

intracellular levels of cyclic-di-GMP. *Yersinia pseudotuberculosis*, the food- and water-borne enteric pathogen from which *Y. pestis* evolved recently, possesses identical *hms* genes and can form biofilm in vitro but not in the flea. The genetic changes in *Y. pestis* that resulted in adapting biofilm-forming capability to the flea gut environment, a critical step in the evolution of vector-borne transmission, have yet to be identified. During a flea bite, *Y. pestis* is regurgitated into the dermis in a unique biofilm phenotype, and this has implications for the initial interaction with the mammalian innate immune response.

1   The Evolution of Arthropod-Borne Transmission of *Yersinia pestis*

1.1   Introduction

The genus *Yersinia* comprises eleven species in the family Enterobacteriaceae of the Gammaproteobacteria. Three of them, *Y. pestis*, *Y. pseudotuberculosis*, and *Yersinia enterocolitica*, are important pathogens of humans and other mammals; one, *Yersinia ruckeri*, is the agent of red mouth disease in rainbow trout and the others are non-pathogens that live in water and soil (Bottone et al. 2005). *Y. pestis*, the causative agent of plague, differs conspicuously from its fellow *Yersinia* species. It is much more invasive and virulent than *Y. pseudotuberculosis* or *Y. enterocolitica*, which cause relatively benign enteric diseases transmitted via contaminated food and water. With a lethal dose to susceptible mammals from an intradermal inoculation site of less than ten cells, *Y. pestis* is one of the most virulent of all microbes and plague one of the most feared diseases of human history (Perry and Fetherston 1997). A second, no less remarkable difference is that *Y. pestis*, uniquely among the enteric group of Gram-negative bacteria, has evolved an arthropod-borne route of transmission. *Y. pestis* is primarily a parasite of rodents that is transmitted by fleas. Permanent enzootic foci exist throughout the world, and plague transmission cycles involve many species of wild rodents and their fleas, making the ecology and epizootiology of plague quite complex (Gage and Kosoy 2005; Pollitzer 1954).

1.2   Y. pestis, *a Recently Emerged Clone of* Y. pseudotuberculosis

Given the radical differences in ecology and natural history between *Y. pestis* and *Y. pseudotuberculosis*, it is surprising that they are in fact very closely related subspecies (Bercovier et al. 1980; Ibrahim et al. 1993). Population genetics analyses indicate that fully virulent *Y. pestis* strains worldwide constitute a clonal group with eight genomic branches separated by minor sequence differences (Achtman et al. 2004). Based on these analyses, it was estimated that *Y. pestis* diverged from its *Y. pseudotuberculosis* ancestor only within the last 10,000-20,000 years (Achtman et al. 1999, 2004). Further comparative genomics analyses confirmed a high degree of genomic identity between the two species (Chain et al. 2004; Hinchliffe et al. 2003; Zhou et al. 2004). The close phylogenetic relationship of *Y. pseudotuberculosis* and *Y. pestis* implies that the change from a comparatively benign food- and water-borne enteric pathogen to a highly virulent, arthropod-borne systemic pathogen occurred quite recently in evolutionary terms and is based on relatively few genetic differences (Hinnebusch 1997; Lorange et al. 2005).

1.3 Arthropod-Borne Transmission Factors of Y. pestis

The most obvious genetic difference between *Y. pestis* and *Y. pseudotuberculosis* is the presence of two *Y. pestis*-unique plasmids (Ferber and Brubaker 1981), presumably acquired sequentially by the *Y. pestis*-progenitor strain via horizontal transfer (Carniel 2003). Each of the two plasmids contains a gene important for transmission by fleas. The 100-kb pFra plasmid encodes *Yersinia* murine toxin (Ymt), a phospholipase D that greatly enhances survival of *Y. pestis* in the flea midgut (Hinnebusch et al. 2002b). The 9.5-kb pPla plasmid encodes the *Y. pestis* plasminogen activator (Pla) (McDonough and Falkow 1989; Sodeinde and Goguen 1988). Although Pla is not required to produce a transmissible infection in the flea or to cause the low-incidence primary septicemic form of plague following flea-bite transmission, it is required for systemic dissemination and the development of bubonic plague following intradermal injection of *Y. pestis* by fleabite or by needle (Brubaker et al. 1965; Hinnebusch et al. 1998; Sebbane et al. 2006; Sodeinde et al. 1992).

As will become evident in the following sections, additional genetic differences between *Y. pestis* and *Y. pseudotuberculosis* that are pertinent to the evolution of arthropod-borne transmission remain to be discovered. Although their chromosomes are nearly identical, *Y. pestis* contains 32 chromosomal genes that so far have not been found in any *Y. pseudotuberculosis* isolate; conversely, 317 genes on the chromosome of *Y. pseudotuberculosis* strain IP32953 were not detected in a panel of *Y. pestis* isolates (Chain et al. 2004). In addition, *Y. pestis* contains many pseudogenes that are intact in *Y. pseudotuberculosis*, most of them metabolic genes that are presumably not needed for a mammal-flea, eukaryotic host-associated life cycle (Chain et al. 2004; Parkhill et al. 2001). It is possible that instances of specific gene loss or change of function as well as gene gain contributed to the evolution of arthropod-borne transmission.

2 Plague Transmission Models

Transmission of *Y. pestis* by fleas can occur by at least three different mechanisms, which may be more or less important among the many different species of flea vectors and at different epizootiologic stages. The first is simple mechanical transmission, which can occur if a flea feeds on a new host shortly after taking a blood meal from a highly septicemic host. Mechanical transmission is akin to dirty-needle transmission, in which the inoculum derives from a residue on the flea mouthparts that remains from a prior infected blood meal (Burroughs 1947). Biological transmission, in contrast, is dependent on bacterial multiplication in the flea midgut and subsequent regurgitation into the bite site. The general mechanism of biological transmission was described by the English medical entomologist A. W. Bacot in two classic papers (Bacot and Martin 1914; Bacot 1915). Bacot observed that *Y. pestis* grew in the form of large aggregates in the midgut of infected fleas and that in some fleas bacterial aggregates also developed in the lumen of the proventriculus, a valve connecting the esophagus and midgut that opens and closes rhythmically during blood feeding. Colonization of the proventriculus was found to be critical for efficient transmission. As the *Y. pestis* aggregates grew in the proventriculus, they interfered with its valvular function, permitting regurgitation of blood, carrying bacteria from the midgut or the proventriculus along with it, back into the bite site. In some fleas, consolidation of the *Y. pestis* aggregate filled the entire lumen of the proventriculus and completely blocked the passage of blood (Fig. 1). However, complete blockage is not essential for efficient transmission; Bacot thought that incompletely blocked fleas might actually be better transmitters (Bacot 1915). This is often overlooked due to the emphasis on blocked fleas, but is an important component of the classic Bacot model because complete proventricular blockage may not develop regularly in all flea vector species (Bacot 1915; Burroughs 1947; Pollitzer 1954). An alternative

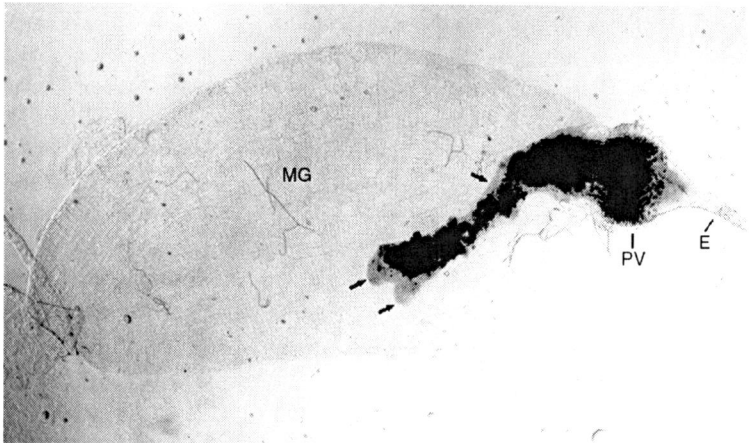

Fig. 1 *Y. pestis* biofilm in the flea. Digestive tract dissected from an *X. cheopis* flea blocked with a dense biofilm consisting of dark masses of *Y. pestis* embedded in a paler, viscous ECM (*arrows*). The contiguous biofilm fills the proventriculus (*PV*) and extends posteriorly into the lumen of the midgut (*MG*). *E*, esophagus

type of regurgitative transmission that may not involve bacterial interference of proventricular function was recently described for the squirrel flea *Oropsylla montana*, which likely contributes to the vector competence of fleas in which proventricular blockage is infrequently observed (Eisen et al. 2006; Webb et al. 2006). Whether this newly described mechanism depends on biofilm formation or is closer to simple mechanical transmission remains to be determined.

2.1   The Coagulase Model of Proventricular Infection

For many years, bacterial aggregation in the flea midgut and proventricular blockage was attributed to a coagulase activity of the pPla-encoded *Y. pestis* plasminogen activator (Pla). Although Pla acts to rapidly degrade fibrin clots at 37°C by activating plasminogen, at lower temperatures typical of the flea gut environment an opposite clot-forming or coagulase activity was observed. According to the coagulase model (Cavanaugh 1971), which is still often cited in textbooks, *Y. pestis* is enveloped and multiplies within a fibrin clot formed from the flea's blood meal by the coagulase activity of Pla. Proventricular colonization and blockage was hypothesized to result from bacteria-laden clots lodging among the proventricular spines. This hypothesis was compelling because it could also explain the decrease in proventricular blockage and flea-borne transmission during hot weather (Cavanaugh and Marshall 1972). The fibrin-dissolving activity of Pla that predominates at higher temperatures was invoked to explain that phenomenon (Cavanaugh 1971). McDonough et al. (1993) subsequently reported that a *Y. pestis pla* mutant caused less mortality to fleas than the isogenic Pla⁺ parent strain, and suggested that this might be due to a decreased ability to cause blockage. However, the mortality was measured at 4 days after infection, before blockage would be expected to occur.

Several lines of investigation contradict the coagulase model, however. The coagulase activity of Pla is weak and has been observed only with rabbit plasma, and not with rodent or human plasma (Beesley et al. 1967; Jawetz and Meyer 1944; Sodeinde et al. 1992). Pla-negative *Y. pestis* strains are able to block the rat flea *Xenopsylla cheopis* as well as wild type strains, with the same inverse relationship between blockage rate and ambient temperature, refuting both predictions of the coagulase model (Hinnebusch et al. 1998). The matrix of the *Y. pestis* aggregates that form in the flea digestive tract is also unaffected by treatment with proteases, including plasmin (Jarrett et al. 2004). More recent work indicates that instead of a role in producing a transmissible infection in the flea vector, the true biological role of Pla occurs after transmission. Plasminogen activation and other proteolytic activities of Pla are required for the development of bubonic plague, because they facilitate dissemination of *Y. pestis* from the skin and lymphatic system (Brubaker et al. 1965; Korhonen et al. 2004; Sebbane et al. 2006; Sodeinde et al. 1992).

2.2 The Biofilm Model of Proventricular Infection

The current model for the autoaggregative growth phenotype first observed by Bacot is that *Y. pestis* grows as a biofilm in the flea digestive tract (Jarrett et al. 2004). According to this model, the decrease in temperature experienced by the bacteria upon leaving the warm-blooded mammal triggers *Y. pestis* to synthesize a stable extracellular biofilm matrix. This extracellular matrix (ECM) surrounds the dense microcolony of *Y. pestis* as they grow and also potentiates bacterial aggregation on the surface of the proventricular spines that leads to transmission. The same general strategy is used by two other arthropod-borne microbial pathogens that are transmitted regurgitatively. The phytopathogen *Xylella fastidiosa* forms a biofilm of polarly attached cells in the foregut of its vectors, which are sap-feeding insects in the leafhopper family (Newman et al. 2004; Purcell et al. 1979); and transmission of *Leishmania* depends on blockage of the anterior midgut of the sandfly vector by parasites embedded in a polysaccharide-containing secretory gel (Rogers et al. 2002; Stierhof et al. 1999). The body of evidence that has accumulated in support of the biofilm model of plague transmission is presented in the following sections.

3 Genetic and Molecular Mechanisms of *Yersinia* Biofilm ECM Synthesis

3.1 The Y. pestis *Pigmentation Phenotype, Hms Locus, and Biofilm ECM*

One of the first of the many temperature-dependent phenotypes described for *Y. pestis* was the formation of darkly pigmented colonies after growth at 28°C or less, but not after growth at 37°C, on agar plates containing hemin or the structurally analogous dye Congo red (Jackson and Burrows 1956; Surgalla and Beesley 1969). This phenotype was termed pigmentation, and is due to avid binding of the chromogens to the outer surface of the bacterial cells (Perry et al. 1993). It was also noted that the pigmentation phenotype correlated with cohesive colonies that came off the agar surface in densely packed masses and with pellicle formation on the surface of glass culture vessels (Jackson and Burrows 1956; Surgalla and Beesley 1969). In retrospect, these findings were the first evidence of *Yersinia* biofilm formation- pellicle growth and Congo red binding to a polysaccharide ECM are now recognized as typical traits of many bacterial biofilms (Heilmann and Götz 1998; Weiner et al. 1999).

Perry et al. (1990) identified a *Y. pestis* chromosomal region required for pigmentation, termed the hemin storage (*hms*) locus; and a four-gene operon in this locus, *hmsHFRS*, was later implicated (Lillard et al. 1997; Pendrak and Perry 1993) (Table 1). Amino acid sequence and domain similarities between the Hms proteins

Table 1 *Yersinia hms* gene products and their predicted functions

Gene	Location of protein[a]	Identified protein domains[b]	Predicted function[b]	Similar genes in other bacteria[b]
Hms locus:				
hmsH	OM	-	ECM synthesis	*pgaA* (*E. coli*)
hmsF	OM	polysaccharide deacetylase, COG1649	ECM synthesis	*pgaB* (*E. coli*) *icaB* (*Staph*)
hmsR	IM	glycosyl transferase	ECM synthesis	*pgaC* (*E. coli*) *icaA* (*Staph*)
hmsS	IM	-	ECM synthesis	*pgaD* (*E. coli*)
Unlinked genes:				
hmsT	IM	GGDEF	Diguanylate cyclase	
hmsP	IM	GGDEF, EAL	Cyclic-di-GMP phosphodiesterase	*yhjK* (*E. coli*)

[a] *OM*, outer membrane; *IM*, inner membrane
[b] See text for references

and the Pga and Ica proteins of *E. coli* and staphylococci that synthesize a poly-β-1,6-*N*-acetyl-*d*-glucosamine ECM required for biofilm formation suggested a link between pigmentation and biofilm phenotypes; and complementation of non-pigmented *Y. pestis hmsR* and *hmsS* mutants with the respective *E. coli pga* homologs restores the pigmentation phenotype (Darby et al. 2002; Heilmann et al. 1996; Jones et al. 1999; Lillard et al. 1997; Mack et al. 1996; Wang et al. 2004).

The ECM of *S. epidermidis* and *Staphylococcus aureus*, called PIA (polysaccharide intercellular adhesin), enables biofilm formation on catheters and other indwelling medical devices, making these bacteria one of the most common causes of nosocomial infections (Vadyvaloo and Otto 2005). A model for PIA biosynthesis by the *ica* (intercellular adhesin) gene products has been developed, based on several genetic and biochemical studies (reviewed in Götz 2002). The glycosyl transferase activity of IcaA, in conjunction with IcaD, first forms intracellular oligomers of 10-20 β-1,6-linked *N*-acetylglucosamine residues (Gerke et al. 1998). The oligomers are then further polymerized and transported through the cell membrane via IcaC, where the polymer is partially deacylated by the polysaccharide deacetylase activity of the cell-surface protein IcaB (Vuong et al. 2004a). Given the membrane localization and similar domains of the Ica, Pga, and Hms gene products, ECM biosynthesis in *E. coli* and *Yersinia* species probably proceeds by an analogous mechanism. Notably, all four of the *Y. pestis hmsHFRS* gene products are required for pigmentation and biofilm phenotype, and site-directed mutagenesis of the periplasmic, deacetylase, and glycosyl transferase domains of HmsH, F, and R, respectively, severely diminished Congo red binding and in vitro biofilm formation (Forman et al. 2006).

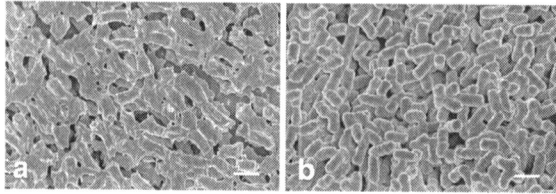

Fig. 2 ECM surrounding Hms⁺ (**a**) but not Hms⁻ (**b**) *Y. pestis* grown on agar plates at 21°C. Scale bar, 1 μm

The ECM structures of staphylococcal and *E. coli* biofilms are biochemically similar, if not identical; both are composed of poly-β-1,6-*N*-acetylglucosamine (Mack et al. 1996; Wang et al. 2004). The *Yersinia* ECM structure has not been determined yet. However, Hms-dependent extracellular material has been observed by electron microscopy techniques specifically designed to preserve relatively labile polysaccharide (Fig. 2), and *Y. pestis* cells expressing the *hms* genes cross-react with anti-PIA antibody (Erickson and Hinnebusch, unpublished data). Furthermore, dispersin B, a β-hexosaminidase from *Actinobacillus actinomycetem-comitans* that disrupts biofilms in that organism (Kaplan et al. 2003), degrades poly-β-1,6-*N*-acetylglucosamine and prevents biofilm formation of *S. epidermidis*, *E. coli*, *Y. pestis*, and other bacteria containing Ica, Pga, and Hms homologs (Itoh et al. 2005). Whereas dispersin B treatment also dispersed preformed *E. coli* and *S. epidermidis* biofilms, however, preformed *Y. pestis* biofilms were unaffected (Itoh et al. 2005). This suggests that the *Y. pestis* ECM is inaccessible to the enzyme or is further modified during maturation of the biofilm, or that other non-ECM components act to stabilize the *Y. pestis* biofilm.

Thus, although it remains to be verified biochemically, there is considerable evidence to support the hypothesis that the *Yersinia* HmsHFRS proteins synthesize a polysaccharide ECM that is structurally related to the Pga- and Ica-dependent ECMs of *E. coli* and staphylococcal biofilms. As in other bacteria, the Hms-dependent ECM is required for Congo red binding (pigmentation), the binding of other ligands with polysaccharide affinity such as calcofluor (Kirillina et al. 2004) and certain lectins (Tan and Darby 2004), as well as for *Yersinia* biofilm formation (Darby et al. 2002; Forman et al. 2006; Jarrett et al. 2004; Joshua et al. 2003).

3.2   Role of the Hms Proteins in Producing In Vitro and In Vivo Biofilms

An essential role for the *hmsHFRS* genes in *Yersinia* biofilm formation has been amply demonstrated using a variety of experimental systems. The *hms* locus is required for *Y. pestis* and *Y. pseudotuberculosis* biofilm growth in 96-well polystyrene

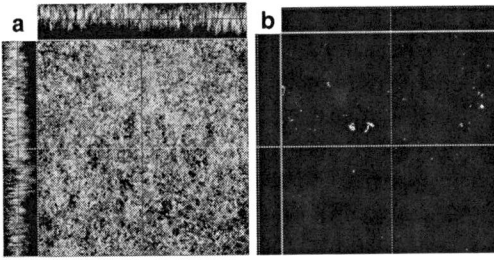

Fig. 3 Biofilm produced on the glass surface of a flow cell after 48 h at 21°C by Hms⁺ (**a**) but not Hms⁻ (**b**) *Y. pestis*

microtiter plates and on the surface of glass continuous-flow (flowcell) chambers (Fig. 3) (Erickson et al. 2006; Forman et al. 2006; Jarrett et al. 2004; Kirillina et al. 2004). Key evidence for the biofilm model of plague transmission came from in vivo studies using the rat flea *X. cheopis* as an infection model, in which the *Y. pestis hms* genes were shown to be required to produce a transmissible infection in the flea proventriculus (Hinnebusch et al. 1996; Jarrett et al. 2004). The first explicit statement of the plague biofilm transmission hypothesis, however, was based on the ability of *Y. pestis* and *Y. pseudotuberculosis* to form Hms-dependent biofilm on the surface of *Caenorhabditis elegans* nematodes, a model that may incorporate aspects of both in vitro and in vivo biofilm formation (Darby et al. 2002; Joshua et al. 2003). In this model, *Yersinia* biofilm aggregates on the mouthparts and blocks the feeding of the nematodes as they crawl across a preformed lawn of ECM-producing bacteria. The in vitro and *C. elegans* models have proven to be useful surrogates to identify genes in addition to *hms* that are important for flea-borne transmission (Darby et al. 2005); however, in vitro, *C. elegans* and flea biofilm phenotypes do not always correlate (Erickson et al. 2006).

3.3 Regulation of Hms-Dependent Biofilm Formation

Proximal regulation of Hms-dependent ECM synthesis in *Y. pestis* appears to depend primarily on intracellular levels of the bacterial second messenger cyclic-di-GMP. *hmsT* and *hmsP*, two recently identified chromosomal genes that are unlinked to the *hmsHFRS* locus and to each other, encode a GGDEF-domain protein with c-di-GMP synthesizing diguanylate cyclase activity and an EAL-domain protein with phosphodiesterase activity, respectively (Bobrov et al. 2005; Hare and McDonough 1999; Jones et al. 1999; Kirillina et al. 2004; Simm et al. 2005). HmsT and HmsP are hypothesized to control Hms-dependent ECM biosynthesis via their opposing c-di-GMP forming and degrading activities. In keeping with this regulatory scheme, copy number of *hmsT* is directly related to c-di-GMP levels and biofilm thickness in *Y. pestis*; conversely, deletion of *hmsT* or disruption of its GGDEF-encoding region result in poor biofilm formation (Kirillina et al. 2004; Simm et al. 2005). As also predicted by the model, elimination of HmsP or its EAL domain result in thicker biofilm (Kirillina et al. 2004). Accordingly, the *Yersinia* Hms system joins a list of systems in several other Gram-negative bacteria, including *Salmonella*, *Gluconacetobacter*, *Rhizobium*, *Pseudomonas*, and *Vibrio*, in which GGDEF- and EAL-protein control of c-di-GMP concentration is involved in the regulation of extracellular polysaccharide and biofilm ECM synthesis (D'Argenio and Miller 2004; Römling and Amikam 2006).

GGDEF and EAL domain proteins are two of the largest superfamilies in eubacteria, with multiple members present in most species (Römling and Amikam 2006). This redundancy suggests that c-di-GMP flux may be influenced by a complex composite of different GGDEF and EAL family member pairs that come into play under different environmental conditions. For example, different GGDEF-domain proteins have been demonstrated to participate in a hierarchical fashion in control of the rdar biofilm phenotype in *Salmonella* (Kader et al. 2006). In addition to *hmsT* and *hmsP*, *Y. pestis* contains seven additional genes that are predicted to encode GGDEF- and EAL-domain proteins (Parkhill et al. 2001) (Table 2). Their role, if any, in *Yersinia* Hms or biofilm regulation is unknown. Interestingly, however, one (YPO3988) is highly similar to *E. coli yhjH*, which inhibits biofilm formation in *Pseudomonas putida* (Gjermansen et al. 2006). Another (YPO0998) is highly similar to *yhdA*, which regulates the Csr small regulatory RNA system known to control the *E. coli* Pga-dependent biofilm ECM (Jackson et al. 2002; Wang et al. 2005).

Many questions remain about the regulation of the Hms system in *Yersinia*. The low-temperature dependence of the Hms phenotype known since the original description (Jackson and Burrows 1956) can be attributed to degradation of HmsT, H, and R at 37°C (Perry et al. 2004), but details of this posttranslational regulation are unknown. The molecular mechanism of c-di-GMP enhancement of biosynthetic Hms protein activity has not been determined. The *hmsT* promoter region contains a binding site for the iron-response regulator protein Fur, suggesting a link between the Hms and low iron response systems that has yet to be fully explored (Jones et al. 1999; Staggs et al. 1994). Answers to these and other questions will provide important

Table 2 Predicted GGDEF- and EAL-domain proteins in *Yersinia pestis* CO92

GGDEF domain		EAL domain		Both GGDEF and EAL domains	
Gene no.	Gene name (homolog)	Gene no.	Gene name (homolog)	Gene no.	Gene name (homolog)
YPO0425	*hmsT*	YPO1274[a]	*rtn*	YPO3996[a]	*hmsP* (*yhjK*)
YPO0449[a]	-	YPO2779	-	YPO3664[a]	(*yhdA*)
YPO1752[b]	-	YPO3988	(*yhjH*)	YPO0998[a]	-

[a] Predicted membrane protein
[b] Pseudogene in CO92 but intact in *Y. pestis* KIM

insight into the biology and evolution of the *Yersinia*. For example, strains of *Y. pestis* and *Y. pseudotuberculosis* that have identical *hmsHFRS*, *hmsT*, and *hmsP* genes differ in their biofilm phenotype in different in vitro conditions (Chain et al. 2004; Darby et al. 2002; Deng et al. 2002; Joshua et al. 2003; Parkhill et al. 2001). Of greatest biological significance, *Y. pseudotuberculosis* can infect the flea midgut but never forms a biofilm in that environment (Erickson et al. 2006). Conversely, *Y. pseudotuberculosis* typically forms thicker biofilms in the *C. elegans* model system than does *Y. pestis* (Tan and Darby 2004).

4   Other Factors Implicated in *Yersinia* Biofilm Formation

4.1   Bacterial Factors

The biofilm growth state is recognized as a complex developmental cycle, involving initial adherence, ECM production, maturation, and eventual dispersion (Stoodley et al. 2002). Initial attachment to a surface can rely on relatively weak and nonspecific physicochemical and electrostatic interactions that are influenced by characteristics of the bacterial outer membrane, the surrounding medium, and the substrate (Beloin et al. 2005). The ECM itself may also promote attachment.

Lipopolysaccharide (LPS), the major component of the Gram-negative outer membrane, has been shown to affect *Yersinia* biofilm. Mutation of *waaA*, *yrbH*, and *gmhA*, which encode enzymes required for the addition of the LPS inner core sugars Kdo (3-deoxy-D-manno-octulosonic acid) and heptose to lipid A, all result in decreased biofilm in *Y. pestis* (Darby et al. 2005; Tan and Darby 2004, 2006). Of these, only the heptose-less *Y. pestis gmhA* mutant has been evaluated in the flea, where it was severely deficient in ability to produce proventricular blockage even though it colonized the flea gut normally (Darby et al. 2005). Deficient biofilm formation of the *Y. pestis waa* and *yrbH* mutant strains (which produce LPS consisting solely of lipid A) on *C. elegans* may be related to their reduced growth rate (Tan and Darby 2005, 2006); alternatively, LPS inner core alteration could conceivably affect initial attachment, or export or stability of the ECM on the outer surface. An unidentified separate activity of YrbH, in addition to its known role in KDO biosynthesis, was also implicated in biofilm formation (Tan and Darby 2006). *Y. pestis*, unlike *Y. pseudotuberculosis*, produces a rough form of LPS lacking O-polysaccharide at temperatures at which the Hms-dependent ECM is made. Loss of O-polysaccharide does not markedly affect the in vitro biofilm forming ability of *Y. pseudotuberculosis*, however (Erickson et al. 2006).

In many bacteria, specific cell-surface adhesins such as certain types of fimbriae or outer membrane proteins are important for initial attachment to the substrate on which a biofilm forms (reviewed in Beloin et al. 2005). The *Y. pestis* genome contains at least ten genetic loci predicted to encode surface fimbriae or adhesins (Parkhill et al. 2001). Two of them, responsible for the F1 protein fibrillar capsule and the pH6 antigen, have known roles in the pathogenesis of plague (Brubaker 1991; Lindler et al. 1990), but functions have not been attributed to the others.

The complex bacterial physiology involved in biofilm development and maturation is indicated by the number of environmental sensing and gene regulatory systems that have been implicated in these processes (reviewed in Beloin et al. 2005; Stoodley et al. 2002). Among them are specific bacterial two-component regulatory systems as well as global regulators of central metabolism. The role of bacterial quorum sensing, a means of intercellular signaling within a community, in biofilm maturation has attracted particular attention (Davies et al. 1998; Parsek and Greenberg 2005; Vuong et al. 2003; Zhu and Mekalanos 2003). Interestingly, a connection between quorum sensing and c-di-GMP regulatory systems has been proposed (Camilli and Bassler 2006). The environmental sensing and subsequent redirection of gene expression pathways that lead to biofilm formation in *Yersinia* remain to be discovered. We have found that a *Y. pestis* strain deleted of all three known quorum-sensing systems is able to form a normal biofilm in the flea (Jarrett et al. 2004). Nevertheless, it is possible that *Y. pestis* quorum-sensing signaling is required for the final step in the biofilm developmental cycle, dispersal of individual cells from the biofilm, which might affect regurgitative transmission from a proventricular biofilm.

Patel et al. recently discovered that the polyamines spermidine and putrescine are important for ECM production and biofilm formation in *Y. pestis,* and suggested possible roles for these cationic molecules in Hms-related signaling pathways or as biosynthetic intermediates or components of ECM (Patel et al. 2006). In *Agrobacterium tumefasciens* and *Vibrio cholerae*, homologs of polyamine transport (Pot) membrane proteins are important for surface attachment (Karatan et al. 2005; Matthysse et al. 1996). Although the *Y. pestis* polyamine transport genes are highly induced in the flea, mutational loss of this system did not affect biofilm formation in the flea, ruling out an essential role for these proteins in adherence, but the polyamine biosynthesis genes of this *pot* mutant were intact (Vadyvaloo et al. 2007).

4.2 Environmental Factors

Besides the many bacterial factors that have been implicated, the biophysical properties of the substrate and surrounding medium can also greatly influence bacterial biofilm formation (Heydorn et al. 2000; Prouty and Gunn 2003). For example, *C. elegans* mutants with an altered surface cuticle do not accumulate *Yersinia* biofilm (Höflich et al. 2004; Joshua et al. 2003). Thus, initiation and development of the biofilm growth phenotype can be multifactorial and environment-specific. Accordingly, for reasons that have yet to be defined but that are likely to be complex, the biofilm growth phenotypes of *Y. pestis* and *Y. pseudotuberculosis* in different in vitro and in vivo environments do not always correlate, and *Y. pseudotuberculosis* never forms biofilm in the digestive tract of *X. cheopis* fleas (Erickson et al. 2006).

5   Characteristics of the *Y. pestis* Transmissible Biofilm Produced in the Flea

During septicemic plague, *Y. pestis* occurs in peripheral blood as individual cells. Immediately after being ingested into the midgut of an *X. cheopis* flea, however, it begins to form multicellular aggregates with evidence of an ECM. Usually, these aggregates are free-floating in the lumen of the midgut, unattached to a substrate. *Y. pestis* does not adhere to or invade the midgut epithelium, putting it at risk of being eliminated in the feces because fleas feed and defecate frequently. In fact, up to half of *X. cheopis* rapidly clear themselves of infection in this way even after feeding on blood containing more than 10^8 *Y. pestis* per milliliter (Engelthaler et al. 2000; Lorange et al. 2005; Pollitzer 1954). Thus, the formation of multicellular aggregates that are too large to pass in the feces may be important to produce a stable infection. For efficient transmission, however, adherence and consolidation of a biofilm to the spines in the proventriculus is required. This usually occurs secondary to the formation of midgut aggregates, although biofilm can occur simultaneously at both sites (Pollitzer 1954). Complete blockage does not usually occur until at least 1-2 weeks after an infectious blood meal, and proventricular colonization does not occur in all fleas with stable midgut infections.

Hms-negative *Y. pestis* strains are able to stably infect the flea midgut in the form of multicellular aggregates, but never colonize the proventriculus (Hinnebusch et al. 1996; Jarrett et al. 2004). This indicates that other bacterial factors besides the ECM are involved in the autoaggregative growth phenotype in the midgut, but that the ECM is essential for attachment and development of a biofilm on the surface of the proventricular spines. The outer coating of the spines is similar or identical to cuticle, the hard, hydrophobic, acellular material that makes up the flea exoskeleton. The Hms-dependent ECM of a *Y. pestis* biofilm may promote aggregation on that surface, in the same way that the biochemically similar Ica-dependent ECM of staphylococcal biofilms promotes aggregation to the hydrophobic, abiotic surface of catheters and other indwelling medical devices (Rupp et al. 1999).

In vivo biofilms of *Y. pestis* in the flea differ in some respects from in vitro biofilms. They are a dark brown color, most likely due to adsorption of hemin derived from red blood cell digestion in the flea midgut; in other words, they exhibit the pigmentation phenotype (Jarrett et al. 2004). The ECM of *Y. pestis* biofilm in the flea gut also appears to be heterogenous and more complex than is observed in vitro. Notably, it has a viscous appearance (Fig. 1) and, unlike in vitro *Y. pestis* biofilm, avidly takes up lipid stains (Jarrett et al. 2004). This suggests that the in vivo ECM consists not only of bacterial exopolysaccharide, but also exogenous material incorporated from the flea gut environment, such as lipid derived from the flea blood meal.

6 Implications of the Biofilm Transmission Phenotype at the Flea-*Y. pestis*-Host Interface

The ability to form an in vivo biofilm is now recognized as an integral part of the infection and disease process of many microbial pathogens (reviewed in Costerton et al. 1999; Parsek and Singh 2003). For *Y. pestis*, an in vivo biofilm type of infection occurs in the invertebrate vector, where it is important for efficient transmission, and not in the vertebrate host. The *Y. pestis* Hms-dependent ECM is not synthesized at mammalian body temperature and is not required for bubonic plague pathogenesis (Lillard et al. 1999). However, *Y. pestis* enters the mammal directly from an infectious biofilm in the flea, and this has implications for the initial interaction with the mammalian innate immune system at the dermal flea bite site.

The ECM that surrounds bacteria in a biofilm has been shown in some cases to protect against phagocytosis and/or intracellular killing, cationic antimicrobial peptides, and antibiotics (Costerton et al. 1999; Jesaitis et al. 2003; Vuong et al. 2004b). It is likely that the infectious inoculum delivered by a flea consists of small clumps of *Y. pestis* that are associated with a complex ECM (Fig. 4) (Jarrett et al. 2004; Lorange et al. 2005). The *Y. pestis* virulence factors known to protect against innate immunity, such as the antiphagocytic F1 capsule and the Type III secretion system that delivers cytotoxic Yop proteins into innate immune cells, are not synthesized in the flea. Thus, it is possible that the ECM may protect *Y. pestis* during the initial encounter with innate immune effector cells immediately after transmission, before the synthesis of specific virulence factors is induced. In addition, if the biofilm ECM made in the flea gut contains lipids derived from mammalian blood, these self components may mask bacterial antigens and avert initial recognition by the immune system. In an initial experiment, we found that both Hms$^+$ and Hms$^-$ *Y. pestis* recovered from flea guts were significantly more resistant to uptake by human polymorphonuclear leukocytes than the same bacteria grown in culture (Jarrett et al. 2004).

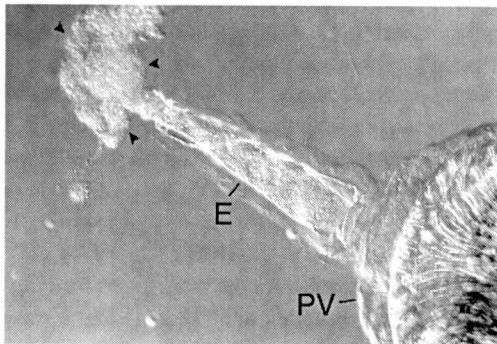

Fig. 4 *Y. pestis* biofilm (*arrowheads*) expelled through the esophagus (*E*) by application of a cover slip to the digestive tract dissected from a blocked *X. cheopis* flea. The array of spines that line the interior of the proventriculus (*PV*) is visible

Y. pestis containing conjugative plasmids that encode high-level resistance to multiple antibiotics have been isolated from human patients in Madagascar (Galimand et al. 1997; Guiyoule et al. 2001). The source of these newly acquired plasmids is unknown, but conjugative plasmid transfer among bacteria can occur at high frequency within a dense, multicellular biofilm (Hausner and Wuertz 1999; Molin and Tolker-Nielsen 2003). High-frequency conjugative transfer to *Y. pestis* within a mixed biofilm in the flea gut has been demonstrated, suggesting that gene transfer from commensal microbial flora in the flea gut that become incorporated into a *Y. pestis* biofilm may be a source of multiple-drug resistant strains (Hinnebusch et al. 2002a).

7 Summary

The genetic changes in *Y. pestis* since it diverged from *Y. pseudotuberculosis* that enabled biofilm development and growth in the digestive tract of the flea was clearly of fundamental importance in the evolution of the new arthropod-borne route of transmission. Because the *hms* genes and the ability to produce environmental biofilm appear to predate this recent divergence, probably relatively few changes were needed; perhaps involving only fine-tuning of environmental sensing and regulatory pathways that induce the Hms system (Erickson et al. 2006). Bacterial biofilm development is multifactorial, however, and much remains to be learned about it in the genus *Yersinia*. Adaptation of biofilm-forming potential to the flea gut environment, along with other discrete changes such as acquisition of two new plasmids, made possible the abrupt (in evolutionary terms) change to flea-borne transmission. Application of the biofilm developmental state to enable arthropod-borne transmission represents a novel function that illustrates the utility of multicellular, adherent growth.

Acknowledgements This work was supported by the Intramural Research Program of the NIH, NIAID.

References

Achtman M, Zurth K, Morelli G, Torrea G, Guiyoule A, Carniel E (1999) *Yersinia pestis*, the cause of plague, is a recently emerged clone of *Yersinia pseudotuberculosis*. Proc Natl Acad Sci U S A 96:14043-14048

Achtman M, Morelli G, Zhu P, Wirth T, Diehl I, Kusecek B, Vogler AJ, Wagner DM, Allender CJ, Easterday WR, Chenal-Francisque V, Worsham P, Thomson NR, Parkhill J, Lindler LE, Carniel E, Keim P (2004) Microevolution and history of the plague bacillus, *Yersinia pestis*. Proc Natl Acad Sci U S A 101:17837-17842

Bacot AW, Martin CJ (1914) Observations on the mechanism of the transmission of plague by fleas. J Hygiene (Plague Suppl 3) 13:423-439

Bacot AW (1915) Further notes on the mechanism of the transmission of plague by fleas. J Hygiene (Plague Suppl 4) 14:774-776

Beesley ED, Brubaker RR, Janssen WA, Surgalla MJ (1967) Pesticins. III. Expression of coagulase and mechanism of fibrinolysis. J Bacteriol 94:19-26

Beloin C, Da Re S, Ghigo J-M (2005) Colonization of abiotic surfaces. In: Böck A, Curtis R III, Kaper JB, Neidhardt FC, Nyström K, Rudd E, Squires CL (eds) EcoSal - *Escherichia coli* and *Salmonella*: cellular and molecular biology. ASM Press, Washington DC

Bercovier H, Mollaret HH, Alonso JM, Brault J, Fanning GR, Steigerwalt A, Brenner DJ (1980) Intra- and interspecies relatedness of *Yersinia pestis* by DNA hybridization and its relationship to *Yersinia pseudotuberculosis*. Curr Microbiol 4:225-229

Bobrov AG, Kirillina O, Perry RD (2005) The phosphodiesterase activity of the HmsP EAL domain is required for negative regulation of biofilm formation in *Yersinia pestis*. FEMS Microbiol Lett 247:123-130

Bottone EJ, Bercovier H, Mollaret HH (2005) Genus XLI. *Yersinia*. In: Brenner DJ, Krieg NR, Staley JT (eds) Bergey's manual of systematic bacteriology, 2nd edn. Springer, Berlin New York Heidelberg

Brubaker RR, Beesley ED, Surgalla MJ (1965) *Pasteurella pestis*: role of pesticin I and iron in experimental plague. Science 149:422-424

Brubaker RR (1991) Factors promoting acute and chronic diseases caused by yersiniae. Clin Microbiol Rev 4:309-324

Burroughs AL (1947) Sylvatic plague studies. The vector efficiency of nine species of fleas compared with *Xenopsylla cheopis*. J Hygiene 45:371-396

Camilli A, Bassler BL (2006) Bacterial small-molecule signaling pathways. Science 311:1113-1116

Carniel E (2003) Evolution of pathogenic *Yersinia*, some lights in the dark. In: Skurnik M, Bengoechea JA, Granfors K (eds) The genus *Yersinia*: entering the functional genomic era. Kluwer Academic, New York, pp 3-11

Cavanaugh DC (1971) Specific effect of temperature upon transmission of the plague bacillus by the oriental rat flea, *Xenopsylla cheopis*. Am J Trop Med Hyg 20:264-273

Cavanaugh DC, Marshall JD (1972) The influence of climate on the seasonal prevalence of plague in the Republic of Vietnam. J Wildl Dis 8:85-94

Chain PSG, Carniel E, Larimer FW, Lamerdin J, Stoutland PO, Regala WM, Georgescu AM, Vergez LM, Land ML, Motin VL, Brubaker RR, Fowler J, Hinnebusch J, Marceau M, Medigue C, Simonet M, Chenal-Francisque V, Souza B, Dacheaux D, Elliot JM, Derbise A, Hauser LJ, Garcia E (2004) Insights into the evolution of *Yersinia pestis* through whole-genome comparison with *Yersinia pseudotuberculosis*. Proc Natl Acad Sci U S A 101:13826-13831

Costerton JW, Stewart PS, Greenberg EP (1999) Bacterial biofilms: a common cause of persistent infections. Science 284:1318-1322

D'Argenio DA, Miller SI (2004) Cyclic di-GMP as a bacterial second messenger. Microbiology 150:2497-2502

Darby C, Hsu JW, Ghori N, Falkow S (2002) *Caenorhabditis elegans*: plague bacteria biofilm blocks food intake. Nature 417:243-244

Darby C, Ananth SL, Tan L, Hinnebusch BJ (2005) Identification of *gmhA*, a *Yersinia pestis* gene required for flea blockage, by using a *Caenorhabditis elegans* biofilm system. Infect Immun 73:7236-7242

Davies DG, Parsek MR, Pearson JP, Iglewski BH, Costerton JW, Greenberg EP (1998) The involvement of cell-to-cell signals in the development of a bacterial biofilm. Science 280:295-298

Deng W, Burland V, Plunkett G, Boutin A, Mayhew GF, Liss P, Perna NT, Rose DJ, Mau B, Zhou S, Schwartz DC, Fetherston JD, Lindler LE, Brubaker RR, Plano GV, Straley SC, McDonough KA, Nilles ML, Matson JS, Blattner FR, Perry RD (2002) Genome sequence of *Yersinia pestis* KIM. J Bacteriol 184:4601-4611

Eisen RJ, Bearden SW, Wilder AP, Montenieri JA, Antolin MF, Gage KL (2006) Early-phase transmission of *Yersinia pestis* by unblocked fleas as a mechanism explaining rapidly spreading plague epizootics. Proc Natl Acad Sci U S A 103:15380-15385

Engelthaler DM, Hinnebusch BJ, Rittner CM, Gage KL (2000) Quantitative competitive PCR as a technique for exploring flea-*Yersina pestis* dynamics. Am J Trop Med Hyg 62:552-560

Erickson DL, Jarrett CO, Wren BW, Hinnebusch BJ (2006) Serotype differences and lack of biofilm formation characterize *Yersinia pseudotuberculosis* infection of the *Xenopsylla cheopis* flea vector of *Yersinia pestis*. J Bacteriol 188:1113-1119

Ferber DM, Brubaker RR (1981) Plasmids in *Yersinia pestis*. Infect Immun 31:839-841

Forman S, Bobrov AG, Kirillina O, Craig SK, Abney J, Fetherston JD, Perry RD (2006) Identification of critical amino acid residues in the plague biofilm Hms proteins. Microbiology 152:3399-3410

Gage KL, Kosoy MY (2005) Natural history of plague: perspectives from more than a century of research. Annu Rev Entomol 50:505-528

Galimand M, Guiyoule A, Gerbaud G, Rasoamanana B, Chanteau S, Carniel E, Courvalin P (1997) Multidrug resistance in *Yersinia pestis* mediated by a transferable plasmid. N Engl J Med 337:677-680

Gerke C, Kraft A, Sussmuth R, Schweitzer O, Götz F (1998) Characterization of the N-acetylglucosaminyltransferase activity involved in the biosynthesis of the *Staphylococcus epidermidis* polysaccharide intercellular adhesin. J Biol Chem 273:18586-18593

Gjermansen M, Ragas P, Tolker-Nielsen T (2006) Proteins with GGDEF and EAL domains regulate *Pseudomonas putida* biofilm formation and dispersal. FEMS Microbiol Lett 265:215-224

Guiyoule A, Gerbaud G, Buchrieser C, Galimand M, Rahalison L, Chanteau S, Courvalin P, Carniel E (2001) Transferable plasmid-mediated resistance to streptomycin in a clinical isolate of *Yersinia pestis*. Emerg Infect Dis 7:43-48

Götz F (2002) *Staphylococcus* and biofilms. Mol Microbiol 43:1367-1378

Hare JM, McDonough KA (1999) High-frequency RecA-dependent and -independent mechanisms of Congo red binding mutations in *Yersinia pestis*. J Bacteriol 181:4896-4904

Hausner M, Wuertz S (1999) High rates of conjugation in bacterial biofilms as determined by quantitative in situ analysis. Appl Environ Microbiol 65:3710-3713

Heilmann C, Schweitzer O, Gerke C, Vanittanakom N, Mack D, Götz F (1996) Molecular basis of intercellular adhesion in the biofilm-forming *Staphylococcus epidermidis*. Mol Microbiol 20:1083-1091

Heilmann C, Götz F (1998) Further characterization of *Staphylococcus epidermidis* transposon mutants deficient in primary attachment or intercellular adhesion. Zentralbl Bakteriol 287:69-83

Heydorn A, Ersbøll BK, Hentzer M, Parsek MR, Givskov M, Molin S (2000) Experimental reproducibility in flow-chamber biofilms. Microbiology 146:2409-2415

Hinchliffe SJ, Isherwood KE, Stabler RA, Prentice MB, Rakin A, Nichols RA, Oyston PC, Hinds J, Titball RW, Wren BW (2003) Application of DNA microarrays to study the evolutionary genomics of *Yersinia pestis* and *Yersinia pseudotuberculosis*. Genome Res 13:2018-2029

Hinnebusch BJ, Perry RD, Schwan TG (1996) Role of the *Yersinia pestis* hemin storage (*hms*) locus in the transmission of plague by fleas. Science 273:367-370

Hinnebusch BJ (1997) Bubonic plague: a molecular genetic case history of the emergence of an infectious disease. J Mol Med 75:645-652

Hinnebusch BJ, Fischer ER, Schwan TG (1998) Evaluation of the role of the *Yersinia pestis* plasminogen activator and other plasmid-encoded factors in temperature-dependent blockage of the flea. J Inf Dis 178:1406-1415

Hinnebusch BJ, Rosso M-L, Schwan TG, Carniel E (2002a) High-frequency conjugative transfer of antibiotic resistance genes to *Yersinia pestis* in the flea midgut. Mol Microbiol 46:349-354

Hinnebusch BJ, Rudolph AE, Cherepanov P, Dixon JE, Schwan TG, Forsberg Å (2002b) Role of Yersinia murine toxin in survival of *Yersinia pestis* in the midgut of the flea vector. Science 296:733-735

Höflich J, Berninsone P, Göbel C, Gravato-Nobre MJ, Libby BJ, Darby C, Politz SM, Hodgkin J, Hirschberg CB, Baumeister R (2004) Loss of *srf-3*-encoded nucleotide sugar transporter activity in *Caenorhabditis elegans* alters surface antigenicity and prevents bacterial adherence. J Biol Chem 279:30440-30448

Ibrahim A, Goebel BM, Liesack W, Griffiths M, Stackebrandt E (1993) The phylogeny of the genus *Yersinia* based on 16S rDNA sequences. FEMS Microbiol Lett 114:173-177

Itoh Y, Wang X, Hinnebusch BJ, Preston JF, Romeo T (2005) Depolymerization of β-1,6-*N*-acetyl-D-glucosamine disrupts the integrity of diverse bacterial biofilms. J Bacteriol 187:382-387

Jackson DW, Suzuki K, Oakford L, Simecka JW, Hart ME, Romeo T (2002) Biofilm formation and dispersal under the influence of the global regulator CsrA of *Escherichia coli*. J Bacteriol 184:290-301

Jackson S, Burrows TW (1956) The pigmentation of *Pasteurella pestis* on a defined medium containing haemin. Br J Exp Pathol 37:570-576

Jarrett CO, Deak E, Isherwood KE, Oyston PC, Fischer ER, Whitney AR, Kobayashi SD, DeLeo FR, Hinnebusch BJ (2004) Transmission of *Yersinia pestis* from an infectious biofilm in the flea vector. J Inf Dis 190:783-792

Jawetz E, Meyer KF (1944) Studies on plague immunity in experimental animals. II. Some factors of the immunity mechanism in bubonic plague. J Immunol 49:15-29

Jesaitis AJ, Franklin MJ, Berglund D, Sasaki M, Lord CI, Bleazard JB, Duffy JE, Beyenal H, Lewandowski Z (2003) Compromised host defense on *Pseudomonas aeruginosa* biofilms: characterization of neutrophil and biofilm interactions. J Immunol 171:4329-4339

Jones HA, Lillard JW, Perry RD (1999) HmsT, a protein essential for expression of the haemin storage (Hms+) phenotype of *Yersinia pestis*. Microbiology 145:2117-2128

Joshua GWP, Karlyshev AV, Smith MP, Isherwood KE, Titball RW, Wren BW (2003) A *Caenorhabditis elegans* model of *Yersinia* infection: biofilm formation on a biotic surface. Microbiology 149:3221-3229

Kader A, Simm R, Gerstel U, Morr M, Römling U (2006) Hierarchical involvement of various GGDEF domain proteins in rdar morphotype development of *Salmonella enterica* serovar Typhimurium. Mol Microbiol 60:602-616

Kaplan JB, Ragunath C, Ramasubbu N, Fine DH (2003) Detachment of *Actinobacillus actinomycetemcomitans* biofilm cells by an endogenous beta-hexosaminidase activity. J Bacteriol 185:4693-4698

Karatan E, Duncan TR, Watnick PI (2005) NspS, a predicted polyamine sensor, mediates activation of *Vibrio cholerae* biofilm formation by norspermidine. J Bacteriol 187:7434-7443

Kirillina O, Fetherston JD, Bobrov AG, Abney J, Perry RD (2004) HmsP, a putative phosphodiesterase, and HmsT, a putative diguanylate cyclase, control Hms-dependent biofilm formation in *Yersinia pestis*. Mol Microbiol 54:75-88

Korhonen TK, Kukkonen M, Virkola R, Lang H, Suomalainen M, Kyllönen P, Lähteenmäki K (2004) The plasminogen activator Pla of *Yersinia pestis*: localized proteolysis and systemic spread. In: Carniel E, Hinnebusch BJ (eds) Yersinia. Molecular and Cellular Biology. Horizon Bioscience, Norfolk, UK, pp 349-362

Lillard JW, Fetherston JD, Pedersen L, Pendrak ML, Perry RD (1997) Sequence and genetic analysis of the hemin storage (*hms*) system of *Yersinia pestis*. Gene 193:13-21

Lillard JW, Bearden SW, Fetherston JD, Perry RD (1999) The haemin storage (Hms+) phenotype of *Yersinia pestis* is not essential for the pathogenesis of bubonic plague in mammals. Microbiology 145:197-209

Lindler LE, Klempner MS, Straley SC (1990) *Yersinia pestis* pH 6 antigen: genetic, biochemical, and virulence characterization of a protein involved in the pathogenesis of bubonic plague. Infect Immun 58:2569-2577

Lorange EA, Race BL, Sebbane F, Hinnebusch BJ (2005) Poor vector competence of fleas and the evolution of hypervirulence in *Yersinia pestis*. J Inf Dis 191:1907-1912

Mack D, Fischer W, Krokotsch A, Leopold K, Hartmann R, Egge H, Laufs R (1996) The intercellular adhesin involved in biofilm accumulation of *Staphylococcus epidermidis* is a linear beta-1,6-linked glucosaminoglycan: purification and structural analysis. J Bacteriol 178:175-183

Matthysse AG, Yarnall HA, Young N (1996) Requirement for genes with homology to ABC transport systems for attachment and virulence of *Agrobacterium tumefaciens*. J Bacteriol 178:5302-5308

McDonough KA, Falkow S (1989) A *Yersinia pestis*-specific DNA fragment encodes temperature-dependent coagulase and fibrinolysin-associated phenotypes. Mol Microbiol 3:767-775

McDonough KA, Barnes AM, Quan TJ, Montenieri J, Falkow S (1993) Mutation in the *pla* gene of *Yersinia pestis* alters the course of the plague bacillus-flea (Siphonaptera: Ceratophyllidae) interaction. J Med Entomol 30:772-780

Molin S, Tolker-Nielsen T (2003) Gene transfer occurs with enhanced efficiency in biofilms and induces enhanced stabilisation of the biofilm structure. Curr Opin Biotechnol 14:255-261

Newman KL, Almeida RP, Purcell AH, Lindow SE (2004) Cell-cell signaling controls *Xylella fastidiosa* interactions with both insects and plants. Proc Natl Acad Sci U S A 101:1737-1742

Parkhill J, Wren BW, Thomson NR, Titball RW, Holden MTG, Prentice MB, Sebhaihia M, James KD, Churcher C, Mungall KL, Baker S, Basham D, Bentley SD, Brooks K, Cerdeño-Tárraga AM, Chillingworth T, Cronin A, Davies RM, Davis P, Dougan G, Feltwell T, Hamlin N, Holroyd S, Jagels K, Karlyshev AV, Leather S, Moule S, Oyston PCF, Quail M, Rutherford K, Simmonds M, Skelton J, Stevens K, Whitehead S, Barrell BG (2001) Genome sequence of *Yersinia pestis*, the causative agent of plague. Nature 413:523-527

Parsek MR, Singh PK (2003) Bacterial biofilms: an emerging link to disease pathogenesis. Ann Rev Micriobiol 57:677-701

Parsek MR, Greenberg EP (2005) Sociomicrobiology: the connections between quorum sensing and biofilms. Trends Microbiol 13:27-33

Patel CN, Wortham BW, Lines JL, Fetherston JD, Perry RD, Oliveira MA (2006) Polyamines are essential for the formation of plague biofilm. J Bacteriol 188:2355-2363

Pendrak ML, Perry RD (1993) Proteins essential for expression of the Hms+ phenotype of *Yersinia pestis*. Mol Microbiol 8:857-864

Perry RD, Pendrak ML, Schuetze P (1990) Identification and cloning of a hemin storage locus involved in the pigmentation phenotype of *Yersinia pestis*. J Bacteriol 172:5929-5937

Perry RD, Lucier TS, Sikkema DJ, Brubaker RR (1993) Storage reservoirs of hemin and inorganic iron in *Yersinia pestis*. Infect Immun 61:32-39

Perry RD, Fetherston JD (1997) *Yersinia pestis* - etiologic agent of plague. Clin Microbiol Rev 10:35-66

Perry RD, Bobrov AG, Kirillina O, Jones HA, Pedersen L, Abney J, Fetherston JD (2004) Temperature regulation of the hemin storage (Hms+) phenotype of *Yersinia pestis* is posttranscriptional. J Bacteriol 186:1638-1647

Pollitzer R (1954) Plague. World Health Organization, Geneva

Prouty AM, Gunn JS (2003) Comparative analysis of *Salmonella enterica* serovar Typhimurium biofilm formation on gallstones and on glass. Infect Immun 71:7154-7158

Purcell AH, Finlay AH, McLean DL (1979) Pierce's disease bacterium: mechanism of transmission by leafhopper vectors. Science 206:839-841

Rogers ME, Chance ML, Bates PA (2002) The role of promastigote secretory gel in the origin and transmission of the infective stage of *Leishmania mexicana* by the sandfly *Lutzomyia longipalpis*. Parasitology 124:495-507

Rupp ME, Ulphani JS, Fey PD, Bartscht K, Mack D (1999) Characterization of the importance of polysaccharide intercellular adhesin/hemagglutinin of *Staphylococcus epidermidis* in the pathogenesis of biomaterial-based infection in a mouse foreign body infection model. Infect Immun 67:2627-2632

Römling U, Amikam D (2006) Cyclic di-GMP as a second messenger. Curr Opin Microbiol 9:218-228

Sebbane F, Jarrett CO, Gardner D, Long D, Hinnebusch BJ (2006) Role of the *Yersinia pestis* plasminogen activator in the incidence of distinct septicemic and bubonic forms of flea-borne plague. Proc Natl Acad Sci U S A 103:5526-5530

Simm R, Fetherston JD, Kader A, Römling U, Perry RD (2005) Phenotypic convergence mediated by GGDEF-domain-containing proteins. J Bacteriol 187:6816-6823

Sodeinde OA, Goguen JD (1988) Genetic analysis of the 9.5-kilobase virulence plasmid of *Yersinia pestis*. Infect Immun 56:2743-2748

Sodeinde OA, Subrahmanyam YV, Stark K, Quan T, Bao Y, Goguen JD (1992) A surface protease and the invasive character of plague. Science 258:1004-1007

Staggs TM, Fetherston JD, Perry RD (1994) Pleiotropic effects of a *Yersinia pestis fur* mutation. J Bacteriol 176:7614-7624

Stierhof YD, Bates PA, Jacobson RL, Rogers ME, Schlein Y, Handman E, Ilg T (1999) Filamentous proteophosphoglycan secreted by *Leishmania* promastigotes forms gel-like three-dimensional networks that obstruct the digestive tract of infected sandfly vectors. Eur J Cell Biol 78:675-689

Stoodley P, Sauer K, Davies DG, Costerton JW (2002) Biofilms as complex differentiated communities. Ann Rev Microbiol 56:187-209

Surgalla MJ, Beesley ED (1969) Congo red agar plating medium for detecting pigmentation in *Pasteurella pestis*. Appl Microbiol 18:834-837

Tan L, Darby C (2004) A movable surface: formation of *Yersinia* sp. biofilms on motile *Caenorhabditis elegans*. J Bacteriol 186:5087-5592

Tan L, Darby C (2005) *Yersinia pestis* is viable with endotoxin composed of only lipid A. J Bacteriol 187:6599-6600

Tan L, Darby C (2006) *Yersinia pestis* YrbH is a multifunctional protein required for both 3-deoxy-D-manno-oct-2-ulosonic acid biosynthesis and biofilm formation. Mol Microbiol 61:861-870

Vadyvaloo V, Otto M (2005) Molecular genetics of *Staphylococcus epidermidis* biofilms on indwelling medical devices. Int J Art Org 28:1069-1078

Vadyvaloo V, Jarrett CO, Sturdevant DE, Sebbane F, Hinnebusch BJ (2007) Analysis of *Yersinia pestis* gene expression in the flea vector. Adv Exp Med Biol 603:192-200

Vuong C, Gerke C, Somerville GA, Fischer ER, Otto M (2003) Quorum-sensing control of biofilm factors in *Staphylococcus epidermidis*. J Infect Dis 188:706-718

Vuong C, Kocianova S, Voyich JM, Yao Y, Fischer ER, DeLeo FR, Otto M (2004a) A crucial role for exopolysaccharide modification in bacterial biofilm formation, immune evasion, and virulence. J Biol Chem 279:54881-54886

Vuong C, Voyich JM, Fischer ER, Braughton KR, Whitney AR, DeLeo FR, Otto M (2004b) Polysaccharide intercellular adhesin (PIA) protects *Staphylococcus epidermidis* against major components of the human innate immune system. Cell Microbiol 6:269-275

Wang X, Preston JF, Romeo T (2004) The *pgaABCD* locus of *Escherichia coli* promotes the synthesis of a polysaccharide adhesin required for biofilm formation. J Bacteriol 186:2442-2449

Wang X, Dubey AK, Suzuki K, Baker CS, Babitzke P, Romeo T (2005) CsrA post-transcriptionally represses *pgaABCD*, responsible for synthesis of a biofilm polysaccharide adhesin of *Escherichia coli*. Mol Microbiol 56:1648-1663

Webb CT, Brooks CP, Gage KL, Antolin MF (2006) Classic flea-borne transmission does not drive plague epizootics in prairie dogs. Proc Natl Acad Sci U S A 103:6236-6241

Weiner R, Seagren E, Arnosti C, Quintero E (1999) Bacterial survival in biofilms: probes for exopolysaccharide and its hydrolysis, and measurements of intra- and interphase mass fluxes. Methods Enzymol 310:403-426

Zhou D, Han Y, Song Y, Huang P, Yang R (2004) Comparative and evolutionary genomics of *Yersinia pestis*. Microbes Infect 6:1226-1234

Zhu J, Mekalanos JJ (2003) Quorum sensing-dependent biofilms enhance colonization in *Vibrio cholerae*. Dev Cell 5:647-656

Escherichia coli Biofilms

C. Beloin, A. Roux, and J.-M. Ghigo(⊡)

Abstract *Escherichia coli* is a predominant species among facultative anaerobic bacteria of the gastrointestinal tract. Both its frequent community lifestyle and the availability of a wide array of genetic tools contributed to establish *E. coli* as a relevant model organism for the study of surface colonization. Several key factors, including different extracellular appendages, are implicated in *E. coli* surface colonization and their expression and activity are finely regulated, both in space and time, to ensure productive events leading to mature biofilm formation. This chapter will present known molecular mechanisms underlying biofilm development in both commensal and pathogenic *E. coli*.

J.-M. Ghigo
Groupe de Génétique des Biofilms, Institut Pasteur, CNRS URA 2172, 25 rue du Dr. Roux, 75724 Paris Cedex 15, France
jmghigo@pasteur.fr

T. Romeo (ed.), *Bacterial Biofilms.*
Current Topics in Microbiology and Immunology 322.
© Springer-Verlag Berlin Heidelberg 2008

1   Introduction

The description of a widespread association between bacteria and surfaces dates back to the dawn of microbiology when, in the seventeenth century, Antone Van Leeuwenhoek observed "animalcules" on his own dental plaque surface. Since then, however, what we now know to be bacteria have most often been studied in artificial but controlled conditions using agitated single-celled planktonic cultures. It is currently admitted that we have explored only a very specific aspect of bacterial biology since, in most ecological niches, bacterial interactions with a surface promote novel behaviors leading to the development of structured and heterogeneous, matrix-encased bacterial communities known as biofilms. Biofilm physiology is characterized by increased tolerance to stress, biocides (including antibiotics) (see the chapter by G.G. Anderson and G.A. O'Toole, this volume) and host immunological defenses, which is at the origin of their resilience in most medical and industrial settings. During the last decade, the negative impact of biofilm development on human activities has stimulated research aimed at providing clues for combatting detrimental biofilms. Most recent studies conducted on genetic requirements underlying bacterial biofilm formation have used a limited selection of model bacteria, including *E. coli*.

Although current knowledge of bacterial biology owes much to work done on planktonic cultures of laboratory strains of *E. coli*, many isolates also have the capacity to form biofilm structures in vivo and in vitro. Indeed, *E. coli* is a predominant species among facultative anaerobic bacteria of the gastrointestinal tract, where it thrives in an environment with structural characteristics of a multispecies biofilm (Costerton et al. 1995; Probert and Gibson 2002).

With over 250 serotypes, *E. coli* is a highly versatile bacterium ranging from harmless gut commensal to intra- or extraintestinal pathogens, including common colonizers of medical devices and the primary causes of recurrent urogenital infections (Kaper et al. 2004).

In this chapter dedicated to *E. coli* biofilms, we will attempt to provide an overview of present knowledge of molecular mechanisms underlying surface colonization by *E. coli*. While particular emphasis will be placed on studies conducted on *E. coli* K-12, in vivo aspects of pathogenic and natural *E. coli* isolate biofilm formation will also be presented.

2   First Contacts with the Surface

2.1   *Approaching the Surface: The role of Motility in* E. coli *Adhesion to Surfaces*

In a liquid environment, bacteria are subjected to hydrodynamic forces, especially when approaching surfaces. Alongside passive movement that is governed by Brownian or gravitational forces, bacteria have developed mechanisms of active motility that enable them to overcome repulsive electrostatic and hydrodynamic forces encountered around surfaces, and consequently to increase their chances of interacting with surfaces (Donlan 2002).

In Gram-negative bacteria such as *E. coli* and *Salmonella*, active motility is dependent on a flagellar apparatus that is necessary for them to swim in liquid or semi-liquid medium. In one of the earliest studies using *E. coli* as a model system, half of the insertion mutants deficient at biofilm formation identified by Pratt and Kolter were found to perturb flagellar functions (Pratt and Kolter 1998). These authors showed that motility itself, and not chemotaxis or direct surface contact by flagella, was required to form a biofilm, and they proposed that, beyond allowing the bacteria to breach repulsive forces, flagella may also allow bacteria to spread along the surface. This analysis was confirmed in *E. coli* by Genevaux and co-workers (Genevaux et al. 1996) and, more recently, by Wood and co-workers, who demonstrated, using a set of different motility mutants, that the biofilm formation capacities of *E. coli* K-12 were directly correlated with its ability to swim (Wood et al. 2006).

Although this suggests that the requirement for force-generating cell-surface organelles is a common theme in biofilm formation, it is not an absolute requirement, and nonmotile bacteria can still form biofilms under certain conditions, as shown for *E. coli* K-12 (Pratt and Kolter 1999), but also for enteroaggregative *E. coli* (EAEC) (Sheikh et al. 2001). In a nonmotile strain that overexpresses the surface adhesins known as curli (see Sect. 2.3.2 below), Prigent-Combaret and co-workers showed that flagellar motility is not required for initial adhesion, nor is it required for biofilm development (Prigent-Combaret et al. 2000). Reisner and co-workers showed that in an *E. coli* strain bearing a conjugative plasmid, known as a strong adhesion factor (Ghigo 2001), the presence of flagellum was dispensable for biofilm formation (Reisner et al. 2003). In these cases, it is possible that the expression of strong adhesion factors may replace force-generating movements during initial interactions between adhering bacteria and the surface (Pratt and Kolter 1999; Geesey 2001; Donlan 2002).

2.2 Primary Adhesion to Surfaces: Reversible Attachment

Whereas flagellar motility helps the bacteria to counteract hydrodynamic and electrostatic forces near the surface, initial bacterial adhesion to abiotic surfaces is likely to be highly dependent on physicochemical and electrostatic interactions between the bacterial envelope itself and the substrate, which is often conditioned by the fluids to which it is exposed (Dunne 2002). Attracting and repulsing forces between the bacteria and the surface lead to reversible attachment of bacteria to the surface, with most of the bacteria leaving the surface to join the planktonic phase either because of mild shear or the bacterium's own mobility (Dunne 2002). This reversible attachment is strongly influenced both by environmental conditions such as pH and the ionic force of the medium or the temperature (Fletcher 1988; Danese et al. 2000a), and by the nature of the surface itself, with rugosity increasing surface adhesion or hydrophobic surfaces such as Teflon or plastic more likely to be colonized by bacteria than hydrophilic surfaces such as glass and metal (Donlan 2002). Furthermore, adsorption and desorption of nutrients at the surface, which compose a so-called conditioning film, are also important factors that may influence bacterial initial attachment both positively and negatively depending on the nature of the organic molecules involved (Zobell 1943; van Loosdrecht et al. 1990).

2.3   Irreversible Adhesion to Surfaces: The Role of Fimbriae

Alongside thermodynamic aspects, the direct contribution of adhesive organelles of the fimbrial family to the irreversible attachment of bacteria to surfaces has been amply demonstrated. Three classes of fimbriae have a role in strengthening the bacteria-to-surface interactions: type 1 fimbriae, curli, and conjugative pili.

2.3.1   Type 1 Fimbriae

Type 1 fimbriae (or pili) are filamentous proteinaceous adhesins commonly expressed both by commensal and pathogenic *E. coli* isolates (Sauer et al. 2000). Bacteria expressing type 1 fimbriae generally present between 100 and 500 fimbriae at their surface. These fimbriae have a tubular structure that is 5-7 nm in diameter and between 0.2 and 2 µm long. Type 1 pili can adhere, in a mannose-dependent manner, to a variety of receptor molecules on eukaryotic cell surfaces (Duncan et al. 2005) and are well documented virulence factors in pathogenic *E. coli* (Kaper et al. 2004). Along with their role in the formation of secreted IgA-mediated biofilm within the gut (Bollinger et al. 2003, 2006; Orndorff et al. 2004), several groups have reported that these adhesins are critical for *E. coli* biofilm formation on abiotic surfaces (Harris et al. 1990; Pratt and Kolter 1998; Cookson et al. 2002; Moreira et al. 2003; Orndorff et al. 2004). Mutants in both *fimA,* the gene encoding the major type 1 pilus subunit, and *fimH*, which codes for the mannose-specific adhesin located at the tip of the pilus, have been reported to reduce *E. coli* initial attachment to polyvinyl chloride and other abiotic surfaces (Pratt and Kolter 1998; Beloin et al. 2004). These findings suggest that the type 1 pili FimH adhesin, besides binding eukaryotic mannose oligosaccharides, may also have nonspecific binding activity at abiotic surfaces (Pratt and Kolter 1998). The expression of type 1 pili is induced by adhesion and biofilm formation at early and late stages (Schembri et al. 2003b; Beloin et al. 2004; Ren et al. 2004a).

While the FimH adhesin itself could be responsible for this nonspecific binding, several studies have indicated that the expression of type 1 pili may affect adhesion of *E. coli* to abiotic surfaces by altering the composition of the outer membrane (Orndorff et al. 2004). Indeed, Otto and co-workers showed that type 1 pili surface contacts mediate a decrease in the abundance of several outer membrane proteins such as BtuB, EF-Tu, OmpA, OmpX, Slp, and TolC (Otto et al. 2001). These changes in the envelope probably affect the general physicochemical characteristics of the bacterial surface and thereby influence adhesion (Otto et al. 2001). Consistently, production of type 1 pili is upregulated in the absence of OmpX. Furthermore, the absence of OmpX also increases exopolysaccharide production as well as decreasing motility of the bacteria. The decrease in the OmpX level upon type 1 pili-mediated surface contacts may therefore serve as a signal leading to physiological adaptation in surface-associated bacteria (Otto and Hermansson 2004).

A recent work by Lacqua and collaborators showed that when *E. coli* K-12 MG1655 is mixed with P1*vir* or λ phages, a phage-tolerant population with increased biofilm formation capacities arose within 24 h, and this appearance was dependent on the presence of type 1 fimbriae. This suggested that type 1 fimbriae-mediated abiotic biofilm formation by K-12 *E. coli* MG1655 strain might represent a strategy to escape bacteriophage attack, thus emphasizing the importance of type 1 fimbriae in *E. coli* physiology (Lacqua et al. 2006).

2.3.2   Curli Fimbriae

Initially identified in *E. coli*, curli fimbriae, also called thin aggregative fimbriae, are produced by other *Enterobacteriaceae* such as *Shigella*, *Citrobacter*, and *Enterobacter* (Smyth et al. 1996). Curli fimbriae aggregate at the cell surface to form 6- to 12-nm-diameter structures whose length varies between 0.5 and 1 μm. Curli have been demonstrated to attach to proteins of the extracellular matrix such as fibronectin, laminin, and plasminogen (Olsen et al. 1989; Ben Nasr et al. 1996), thus promoting adhesion of the bacteria to different human cells. In addition, curli adhesive fibers also promote biofilm formation to abiotic surfaces both by facilitating initial cell-surface interactions and subsequent cell-cell interactions (Vidal et al. 1998; Cookson et al. 2002; Uhlich et al. 2006). Genes involved in curli production are clustered in two divergently transcribed operons: the *csgBA* operon, encoding the structural components of curli, the *csgDEFG* operon, encoding a transcriptional regulator (CsgD) and the curli export machinery (CsgE-G). In environmental and clinical isolates of *E. coli*, the synthesis of curli is subject to tight and complex regulation allowing curli production notably at 37°C and/or 28°C depending on the isolates, whereas the expression of curli is cryptic in most *E. coli* laboratory strains. However, Vidal and co-workers isolated, in this type of domesticated strain, a gain-of-function mutant with increased capacity for surface adhesion and cell-to-cell interactions due to the constitutive expression of the cell-surface adhesin curli (Vidal et al. 1998). They determined that the hyperadhesive phenotype was the result of a point mutation in the OmpR protein that constitutes, with the EnvZ protein, a two-component regulatory system that senses variations in osmolarity. This *ompR* allele (*ompR*234) leads to more efficient OmpR-dependent activation of the *csgD* promoter, which stimulates curli production and biofilm formation in laboratory strains (Vidal et al. 1998; Prigent-Combaret et al. 2000; Prigent-Combaret et al. 2001).

Along with the EnvZ/OmpR two-component system, several transcriptional regulators (CpxR, RcsCDB, RpoS, H-NS, IHF, Crl, MlrA) responding to different environmental and stress conditions such as temperature, osmolarity, pH and oxygenation are involved in regulation of curli expression through a network of interactions between transcription factors and the *csg* regulatory region (Dorel et al. 1999; Prigent-Combaret et al. 2001; Brombacher et al. 2003; Gerstel et al. 2003; Jubelin et al. 2005; Vianney et al. 2005). This complex regulatory network is presumed to allow fine-tuning of curli expression that may play a role in colonization of specific niches by *E. coli,* especially in the human body (Prigent-Combaret et al. 2001; Kikuchi et al. 2005).

2.3.3 Conjugative Pili

Although most laboratory *E. coli* K-12 strains are poor biofilm formers, the introduction, either artificially or naturally, in mixed *E. coli* communities of a conjugative plasmid in these strains induces formation of a thick mature biofilm (Ghigo 2001; Reisner et al. 2003, 2006). Mutational analysis of the conjugative transfer apparatus genes of the F plasmid demonstrated that this phenotype does not depend on the ability of the plasmids to mediate DNA transfer, but it does require a functional conjugative pilus (Ghigo 2001). The F-pilus promotes both initial adhesion and biofilm maturation through nonspecific attachment to abiotic surfaces and subsequent cell-to-cell contacts, which stabilize the structure of the biofilm (Ghigo 2001; Molin and Tolker-Nielsen 2003; Reisner et al. 2003). Reisner and co-workers also showed that expression of the F conjugative pilus could functionally substitute for other known adhesion factors such as type1 pili, Ag43 or curli (Reisner et al. 2003). Plasmid-mediated biofilm production is not restricted to the F plasmids, and most tested conjugative plasmids directly contribute, upon derepression of their conjugative function, to bacterial host capacity to form a biofilm (Ghigo 2001). This general connection between conjugation and biofilm formation is consistent with early observations showing that surface contacts positively affect the dynamics of plasmid transfer (Simonsen 1990). Since then, numerous studies investigating transfer of both conjugative and nonconjugative plasmids indicate that physical contact between donor and recipient cells is highly favored within monospecific and mixed *E. coli* biofilms, where efficient horizontal transfer of genetic material has been demonstrated (Lebaron et al. 1997; Hausner and Wuertz 1999; Licht et al. 1999; Dionisio et al. 2002; Molin and Tolker-Nielsen 2003; Maeda et al. 2006). In addition, those studies showed that both conjugative and nonconjugative plasmids are likely to carry determinants for biofilm initiation and architecture, thus promoting biofilm development, which in turn affects the extent of plasmid-mediated horizontal gene transfer within the biofilm (Wuertz et al. 2004).

Several cell adhesins encoded by genes on plasmids or mobile genetic elements have been characterized in pathogenic *E. coli* (Henderson et al. 1998; Kaper et al. 2004; Dudley et al. 2006), suggesting that the extrachromosomal gene pool (plasmids and other mobile genetic elements) may also constitute an important source of adhesion factors leading to biofilm formation, influencing both the probability of biofilm-related infection and of conjugational spread of plasmid-borne virulence factors (Amabile-Cuevas and Chicurel 1996; Ghigo 2001; Molin and Tolker-Nielsen 2003).

This section has presented both environmental and bacterial structures leading to irreversible surface attachment. The section that follows will address the question of biofilm maturation, whereby interbacterial adhesion, rather than direct contact with the substrate, leads to progressive buildup of the mature biofilm.

3   Building the Mature Biofilm

Biofilm maturation corresponds to the three-dimensional growth of the biofilm that occurs after initial attachment to the surface. Mostly due to bacterium-bacterium interactions, this process leads to the formation of a heterogeneous physicochemical environment in which biofilm bacteria display characteristic physiological traits that distinguish them from their planktonic counterparts. Whereas the delimitation between initial attachment and maturation is gradual - both processes involve some of the structures described in the previous section - we will now describe the surface proteins and extracellular matrix components particularly involved in bacterial interadhesion and biofilm architecture.

3.1   Surface Adhesins that Contribute to Biofilm Structure

3.1.1   Autotransporter Adhesins

In most Gram-negative bacteria, the translocation of proteins toward the extracellular medium requires crossing of the cytoplasmic membrane, the periplasm, and the outer membrane. This translocation can be achieved through at least six different secretion pathways (Economou et al. 2006). Among them, the type V secretion pathway enables a family of proteins to reach the surface with a very limited number of accessory secretion factors because most information necessary to the translocation process is contained within the secreted protein itself. These proteins, which can therefore carry out their own transport to the outer membrane, are called autotransported or autotransporter proteins.

3.1.1.1   The Type V Secretion Pathway

Pohlner and collaborators were the first to describe a model of translocation of the *Neisseria gonorrhoeae* IgA1 protease by the type V secretion pathway (Pohlner et al. 1987). Since then, many autotransporter proteins have been described. Each of these proteins displays a modular structure with four characteristic domains:

(1) an N-terminal signal peptide that allows translocation through the cytoplasmic membrane via the Sec general pathway; (2) a passenger alpha domain that provides functionality to the secreted proteins and is exposed to the cell surface or released in the extracellular medium; (3) a linker necessary for translocation of the passenger domain through the outer membrane; and (4) a C-terminal beta-domain forming a transmembrane pore (Henderson et al. 2004). The study of the mechanisms of type V secretion showed that, once exported through the cytoplasmic membrane via the mechanism of Sec secretion, the signal peptide is then cleaved by a peptidase that releases a mature protein into the periplasm. The beta-domain of the now periplasmic protein is then thought to insert spontaneously and form a transmembrane beta-barrel pore, enabling translocation of the passenger domain onto the surface of the cell. Though this was initially thought to be an autonomous process, some accessory factors such as the Omp85 protein are nevertheless involved in insertion of autotransporter proteins into the outer membrane (Voulhoux et al. 2003; Oomen et al. 2004). However, Omp85 also seems to be required for other outer membrane protein insertions and is therefore not specific to the type V secretion pathway. The passenger domain can then either undergo self autocatalytic cleavage or remain attached to the external membrane (Henderson et al. 1998). Autotransporter proteins were identified in most Gram-negative bacteria and were classified into three families according to the function carried by their passenger alpha domain: proteases such as the IgA1 protein, esterases, such as ApeE of *Salmonella typhimurium* and adhesins (Henderson et al. 1998).

3.1.1.2   Antigen 43

In 1980, B. Diderichsen identified an *E. coli* gene whose mutation affects various phenotypes associated with the surface properties of the bacteria. He showed that a change in this gene, called *flu*, prevented flocculation of bacteria at the bottom of a tube and modified the morphology of colonies on agar plates (Diderichsen 1980). The *flu* gene encodes antigen 43 (Ag43), a major outer membrane protein found in most commensal and pathogenic *E. coli*. Although *E. coli* K-12 has only one copy of *flu*, most other strains of *E. coli* have several copies of this gene.

Ag43 is a self-recognizing surface autotransporter protein that does not seem to be involved in non-specific initial adhesion to abiotic surfaces, but rather, promotes cell-to-cell adhesion (Kjaergaard et al. 2000a). While, in liquid culture, this property leads to autoaggregation and clump formation rapidly followed by bacterial sedimentation, it also facilitates bacteria-bacteria adhesion and leads to the three-dimensional development of the biofilm (Owen et al. 1996; Henderson et al. 1997a; Hasman et al. 1999; Kjaergaard et al. 2000a; Schembri et al. 2003a). When expressed in different species, Ag43 can also be used to promote mixed biofilm formation between different bacteria, for example, between *E. coli* and *Pseudomonas aeruginosa* (Kjaergaard et al. 2000a, 2000b).

Hence, Ag43 seems to play a key role in biofilm maturation on abiotic surfaces, but also in eukaryotic cells, where a recently discovered glycosylation reaction on

Ag43 could play a role, suggesting a link between expression of Ag43 and the ability of pathogenic *E. coli* to adhere to and form biofilm-like structures on epithelial cells (Anderson et al. 2003; Justice et al. 2004; Sherlock et al. 2006). Several groups have consistently demonstrated that there is a correlation between an increased level of *flu* expression and *E. coli* capacity to form a biofilm on different abiotic surfaces (Danese et al. 2000a; Kjaergaard et al. 2000a, 2000b; Beloin et al. 2006). Moreover, global gene expression studies showed that the biofilm lifestyle is often associated with increased expression of *flu* as compared to planktonic growth (Schembri et al. 2003b).

3.1.1.3   AidA and TibA Proteins

AidA adhesin from diarrhea-causing *E. coli* and TibA adhesin/invasin associated with some enterotoxigenic *E. coli* are two glycosylated surface proteins involved in bacterial adhesion to a variety of eukaryotic cells. Both AidA and TibA are autotransporter proteins sharing approximately 25% identity at the sequence level with Ag43, and they play a role in the virulence of different pathogenic *E. coli* strains. Recent studies have shown that, in addition to autoaggregation, expression of these proteins also promotes biofilm formation on abiotic surfaces (Sherlock et al. 2004, 2005). Though they are quite different with respect to size, glycosylation, and processing, these three proteins share common properties: all are self-associating proteins that cause bacterial aggregation and enhance biofilm formation. They can also interact with each other via heterologous interactions, promoting the formation of mixed bacterial aggregates. Based on these properties, Klemm and co-workers proposed that they be classified together in a subgroup termed SAAT, for self-associating autotransporters (Klemm et al. 2006).

3.1.2   Exploring the Adhesin Potential of *E. coli*

Genetic analyses have revealed the diversity of *E. coli* adhesins contributing either to colonization or biofilm maturation. Few, if any, of these adhesion factors are absolutely required for biofilm formation; instead, they can be replaced by alternative adhesion factors. A recent study demonstrated that four previously uncharacterized *E. coli* genes (*yfaL*, *yeeJ*, *ypjA*, and *ycgV*), sharing homologies with the autotransporter adhesin Ag43 and biofilm-associated proteins (Bap) (Cucarella et al. 2001; Latasa et al. 2005), lead to clear adhesion and a biofilm phenotype when expressed from a chromosomally introduced inducible promoter. Deletion of genes coding for these putative adhesins does not significantly alter the adhesion phenotype of the wild type MG1655 *E. coli* K-12 strain under laboratory conditions, indicating that these genes may be cryptic (Roux et al. 2005).

Those studies demonstrated that *E. coli* K-12 probably possesses a large and partly unexplored arsenal of surface adhesins with different binding specificities that are expressed under specific physiological conditions, possibly in response to different environmental cues. This adhesion potential is likely to be even greater in some pathogenic *E. coli* isolates, which not only often have a larger genome than *E. coli* K-12 and therefore express new types of adhesin or fimbrial structures (Buckles et al. 2004), but also carry several plasmids that can contribute to adherence to cells and abiotic surfaces (Perna et al. 2001; Dobrindt et al. 2002; Welch et al. 2002; Dudley et al. 2006).

3.2   Biofilm Matrix Polysaccharides

One of the most distinctive features that distinguishes biofilms from planktonic populations is the presence of an extracellular matrix embedding the biofilm bacteria and determining mature biofilm architecture (Sutherland 2001; Starkey et al. 2004). Along with expression of proteinaceous adhesins, production of this matrix is essential for maturation of the biofilm structure. The biofilm matrix is a complex milieu essentially composed of water (97%), but it also includes exopolysaccharide polymers, proteins, nucleic acids, lipids/phospholipids, absorbed nutrients, and metabolites (Ghannoum and O'Toole 2001).

3.2.1   Role of the Biofilm Matrix

Although the matrix is a hallmark of bacterial biofilms, its role is not fully understood. The biofilm matrix offers a constantly hydrated viscous layer protecting embedded bacteria from desiccation or from host defenses by preventing recognition of biofilm bacteria by the immune system. The matrix may also play a significant protective role as a diffusion barrier and a sink for toxic molecules (antimicrobials, hydroxyl radicals, and superoxide anions). The biofilm matrix could also inhibit wash-out of enzymes, nutrients, or even signaling molecules that could then accumulate locally and create more favorable microenvironments within the biofilm (Redfield 2002; Welch et al. 2002; Starkey et al. 2004). All these aspects of the putative roles of the matrix could contribute to development of phenotypic resistance of pathogenic *E. coli* biofilms and lead to persistent infections (Anderson et al. 2003; Justice et al. 2004).

In addition to its protective role, one of the main functions of the matrix is probably also a structural one. The adhesive properties of the matrix enable the bacteria to remain in proximity to the surface and to adhere to each other. Moreover, the interactions between polysaccharides and the other components of the matrix, such as those between cellulose and curli, may participate in three-dimensional growth of the biofilm (White et al. 2003).

Due to biofilm heterogeneity, analysis of the extracellular polymeric substance (EPS) is progressing slowly, and little is yet known about the composition of the biofilm matrix (Sutherland 2001). Several exopolysaccharides found in the *E. coli* biofilm matrix (cellulose, PGA, colanic acid) are key components of the biofilm matrix, while others such are lipopolysaccharides and capsular polysaccharides may not accumulate significantly in the matrix, but still play an important indirect role in biofilm formation. While these components may coexist in the matrix, our current knowledge seems to indicate that they are subject to very distinct regulatory pathways, the coordinated expression of which remains to be clarified.

3.2.2 Polysaccharides Secreted in the Biofilm Matrix

Secreted polysaccharides have been recognized as key elements that shape and provide structural support for the biofilm (Sutherland 2001). These polymers are very diverse and are often involved in the establishment of productive cell-to-cell contacts that contribute to the formation of biofilms at liquid-solid interfaces, pellicles at air-liquid interfaces, cell aggregates and clumps in liquid cultures, and wrinkled colony morphology on agar plates. Evidence for a structural role of some of these matrix polysaccharides is accumulating, and the regulation of production of these exopolysaccharides is now actively being investigated in different bacteria (Kirillina et al. 2004; Branda et al. 2005; Simm et al. 2005). To date, three exopolysaccharides, β-1,6-*N*-acetyl-D-glucosamine polymer (PGA), colanic acid, and cellulose, have been detected in the biofilm matrix of *E. coli* and have been shown to be important for biofilm formation.

3.2.2.1 Poly-β-1,6-*N*-acetyl-glucosamine

β-1,6-*N*-acetylglucosamine (β-1,6-GlcNAc) is a polysaccharide polymer known to participate in biofilm formation in *Staphylococcus aureus* and *Staphylococcus epidermidis*, where it contributes to their virulence (Mack et al. 1996; Rupp et al. 2001; Gotz 2002; Maira-Litran et al. 2002). β-1,6-GlcNAc, or PGA, was recently identified in *E. coli* K-12, where the expression of β-1,6-GlcNAc exopolysaccharide polymer is involved in both cell-cell adhesion and attachment to surfaces (Agladze et al. 2005). Moreover, PGA depolymerization by treatment with metaperiodate or a β-hexosaminidase from *Actinobacillus actinomycetemcomitans* (DspB), which degrade the β-1,6-GlcNAc, leads to nearly complete disruption and dispersion of the biofilm (Wang et al. 2004; Itoh et al. 2005). PGA production depends on the *pgaABCD* locus (Wang et al. 2004). The *E. coli* *pgaABCD* (or *ycdSRQP*) operon encodes proteins involved in the synthesis (the PgaC glycosyltransferase), export and localization of the PGA polymer. The *pgaABCD* operon exhibits features of a horizontally transferred locus and is present in a variety of eubacteria. Therefore, it has been proposed that β-1,6-GlcNAc serves as an adhesin that stabilizes biofilms of *E. coli* and other bacteria such as *A. actinomycetemcomitans* and *Actinobacillus pleuropneumoniae* (Kaplan et al. 2004; Wang et al. 2004).

3.2.2.2   Cellulose

Cellulose, the main component of plant cell wall, is a homopolysaccharide composed of D-glucopyranose units linked by β-1→4 glycosidic bonds. Outside of the
plant kingdom, cellulose has primarily been thought to be produced only by a few
bacterial species such as the model organism *Gluconacetobacter xylinum* (Czaja
et al. 2006). The ability of cellulose to bind fluorescent chemical dyes such as calcofluor has provided a convenient screen for cellulose-producing bacteria, showing
that cellulose production is a widespread phenomenon in Enterobacteriaceae,
including *Salmonella enterica* serovar Typhimurium, *S. enterica* subsp. *Enterica*
serovar Enteritidis, and commensal and pathogenic strains of *E. coli*, *Citrobacter
spp.* and *Enterobacter spp.* (Zogaj et al. 2001; Solano et al. 2002; Romling et al.
2003; Zogaj et al. 2003; Romling 2005; Da Re and Ghigo 2006; Uhlich et al. 2006).
In these bacteria, cellulose production is clearly associated with the ability to form
a rigid biofilm at the air-liquid interface; however, these characteristics vary between
strains and serovars and are highly dependent on environmental conditions.

Genetic analysis performed in *Salmonella* serovar Typhimurium and *Salmonella*
serovar Enteritidis showed that cellulose synthesis genes are organized as two
divergently transcribed operons, *bcsABZC* and *bcsEFG*, which are constitutively
expressed and composed of genes sharing homologies with genes of the bacterial
cellulose operon of *G. xylinum* (Gerstel et al. 2003; Romling 2005).

Although these genes are present in most enterobacterial genomes, including
Salmonella, E. coli, Shigella, Enterobacter, and *Citrobacter* (Zogaj et al. 2003),
little is known about the function and localization of the bacterial cellulosome.
BcsA is a cytoplasmic membrane protein whose cellulose synthase activity is allosterically controlled upon binding of a small molecule called cyclic-di-GMP
(c-di-GMP), a ubiquitous second messenger produced and degraded by diguanylate
cyclase and phosphodiesterases, respectively. In *Gluconacetobacter xylinus*, it has
been suggested that c-di-GMP binds to BcsB, promoting an allosteric change in the
protein conformation that leads to its activation (Mayer et al. 1991). Recently, a
PilZ domain believed to be part of the c-di-GMP binding protein has been identified
in several bacterial cellulose synthases, including BscA, although direct evidence
for c-di-GMP binding is still missing (Amikam and Galperin 2006). c-di-GMP is
now known to antagonistically control the motility and virulence of single, planktonic cells, on the one hand, and cell adhesion and persistence of multicellular
communities on the other (Jenal and Malone 2006; Romling and Amikam 2006).

Genetic analyses performed mostly in *S. Typhimurium* revealed that the combined
and coregulated syntheses of cellulose and curli fimbriae lead to a distinctive phenotype on Congo red agar plates, the red dry and rough (rdar) morphotype (Zogaj et al.
2001; Solano et al. 2002; Romling et al. 2003; Da Re and Ghigo 2006). While most
of what is known of the regulation of cellulose production has been learned from
Salmonella, cellulose has also been found in *E. coli*, where cellulose synthesis is
correlated with biofilm formation and expression of multicellular behavior (rdar
morphotype), and where treatment with cellulase totally disperses existing biofilms
(Zogaj et al. 2001, 2003; Romling 2002; Da Re and Ghigo 2006).

3.2.2.3   Colanic Acid

Colanic acid is a negatively charged polymer of glucose, galactose, fucose, and glucuronic acid that forms a protective capsule around the bacterial cell under specific growth and environmental conditions (for example, colanic acid is not produced in rich medium at 37°C). Colanic acid has a structure and assembly pathway very similar to that of the group I capsule and is therefore often included in that category. However, in contrast to most capsular types, a significant portion of the colanic acid produced is released into the extracellular medium.

Colanic acid synthesis involves 19 genes located in the same cluster, named *wca* (formerly known as *cps*) (Stevenson et al. 1996). It is induced by the three-component system RcsC/RcsD/RcsB and requires an auxiliary positive transcription regulator RcsA (Majdalani and Gottesman 2005). Although the signal for sensor kinase RcsC remains uncharacterized, RcsC seems to respond to complex cues such as desiccation, osmotic stress, the level of periplasmic glucans, and growth on a solid surface (Ophir and Gutnick 1994; Sledjeski and Gottesman 1996; Ferrieres and Clarke 2003). A recent observation also indicates that colanic acid is induced by near-lethal levels of a subset of β-lactam antibiotics that may exacerbate the formation and persistence of a biofilm (Sailer et al. 2003). Although colanic acid has been reported to impair initial bacterial attachment, its synthesis is consistently upregulated within biofilms, and its production plays a role in the development of the mature biofilm architecture (Prigent-Combaret and Lejeune 1999; Prigent-Combaret et al. 1999; Danese et al. 2000b; Hanna et al. 2003). Interestingly, several groups have reported that expression of the colanic acid capsule may also have an inhibitory effect upon the biofilm ability of *E. coli* strains by masking autotransporter adhesins such as Ag43 and AidA (Hanna et al. 2003; Schembri et al. 2004).

3.2.3   Cell Surface Polysaccharides

Cell-surface glycoconjugates play a critical role in interactions between bacteria and their immediate environment. Besides released polysaccharides that have been identified as part of the biofilm matrix, surface polysaccharides can also contribute to the biofilm phenotype. Most *E. coli* isolates produce a complex layer of serotype-specific surface polysaccharides: the lipopolysaccharide (LPS) O antigen and capsular polysaccharide K antigen. Variations in the structure of these polysaccharides give rise to 170 different O antigens and 80 K antigens that enable typing of most enterobacteria.

3.2.3.1   Lipopolysaccharides

The lipopolysaccharide (LPSs), also known as endotoxin, is a glycolipidic polymer that constitutes the main component of the outer leaflet of the outer membrane of Gram-negative bacterium. Constitutively expressed, LPS consists of three parts:

lipid A, which is the toxic component and to which the core region is attached, which can be divided into an inner and an outer part; and finally the O-antigen polysaccharide, which is specific to each of the 170 E. coli serogroups.

The sugar residues in lipid A and the core region are decorated to varying extents with phosphate groups and phosphodiester-linked derivatives, ensuring micro-heterogeneity in each strain. The lipid A part is highly conserved in E. coli, while the core contains six different basic structures, denoted R1-R6. The O-polysaccha-ride is linked to a sugar in the outer core. The O-antigen, absent in rough strains such as E. coli K-12, usually consists of 10-25 repeating units containing two to seven sugar residues. Thus, the molecular mass of the LPS present in smooth strains will be up to 25 kDa. Finally, in some serotypes, the core can be bound to group 1 capsule, forming K_{LPS} (Raetz 1996).

More than 50 genes are required to synthesize LPS and assemble it at the cell surface. Some of these genes are clustered in large operons or are isolated on the E. coli chromosome. Mutations affecting LPS synthesis have been shown to affect E. coli ability to adhere to abiotic surfaces (Genevaux et al. 1999), suggesting a role of LPS in adhesion processes. Studies investigating E. coli mature biofilm formation also reported that LPS mutation could lead to a significant decrease in biofilm capacity. Inactivation of waaG, coding for a LPS core glycosyltransferase that resulted in a truncated LPS core structure and a deep rough phenotype, com-pletely abolished biofilm formation of uropathogenic strain 536 without affecting its growth rate in liquid culture, suggesting that an intact LPS core is a major factor in adhesion to abiotic surfaces in E. coli strain 536 (Landini and Zehnder 2002; Beloin et al. 2006). However, since alteration of LPS synthesis can also impair Type 1 pili and colanic acid expression as well as bacterial motility, the phenotype of LPS mutants could still be attributed to indirect effects.

In contrast, like capsule masks the function of short membrane adhesins, it has recently been reported that the reduction in LPS expression caused by an rfaH mutation could unmask E. coli adhesins and therefore allow initial adherence and/ or biofilm formation (Beloin et al. 2006).

These results suggest two distinct mechanisms by which LPS either promotes or inhibits biofilm formation, mainly by interacting with cell-surface-exposed adhesion factors.

3.2.3.2   Capsules

E. coli capsules are surface-enveloping structures comprising high-molecular-weight capsular polysaccharides that are firmly attached to the cell (see, however, the dis-cussion in this section below). They are well-established virulence factors, often acting by protecting the cell from opsonophagocytosis and complement-mediated killing (Whitfield 2006). The 80 different capsular serotypes in E. coli were origi-nally divided into more than 80 groups based on serological properties. Despite the diversity of bacterial capsule glycoconjugates and the complexity of their synthesis and assembly processes, later revisions classified capsules into four groups. E. coli group 1 and 4 capsules share a common assembly system, and this is fundamentally different from that used for group 2 and 3 capsules (Whitfield and Roberts 1999; Whitfield 2006).

As seen for colanic acid and LPS, the *E. coli* capsule has also been shown to play an indirect role in biofilms by shielding of bacterial surface adhesin (Schembri et al. 2004). While capsular polysaccharides are linked to the cell surface of the bacterium via covalent attachments, capsule can be released into the growth medium as a consequence of the instability of phosphodiester linkage between the polysaccharide and the phospholipid membrane anchor (Roberts 1996; Whitfield 2006). Recently, group II capsular polysaccharides were shown to be significantly released into the culture supernatant and to display antiadhesion activities toward both Gram-positive and Gram-negative bacteria, therefore antagonizing biofilm formation by a mechanism distinct from steric hindrance of surface adhesin (Valle et al. 2006). Capsule-mediated biofilm inhibition is widespread in extraintestinal *E. coli*, suggesting that the anti-biofilm property of group II capsular polysaccharides could also play a role in the biology of these pathogens. Group II capsule may contribute to competitive interactions (bacterial interference) within bacterial communities, or to modulating *E. coli*'s own adhesion to surfaces encountered during the intestinal or urinary tract colonization process. Analyses showed that group II capsular polysaccharides affect biofilm formation by weakening cell-surface contacts (initial adhesion), but also by reducing cell-cell interactions (biofilm maturation). Interestingly, direct treatment of abiotic surfaces with group II capsular polysaccharides drastically reduces both initial adhesion and biofilm development by important nosocomial pathogens, which may be used in the design of new anti-biofilm strategies (Valle et al. 2006).

We have seen that *E. coli* biofilm initiation and maturation can involve many different factors. None of them, however, are strictly required. Indeed, *E. coli*'s ability to form a biofilm depends considerably on environmental conditions, and even well-demonstrated adhesion factors can be replaced by others. For instance, expression of conjugative pili totally overcomes the need for curli, type 1 pili of flagellar expression (Reisner et al. 2003). These results not only suggest that many different pathways can be used during *E. coli* biofilm formation, but also that regulatory mechanisms could coordinate the biofilm adhesion and maturation processes.

4   Regulatory Events During Biofilm Development

4.1   Predisposition to Surface Adhesion

4.1.1   Phase Variation and Coordinated Expression of Surface Components

In bacteria, the expression of many surface components is governed by a specific regulation mode called phase variation. This process induces the differential expression of one or several genes and leads to the emergence of two subpopulations within a clonal population either expressing (ON situation) or not expressing (OFF situation) factor(s) suggested to cause phase variation. In addition to its known role in protecting bacteria from the immune system, this mechanism of phase variation is now recognized to modulate exposure of cell surface components and especially adhesin molecules (van der Woude 2006). In *E. coli*, two major adhesins implicated in biofilm formation are subjected to this phase variation mechanism: type 1 fimbriae and the autotransported Ag43 adhesin.

The promoter for the fimbrial subunit gene, *fimA*, lies within a short segment of invertible DNA known as the *fim* switch (*fimS*), and the orientation of the switch in the chromosome determines whether *fimA* is transcribed or not (Abraham et al. 1985). Inversion is catalyzed by two site-specific recombinases, the FimB and FimE proteins. The FimB protein inverts the switch in either direction, while FimE inverts it predominantly to the OFF orientation. When FimB and FimE are coexpressed, FimE activity dominates and the switch turns to the OFF phase (Klemm 1986; Dorman and Higgins 1987; Eisenstein et al. 1987; Gally et al. 1996; Blomfield et al. 1997). This dominance can be environmentally modulated, notably via DNA supercoiling modulation that modifies the formation and stabilization of a recombination-proficient protein nucleocomplex encompassing several nucleoid-associated proteins such as IHF, H-NS, and Lrp (reviewed in van der Woude and Baumler 2004). Other regulators such as the LrhA protein, a regulator of the LysR family previously described as a repressor of flagellar motility (Lehnen et al. 2002), also influence phase variation of type 1 fimbriae. LrhA appeared to be a repressor of type 1 fimbriae both in K-12 and UPEC strains, mainly via its activation of the *fimE* recombinase gene (Blumer et al. 2005). Consequently, LrhA acts as a repressor of the initial phase of biofilm formation in *E. coli* (Blumer et al. 2005). Recently, *fimB*- and *fimE*-independent *fimS* phase variation of type 1 fimbriae has been identified both in meningitis-causing *E. coli* K1, by tyrosine site-specific recombinase HbiF (Xie et al. 2006), and in uropathogenic *E. coli* CFT073 by two recombinases encoded by *ipuA* and *ibpA* (Bryan et al. 2006).

The expression of Ag43 is also phase-variable and represents a classical example of bacterial bi-stability. The shift from Ag43$^+$ to Ag43$^-$ cells is governed by a mechanism involving the concerted action of both the Dam GATC DNA-methylating enzyme deoxyadenosine methylase (activation) and the transcriptional regulator OxyR (repression) (Owen et al. 1996; Henderson et al. 1997b; Wallecha et al. 2002). Three GATC sites are present in the promoter region of the *flu* gene, two of which overlap with an OxyR binding site centered at position +37 from the +1 of transcription. In a wild type situation, each bacterium is either in an Ag43 OFF situation, if, after DNA replication, the OxyR protein manages to bind to its consensus site before DNA methylation, thus stopping RNA polymerase progression, or in an Ag43 ON situation, if DNA methylation occurs before OxyR can bind to DNA. This ON or OFF state of each bacterium is reflected by, respectively, the presence or absence of Ag43 at the cell surface.

The studies presented above clearly indicate that complex regulation mechanisms are used by bacteria to modulate expression of different factors required for biofilm formation, especially when these factors are required at different times in biofilm development. For example, type 1 fimbriae are important in the initial steps of bacteria-to-surface interactions, whereas Ag43 is required later to promote bacteria-to-bacteria interactions. In addition to the individual mechanisms of phase variation, several mechanisms of coordinated regulation between different surface components involved in adhesion have been reported (Holden and Gally 2004). Among coordinated regulation implicating different adhesins, flagella, and capsules, the best characterized is the coordination between type 1 fimbriae and Ag43 production. Schembri and co-workers first showed that overexpression of type 1 fimbriae (and also of P or F1C fimbriae) abrogates exposure of Ag43 to cell surfaces (Schembri and Klemm 2001). This effect is mediated by a regulatory effect at the Ag43-encoding gene, *flu*, whose quantity of RNA transcript is enhanced 20-fold when genes encoding type 1 fimbriae are absent (Schembri et al. 2002). Fimbriation does not affect Ag43 production in an *oxyR* background in which the *flu* promoter is in an ON situation (Schembri and Klemm 2001). Until recently, it was presumed that *flu* was only repressed by the reduced form of OxyR (Henderson et al. 1999; Haagmans and van der Woude 2000). Two results suggest that a modification in the redox status of OxyR could explain the effect of type 1 fimbriae expression on Ag43 production: first, the addition of the reducing agent DTT counteracts the effect of deletion of type 1 fimbriae; second, overproduction of flagella, which do not contain any disulfide bonds, has no effect on Ag43 exposure at the cell surface (Schembri and Klemm 2001). This suggests that expression of organelles containing disulfide bonds, such as type 1, P or F1C fimbriae, could affect cellular thiodisulfide status and thus modify the redox status of OxyR and the expression of Ag43. However, this hypothesis is not consistent with recent indications that phase variation in Ag43 is independent of the oxidation status of OxyR (Wallecha et al. 2003). However, in addition to its role in biofilm formation, Ag43-mediated autoaggregation seems to protect cells from oxidizing agents (Schembri et al. 2003a). Moreover, independently of Ag43 expression, the presence of fimbriae on the cell surface seems to abrogate the intercellular Ag43-Ag43 interaction that is required for autoaggregation to occur (Hasman et al. 1999). The reciprocal is not true, and Ag43 or OxyR status does not appear to influence fimbriae expression (Hasman et al. 1999). Hence, the correlation between Ag43 expression, type 1 pili, and OxyR status remains to be clarified.

The presence of type 1 pili is not the only extracellular component to interfere with Ag43 activity. A recent study demonstrated that the presence of capsules such as K1 or K5 capsules could block Ag43 functionality (Schembri et al. 2004). Whereas type 1 pili production interferes with both Ag43 expression and interaction of Ag43 molecules, capsule production appears solely to sterically shield the Ag43-Ag43 interaction. As a consequence, encapsulated cells expressing both capsule and Ag43 are impaired in biofilm formation on a polystyrene abiotic surface compared with nonencapsulated cells.

A recent study by Ulett and co-workers demonstrated that Ag43-mediated auto-aggregation impaired motility (Ulett et al. 2006), thus reinforcing the idea that *E. coli* has developed specific mechanisms to coordinately express different surface components required at different stages of biofilm formation.

4.1.2 Fertility Inhibition System

Most of the genes required for conjugative transfer of F plasmid DNA are encoded by the 33-kb transfer (tra) operon of F-like conjugative plasmids (Frost et al. 1994). Transcription of the *tra* operon is positively regulated by the TraJ transcriptional activator which, in turn, is negatively regulated by the FinOP fertility inhibition system. The FinOP system consists of an antisense RNA, FinP, and a 21.2-kDa protein, FinO, which together inhibit TraJ expression. Except for the F-plasmid itself, whose *finO* gene is interrupted by an IS3 insertion element, in F-like plasmids the regulatory activity of FinP depends upon the action of the plasmid-encoded protein, FinO (Yoshioka et al. 1987; van Biesen and Frost 1992). However, this fertility inhibition mechanism is not completely controlled, thus leaving in a planktonic population of F-like-bearing bacteria a low proportion of bacteria expressing conjugation function. These piliated bacteria are therefore prompted to easily integrate into a preformed biofilm and to transmit their conjugative plasmid to the recipient population, thus promoting maturation of the biofilm. Hence, this fertility inhibition mechanism may help bacteria to switch from a planktonic to a sessile mode of growth.

4.2 Regulatory Events upon Initial Interaction with Surfaces

Environmental conditions existing on immerged surfaces are different from those found in the surrounding medium. For example, the formation of a conditioning film on the surface modifies nutriment concentrations as well as other physico-chemical factors such as pH, oxygenation, and osmolarity. How bacteria know that they are on a surface is still poorly understood. Among possible inducing cues are direct physical contact, perception of extracellular signals, or gradients and bacteria-to-bacteria interactions (Harshey and Toguchi 1996) that allow the bacteria to produce productive cell-surface interactions and therefore lead to the formation of stable biofilm. *E. coli* has proven to be a valuable model for investigating some aspects of these questions.

4.2.1 Sensing the Surface: Regulation of Genes upon Direct Contact of Bacteria with Surfaces

4.2.1.1 The *cpx* System Senses the Surface and Neighboring Bacteria

The two-component regulatory *cpxRA* system is composed of the sensor membrane protein CpxA and of the cytoplasmic regulator CpxR. This system is known to respond to envelope stresses such as overproduction and misfolding of membrane proteins and elevated pH (Raivio and Silhavy 2001) and notably to activate the expression of genes encoding protein chaperones or proteases such as *dsbA* and *degP*. Activation of the *cpx* pathway therefore participates in the adaptation of bacteria to environmental stresses. Early adhesion of *E. coli* cells to abiotic surfaces, probably by inducing membrane perturbation, has been shown to activate the *cpx* system through a process called surface sensing (Otto and Silhavy 2002). This abiotic surface contact induction depends on CpxR, the cognate sensor of the system, but also on NlpE, an outer membrane lipoprotein previously shown to induce the *cpx* system when overproduced (Snyder et al. 1995). A mutation in the *cpxR* gene alters cell-surface interactions (Otto and Silhavy 2002). This suggests that the *cpx* system is required for an adaptive response necessary for stabilizing contact of attached cells with the surface. Among responses of the *cpx* system to surface contact are modifications in cell-surface composition through regulation of cell-surface protein expression (i.e., OmpC) (De Wulf et al. 2002) and modulation of flagellar gene expression, since *cpxR* is a repressor of motility and chemotaxis genes (De Wulf et al. 2002), two pathways that must be switched off to optimize initial attachment and to avoid leaving the recently colonized surface. Flagellar encoding genes are consistently repressed early after bacteria reach the surface (Prigent-Combaret and Lejeune 1999; Ren et al. 2004a), and overexpression of flagellar genes, either directly or indirectly via disruption of flagellar gene repressors such as *hha / ybaJ* genes, reduces biofilm formation by *E. coli* (Tenorio et al. 2003; Barrios et al. 2006). In addition to sensing the surfaces, the *cpx* system may also sense neighboring bacteria. Components of the *cpx* system are indeed induced in mature *E. coli* biofilms, where most of the bacteria are in contact with one another rather than with the surface (Beloin et al. 2004). Induction of the *cpx* pathway may therefore play an important role in maturation of biofilm by affecting both initial adhesion between bacteria and the surface and subsequent interactions between the bacteria themselves (Otto and Silhavy 2002; Beloin et al. 2004). Since the *cpx* pathway, when activated, also impairs biogenesis of proteinaceous adhering appendages such as curli or F-pili, Dorel and co-workers recently proposed that the role of this system in each newly divided cell in a biofilm might well be to turn down expression of such energy-costly adhering molecules right after productive interactions are secured between the bacteria and surfaces, and between bacteria themselves (Dorel et al. 2006).

4.2.1.2 The *rcs* System Mediates Biofilm Maturation

The *rcs* two-component regulatory pathway is composed of membrane-associated proteins RcsC and RcsD and the cytoplasmic response regulator RcsB. Rcs phosphorelay in *E. coli* was originally described as being a regulator of *cps* operon expression, encoding proteins required for production of capsular polysaccharide colanic acid (for review see Majdalani and Gottesman 2005; Huang et al. 2006). Although the exact signals sensed by RcsC remain incompletely characterized, several studies have shown that this sensor kinase responds to complex signals, including desiccation and changes in osmolarity (Ophir and Gutnick 1994; Sledjeski and Gottesman 1996). More recently, the RcsC sensor kinase has been shown to respond to growth on a solid surface by sensing membrane perturbations and therefore to be required for normal biofilm development in *E. coli* (Ferrieres and Clarke 2003). Ferrières and co-workers demonstrated that surface induction of colanic acid genes observed in *E. coli* (Prigent-Combaret et al. 1999, 2000) depended on the *rcs* pathway, which also represses expression of surface appendages such as curli, fimbrial proteins, and Ag43 (Ferrieres and Clarke 2003). Considering that flagellar gene expression in *E. coli* is repressed by the *rcs* two-component signaling pathway as well (Francez-Charlot et al. 2003), induction of the *rcs* system is likely to cause remodeling of the bacterial surface. Moreover, the fact that initial attachment of cells is not decreased in an *rscC* or *rcsB* mutant (Danese et al. 2000b) suggests that this regulon plays a role in a later step, i.e., biofilm maturation, where colanic acid production, in particular, is necessary for proper architecture of the biofilm, whereas flagella need to be repressed.

Recent studies on modifications in the composition of outer membrane proteins during fimbriae-mediated adhesion also demonstrated that such cell-surface interactions indeed induce remodeling of the bacterial envelope (Otto et al. 2001). Among proteins whose the quantity is reduced upon attachment to abiotic surfaces is the OmpX protein (see Sect. 2.3.1) (Otto and Hermansson 2004).

Consequently, it appears that contact with a solid surface induces an adaptive response in *E. coli* cells leading to stable adhesion. This surface-sensing mechanism appears to involve several pathways, whose complex overlapping and interplay need to be elucidated.

4.2.2   Sensing Microenvironments at an Abiotic Surface: The EnvZ/OmpR Two-Component Pathway Senses Surface Osmolarity

As stated earlier in this chapter, surfaces tend to adsorb organic molecules and to acquire a conditioning film that locally modifies osmolarity compared with the surrounding medium. Bacteria entering into contact with surfaces will therefore face a favorable growth environment, especially if the surrounding medium contains low nutrient concentrations. It has been proposed that this growth advantage may be a significant selective force driving growth on abiotic surfaces (Costerton et al. 1987). The EnvZ/OmpR two-component pathway is known to respond to external osmolarity by regulating transcription of the *ompF* and *ompC* porin-encoding genes; moreover, it increases surface adhesion in response to moderate increases in osmolarity (Prigent-Combaret et al. 2001). This EnvZ/OmpR activity may thus favor adhesion in zones supporting high metabolic activity, both by repressing flagellar gene expression (Shin and Park 1995; Oshima et al. 2002) and by activating curli expression upon initial adhesion. The OmpR effect on curli expression is mediated by activation of the regulator CsgD by phosphorylated OmpR. Interestingly, *csgD* encodes a key regulator of the FixJ family that positively regulates production of curli and cellulose in *E. coli* and *Salmonella*, and these two factors promote biofilm formation (see Zogaj et al. 2001; Solano et al. 2002). Recently, modulation of *csgD* expression by modification of osmolarity was shown to be essentially dependent on the interplay between the two-component systems EnvZ/OmpR and CpxA/CpxR and the histone-like protein H-NS (Jubelin et al. 2005).

A recent study identified the gene *bolA* to be a potential regulator of biofilm formation. *bolA* has been described as being negatively regulated by OmpR (Yamamoto et al. 2000). It is involved in *E. coli* morphogenesis and was shown, when overexpressed, to cause round-cell morphology (Aldea et al. 1988) and to increase the ratio of OmpC/OmpF porins, leading to a subsequent decrease in cell membrane porin-mediated permeability (Freire et al. 2006). Furthermore, the morphological effect of *bolA* overexpression depends on an active *ftsZ* gene product (Aldea et al. 1988). *bolA* has been described as a regulator of cell-wall biosynthetic and biosynthesis enzymes (Santos et al. 2002). In minimal medium, mutation of *bolA* slightly reduces *E. coli* biofilm formation in 96-well polystyrene microtiter plates, whereas overexpression of *bolA* strongly induces biofilm formation (Vieira et al. 2004). These results hint at a physiological connection between cell morphology, cell division, and biofilm formation.

4.2.3   A Role for Small Molecules in the Switch from Planktonic to Sessile Life

4.2.3.1   Cyclic di-GMP

Cyclic di-GMP (c-di-GMP) is a second messenger whose synthesis is performed by diguanylate cyclases, proteins bearing GGDEF motifs, whereas its degradation is ensured by proteins containing EAL domains and with phosphodiesterase activity (Romling et al. 2005). *E. coli*, like most bacteria, contains multiple diguanylate cyclases and phosphodiesterases with as many as 19 GGDEF proteins and 13 EAL proteins detected in its genome (Romling et al. 2005). It is likely, therefore, that localization of these proteins in the bacterial cell is a key factor enabling fine-tuning of c-di-GMP concentrations in different areas of the cell. c-di-GMP appears to act posttranscriptionally by activation of specific proteins through direct interactions, as has been shown for the protein complex containing the cellulose synthase BcsA of *G. xylinus* (Amikam and Benziman 1989). A domain called PilZ was recently bioinformatically identified in several GGDEF and/or EAL proteins, and notably in the BcsA and YcgR proteins in *E. coli*, as being a possible domain of interaction with c-di-GMP (Amikam and Galperin 2006). Among numerous cellular functions linked to c-di-GMP is biofilm formation. In *E. coli*, as in other bacteria, a correlation has been demonstrated between a high c-di-GMP concentration and biofilm formation vs a low c-di-GMP concentration and motility (Simm et al. 2004). Cellulose synthesis and rotation of flagella are the main functions responsible for such correlations. Ko and Park showed that expression of the EAL domain protein YhjH is required to stimulate swimming motility in a *hns* mutant of *E. coli*, whereas swimming motility was stimulated by deletion of *ycgR* that encodes a PilZ domain protein (Ko and Park 2000). One hypothesis is that a YcgR-c-di-GMP complex inhibits the motor function of the flagella, whereas degradation of cylic-di-GMP by YhjH relieves inhibition of motility by YcgR (Romling and Amikam 2006). Extrapolating from the work done in *Salmonella* on regulation of cellulose production (Zogaj et al. 2001, 2003; Romling 2002; Solano et al. 2002), activation by the regulator CsgD of expression of *adrA*, a diguanylate cyclase-encoding gene, was thought to be the unique pathway employed by *E. coli* to activate cellulose production. However, recent studies demonstrate that in *E. coli* the mechanism of regulation of cellulose synthesis is more complex than previously thought. Brombacher and co-workers showed that in *E. coli* K-12 MG1655, in addition to *adrA*, CsgD also activated *yoaD*, a putative phosphodiesterase-encoding gene whose mutation consistently activates cellulose production. Consequently, in this strain, cellulose production might be regulated by subtle modifications in c-di-GMP concentrations resulting from AdrA- and YoaD *csgD*-dependent activation (Brombacher et al. 2006). A recent study also demonstrated that, in addition to being regulated by RpoS through its action on MlrA (whose mechanism of function is unknown) (Gerstel et al. 2003), two RpoS-regulated GGDEF and/or EAL proteins, YdaM and YciR, inversely regulate *csgD* expression, thus influencing both curli and cellulose production (Weber et al. 2006). The expression of genes encoding these two proteins was also inversely regulated by H-NS, again increasing the complexity of the mechanisms governing *csgD* regulation (Weber et al. 2006). Moreover, Da Re and Ghigo showed that, alongside a *csgD*- and *adrA*-dependent mechanism of cellulose synthesis identified in the enteroaggregative strain 55989, a *csgD*-independent pathway leads to cellulose synthesis in the commensal 1094 and probiotic Nissle 1917 (DSM6601) strains, whereas *csgD* activation of curli production is maintained (Da Re and Ghigo 2006). This pathway, instead of using *adrA*, passes through expression of another putative diguanylate cyclase called YedQ (Da Re and Ghigo 2006). Furthermore, in another *E. coli* commensal isolate 1125, cellulose production appeared to be independent of *csgD*, *adrA* and *yedQ*, thus suggesting that an as yet uncharacterized GGDEF protein distinct from both AdrA and YedQ may have acquired this cellulose regulatory function.

4.2.3.2 Acetyl Phosphate

Among small molecules that coordinate gene expression in response to environmental stimuli, two molecules have recently been identified as signals linking nutrient status to biofilm formation in *E. coli*: acetyl phosphate (AcP), which accumulates intracellularly in the presence of an abundant carbon source and/or a low oxygen concentration in the medium (Wolfe et al. 2003), and ppGpp, the molecule of the stringent response that accumulates upon nutrient starvation conditions (Balzer and McLean 2002) (see Sect. 4.2.3 below). Local depletion of oxygen occurring when bacteria reach a surface is hypothesized to be the signal causing intracellular AcP levels to rise. An increase in AcP levels correlates notably with an elevated level of type 1 pili and colanic acid gene expression, and with a decreased level of flagellar gene expression, favoring maturation of the biofilm

(Pruss and Wolfe 1994; Wolfe et al. 2003). One hypothesis is that AcP could influence biofilm formation by acting as a phosphodonor for response regulators such as FimZ, OmpR, and RcsB, known to control biofilm-associated genes. Fredericks and co-workers determined that AcP indeed exerts its effect on capsular and flagellar gene expression completely or partially by means of the Rcs phosphorelay (Fredericks et al. 2006). In addition, as in the surface-sensing mechanisms described in Sect. 4.2.1above, modifications in the intracellular AcP pool could also lead to bacterial surface modifications. Indeed, genes that respond negatively to high AcP levels include these encoding outer membrane porins (*ompF, ompC*) and other proteins associated with, or predicted to be associated with, the envelope (*rbsB, b1996, yqiH, glpD, rfbX, rbsD*) (Wolfe et al. 2003).

4.2.3.3 The Alarmone ppGpp

The effect of ppGpp on biofilm formation was assessed by comparing wild type *E. coli* K-12 MG1655 with an isogenic *relA spoT* mutant that does not produce ppGpp (Balzer and McLean 2002). Clearly, in nutrient-limited conditions, the absence of ppGpp production caused a decrease in biofilm cell density, therefore signifying that the stringent response is necessary for normal development of *E. coli* biofilm. However, in rich LB medium where the ability to synthesize ppGpp is also crucial, the reverse was observed, i.e., the absence of ppGpp production caused an increase in biofilm cell density (Balzer and McLean 2002). A recent work by Aberg and co-workers seems to contradict this result, since they showed that in LB broth, the absence of ppGpp reduced biofilm formation in uropathogenic and K-12 *E. coli* strains (Aberg et al. 2006). Furthermore, Aberg and co-workers showed that ppGpp promoted biofilm formation via increased type 1 fimbriae expression due to activation of expression of the *fimB* recombinase gene. This ppGpp effect is, in fact, independent of RpoS, H-NS, and NanR, three previously described regulators of *fimB* expression (Aberg et al. 2006). The authors proposed that induction of intracellular ppGpp levels can be considered as an alert signal that promotes expression of genes, including those for type 1 fimbriae, which could both adapt cells to slow growth and increase the probability of surviving stressful environments, two situations encountered by bacteria within biofilm. It can also be envisaged that ppGpp signaling in planktonic bacteria growing in nutrient-depleted media, and via an increase in type 1 fimbriae production, may favor colonization of surfaces in which conditions are more advantageous.

4.2.3.4 *N*-Acetyl-Glucosamine and *N*-Acetylglucosamine-6-P

E. coli can use *N*-acetylglucosamine-6-P (GlcNAc-6P) as a carbon source and/or as a precursor for peptidoglycan and lipopolysaccharide biosynthesis. GlcNAc-6-P is obtained either by phosphorylation of *N*-acetyl-glucosamine (GlcNAc) acquired from the environment or by de novo synthesis. When high levels of

GlcNAc-6P accumulate in the cell, the repression exerted by the protein NagC on *nag* genes (*nagABCDE*) is relieved and these genes then support utilization of GlcNAc as a carbon source. Barhnart and co-workers found that curli gene expression is repressed in response to artificially elevated levels of intracellular GlcNAc-6-P created by deletion of *nagC* or *nagA*, a deacetylase that converts GlcNAc-6-P to glucosamine-6-phosphate (GlcN-6P) (Barnhart et al. 2006). Production of another adhesive fiber, type 1 fimbriae, was previously shown to be activated by GlcNAc-6-P (Sohanpal et al. 2004). This later effect appeared to be NagC-dependent through activation of *fimB* recombinase gene expression that switches the expression of type 1 fimbriae from OFF to ON. In addition to its role as a carbon source, GlcNAc-6-P might well be a signal for the cell to produce adhesive molecules. Interestingly, *E. coli* produces an extracellular polysaccharide composed of the GlcNAc subunit (the PGA) that promotes attachment to solid surfaces, cell-cell adherence, and stabilization of biofilm structure (Wang et al. 2004; Itoh et al. 2005). However, a possible link between PGA production and regulation of cell surface molecules promoting adhesion remains to be elucidated.

4.3   Genetic Regulation Within E. coli *Biofilms*

4.3.1   Gene Expression Within Biofilm: Cause or Consequence?

Planktonic and surface-attached growth modes are simple to distinguish phenotypically. These two lifestyles are thought to require or involve a different gene expression setup, leading to the expression of some of the phenotypic characteristics of the biofilm phenotype. The existence of changes in gene expression within biofilm compared with a non-biofilm mode of growth was recognized early on. Gene fusion studies suggest that the expression of up to 38% of the *E. coli* genome is affected by biofilm formation (Prigent-Combaret et al. 1999). Such evidence for differential gene expression within bacterial biofilms has been provided by recent studies using DNA arrays. These studies indicate that, in fact, a lower proportion of the *E. coli* genome (5%-12%) is subject to differential expression in sessile vs planktonic life (Schembri et al. 2003b; Beloin et al. 2004; Ren et al. 2004a). While these studies (along with others conducted in other microorganisms) suggest the existence of a common pattern of gene expression in *E. coli* biofilms and have indeed identified some genes required for biofilm formation, detailed comparison of genes discovered reveal only a very modest overlap between the different studies. This underscores the difficulty in comparing analyses carried out with different strains, different experimental setups (biofilm device, medium, presence or absence of flow) and different time scales (i.e., with different *E. coli* biofilms), but it also raises the possibility that global analyses are not really appropriate for dealing with the extreme complexity in time and space that resides within a biofilm (Beloin and Ghigo 2005).

Moreover, it is important to distinguish between factors required for biofilm formation and factors induced by particular biofilm conditions. Indeed, although biofilm formation requires the expression of specific factors (see Sect. 4.3.1), major modifications in gene expression patterns within biofilms could also be induced by the drastic environmental changes occurring during biofilm formation. Biochemical and genetic evidence support the hypothesis that bacteria face different conditions within a biofilm as compared with planktonic growth (Huang et al. 1998; Prigent-Combaret et al. 1999, 2001). Indeed, biofilm bacteria are likely to be subjected to progressive microaerobic conditions, increased osmotic pressure, pH variation and decreased nutrient accessibility. These biofilm conditions often have strong similarities with conditions that prevail in stationary phase (planktonic) cultures, and when the stationary phase character of the bacterial lifestyle within biofilm has been investigated, it has generally been shown that a significant part of the *E. coli* K-12 biofilm response involves stationary-phase-induced genes (Schembri et al. 2003b; Beloin et al. 2004). Since many changes observed in biofilm gene expression are potentially a consequence rather than a cause of biofilm formation, the question as to whether these genes encode functions that are required by, or that are induced by, biofilm conditions remains to be determined.

4.3.2 Regulation of Biofilm Formation by Central Carbon Flux

Catabolite repression has recently been recognized to be a regulatory signal controlling *E. coli* biofilm formation (Jackson et al. 2002a). The presence of 0.2% glucose in rich medium appears to decrease biofilm biomass. However, this effect is more pronounced when glucose is added during the initial steps of biofilm formation rather than in later stages of biofilm maturation, suggesting that catabolite repression preferentially affects components required in the early stages of bacterial adhesion (Jackson et al. 2002a). Glucose repression is partially mediated by the cAMP receptor protein CRP. Indeed, a *crp* mutant displays decreased biofilm formation abilities compared with the wild type (Jackson et al. 2002a). Recently, Domka and co-workers identified two biofilm-induced genes, *yceP* and *yliH* (renamed *bssS* and *bssR* for regulator of biofilm through signal secretion) (Schembri et al. 2003b; Beloin et al. 2004; Ren et al. 2004a), which appear to be key regulators of several genes involved in catabolite repression and could participate in the negative effect of glucose on *E. coli* biofilm formation (Domka et al. 2006). Mutations in these two genes increased biofilm formation only in the presence of glucose and could possibly reduce both phosphorylation and transport of glucose (Domka et al. 2006). BssS (YceP) and BssR (YliH) could notably repress biofilm formation via different systems modulated by glucose, implicating regulators such as RpoS, CRP, CreC, and CsrA (Domka et al. 2006). These two genes were also recently identified as biofilm-induced when two asymptomatic uropathogenic *E. coli* strains (83972 and VR50) were grown in urine (Hancock and Klemm 2006). As opposed to results by Domka and co-workers (2006), the mutation of *yceP* in this study led to decreased biofilm formation, as previously observed in a study on biofilm formation of the F plasmid-bearing strain TG1 (Beloin et al. 2004). Therefore, these two genes, and especially *yceP*, appear to play differential roles in biofilm formation depending on the growth conditions and strains used.

The CsrA protein has been extensively studied in recent years. This protein has been shown to repress biofilm formation (Jackson et al. 2002b). Until recently, CsrA was thought to affect biofilm formation only through repression of glycogen metabolism and its regulatory effect on the swimming-motility master regulator *flhDC* (Wei et al. 2001). Expression of *csrA* is indeed sharply decreased a few hours after initiation of growth on surfaces, a profile that is compatible with the decrease in flagellar gene expression upon attachment. On the other hand, *csrA* expression is reactivated after maturation of the biofilm (2-day-old biofilm). An increase in *csrA* expression in mature *E. coli* biofilm might also lead to resumption of swimming motility. Therefore, there may be a link between increased flagellar gene expression and biofilm detachment after reinitiation of swimming motility, a hypothesis found in early work on motility and biofilm (Pratt and Kolter 1998). In line with this, Jackson and co-workers showed that overexpression of *csrA* was responsible for biofilm dispersal (Jackson et al. 2002b). Lately, Wang and co-workers have shown that the biofilm effect exerted by CsrA is, in fact, results essentially from its effect on production of the polysaccharide adhesin PGA (*pgaABCD*), with a deletion of *csrA* having no effect on biofilm formation in a *pgaC* mutant (Wang et al. 2004, 2005). CsrA directly affects PGA production at a posttranscriptional level and may indirectly affect *pgaABCD* expression via its effect upon an as yet unidentified regulator of the *pgaABCD* operon (Wang et al. 2005). Two untranslated RNAs, CsrB and CsrC, antagonize CsrA activity by sequestering this protein; consequently, deletion of either *csrB* or *csrC* represses biofilm formation (Wang et al. 2005). The Csr (carbon storage regulatory) system also involves the UvrY protein, the cognate regulator of the two-component system BarA/UvrY (Suzuki et al. 2002; Sahu et al. 2003). Whereas CsrA is necessary for UvrY activity, UvrY in turn activates *csrB* expression, thus implementing expression of the CsrA/CsrB/CsrC negative regulatory loop (Suzuki et al. 2002). Mutation of either *barA* or *uvrY* attenuates biofilm formation, suggesting that BarA and UvrY are necessary for development of a biofilm (Suzuki et al. 2002; Sahu et al. 2003). Besides the high number of genes regulated by UvrY (Oshima et al. 2002), the effect of UvrY on biofilm formation is directly linked to its role on CsrB/CsrC and disappears in a mutant that is unable to synthesize PGA (Wang et al. 2005). Recently, a new activator of *pgaABCD*, and thus biofilm formation, was discovered: the LysR-type-positive regulator NhaR that seems to activate this operon specifically in response to increased Na+ concentration or pH (Goller et al. 2006).

4.3.3 Quorum-Sensing Molecules Regulate *E. coli* Biofilm Formation

4.3.3.1 SdiA, a Homoserine Lactone, Activates UvrY and Biofilm Formation

While *E. coli* is not known to synthesize *N*-acylated homoserine lactones (AHL) (Ahmer 2004) and has no apparent AHL synthase in its genome, it contains the *sdiA* gene that encodes a protein of the LuxR family. LuxR proteins possess one domain for binding *N*-acylated homoserine lactones and a second domain for binding DNA. An *sdiA* mutant has been shown to produce threefold less biofilm than a wild type *E. coli* strain (Suzuki et al. 2002). This effect appears to be mediated by SdiA activation of the *uvrY* gene (Suzuki et al. 2002), and consequently, predominantly by the CsrB/CsrC untranslated RNA effect on CsrA. While the environmental signal that permits SdiA of *E. coli* to regulate *uvrY* expression remains to be determined, a study by van Houdt and co-workers showed that, at 30°C, *E. coli* responds in an SdiA-dependent manner to the addition of AHL by modification of expression of 15 genes, including upregulation of *uvrY* (Van Houdt et al. 2006). Also consistent with these results is the increased expression of *sdiA* in mutants of either *yceP* or *yliH*, whose deletion caused an increase in *E. coli* biofilm formation (Domka et al. 2006). This leaves us with the possibility that the biofilm formation abilities of *E. coli* can potentially be modulated by quorum-sensing AHL signaling molecules from other species, eventually interacting with *E. coli* in natural environments.

4.3.3.2   The AI-2 Signaling Molecule Modulates *E. coli* Biofilm Formation

E. coli strains do secrete the autoinducer-2 (AI-2) quorum-signaling molecule that is encoded by genes of the *luxS* family and that has been regarded as a universal cell-cell communication signal (Xavier and Bassler 2003). An *lsr*-like transporter system has been described recently in *E. coli* K-12 and this could serve to internalize AI-2 molecules (Xavier and Bassler 2005). Nor does disruption of the AI-2 signaling system of *E. coli* appear to modify biofilm maturation mediated by derepressed IncF plasmids in a flow-cell system (Reisner et al. 2003). A *luxS* mutation either did not affect or only moderately reduced the initial adhesion steps in biofilm formation on a microtiter plate (Colon-Gonzalez et al. 2004; Gonzalez Barrios et al. 2006). However, a furanone-based molecule that inhibits the *E. coli* AI-2 signaling system has been shown to decrease the thickness of *E. coli* biofilm formed on steel coupons or on air-liquid interfaces, and to increase the percentage of dead cells within the same biofilm (Ren et al. 2001, 2004b). Consistently, Gonzalez Barrios and co-workers showed that the addition of AI-2 activated *E. coli* biofilm formation through a complex regulatory cascade where MqsR (motility quorum-sensing regulator), encoded by a biofilm-induced gene (Ren et al. 2004a), induced the flagellar operon activator two-component system QseBC (Sperandio et al. 2002) that in turn activated *E. coli* swimming motility (Gonzalez Barrios et al. 2006). In the presence of glucose, YdgG, another protein encoded by a biofilm-induced gene (Ren et al. 2004a) and renamed TqsA (transport quorum-sensing A), may participate in this regulatory cascade by exporting AI-2 molecules outside the cells, as well as the two genes *bssS* (*yceP*) and *bssR* (*yliH*) that are implicated in catabolite repression of the *lsr* operon that imports AI-2 into the cells (Domka et al. 2006). Indeed, a deletion of *ydgG* leads, in LB + glucose, to increased biofilm formation and, in different media, to an increase in intracellular levels of AI-2 (Herzberg et al. 2006). Furthermore, YdgG was found to repress cell surface determinants (genes related to flagellum, type 1 fimbriae, Ag43, curli, and polysaccharide production), as well as 10 genes newly recognized as important for *E. coli* biofilm formation (*yfjR, bioF, yccW, yjbE, yceO, ttdA, fumB, yjiP, gutQ,* and *yihR*), and it appears to control these genes through AI-2 transport (Herzberg et al. 2006).

4.3.3.3 Indole

Indole production is a phenotypic trait displayed by several Gram-negative bacteria including *E. coli*. Indole is produced by the degradation of tryptophane, a reaction performed by tryptophanase encoded by the *tnaA* gene (Newton and Snell 1964). Indole has been described as a potential extracellular signal (Wang et al. 2001). Genes necessary for indole production (including *tnaA*) have been shown to be induced by addition of *E. coli* stationary-phase supernatant (Ren et al. 2004c), suggesting the existence of complex cross-talk between different extracellular signaling pathways. A mutant of *E. coli* K-12 S17-1 for gene *tnaA* is unable to produce a biofilm in 96-well polystyrene microtiter plates in LB medium (Di Martino et al. 2002). Addition of exogenous indole has no effect on biofilm formation of S17-1 itself (as shown also for *E. coli* K-12 MG1655 strains; Bianco et al. 2006), but restores normal biofilm formation in S17-1 *tnaA* (Di Martino et al. 2003). Moreover, whereas oxindolyl-L-alanine, a specific inhibitor of tryptophanase, has no effect on biofilm development of *Klebsiella pneumoniae*, an indole nonproducing species, it has a dose-dependent inhibitory effect on biofilm development of S17-1 and also of other indole-producing species such as urinary isolates of *E. coli, Klebsiella oxytoca, Citrobacter koseri, Providencia stuartii,* and *Morganella morganii* grown in LB or synthetic urine (Di Martino et al. 2003). A link between indole and AI-2 pathways has been recently pointed out by Herzberg and co-workers and Domka and co-workers (Domka et al. 2006; Herzberg et al. 2006). In LB + glucose, a mutation of *tqsA* (*ydgG*) that exports AI-2 seems to activate the expression of the AcrEF multidrug efflux pump that exports indole outside the cells (Herzberg et al. 2006). Moreover, in LB + glucose and not in LB, mutation of *bssS* (*yceP*) or *bssR* (*yliH*), which also regulates AI-2 concentration, strongly reduces both intra- and extracellular concentrations of indole in *E. coli* K-12 BW25113, and at the same time is responsible for induction of expression of *acrE* and *acrF* and for repression of *mtr*, encoding pumps that both export and import indole (Domka et al. 2006). These authors conclude that *bssR* (*yliH*) and *bssS* (*yceP*) mutants increase biofilm formation by repressing indole concentrations through a catabolite repression-related process; they infer that indole represses *E. coli* biofilm formation, a conclusion that appears to be in conflict with results of Di Martino and co-workers (Di Martino et al. 2002, 2003). Given the differences in the media (LB or synthetic urine versus LB + glucose) and strains used (S17-1 and other Gram-negative bacteria vs BW25113), as well as the types of mutants analyzed (*tnaA* vs *bssR* and *bssS*), the role of indole in biofilm development appears to depend considerably on the conditions used (as does the role of *yceP* and *yliH*), and therefore remains to be elucidated.

4.3.3.4 *O*-Acetyl-l-Serine, Another Extracellular Signal, Regulates *E. coli* Biofilm Formation

Another diffusible molecule, *O*-acetyl-l-serine (OAS), appears to modulate *E. coli* biofilm formation. A mutation in the gene coding for a serine acetyltransferase *cysE*, which catalyzes the conversion of serine to *O*-acetyl-l-serine, was shown to enhance biofilm formation through reduction of the amount of an extracellular signal molecule. The authors suggest that OAS or other cysteine metabolites may play a physiological role, possibly by activating genes whose expression leads to inhibition of biofilm formation (Sturgill et al. 2004).

4.3.4 Regulation of Biofilm and Virulence in *E. coli*

As stated earlier in this chapter, another regulator, the virulence activator RfaH, has recently been linked to biofilm formation in *E. coli* (Beloin et al. 2006). A mutation in *rfaH* was shown to derepress biofilm formation in several *E. coli* strains, including uropathogenic strain 536. Since expression of several *E. coli* virulence-associated genes depends on RfaH, the increased biofilm phenotype of the nonvirulent *rfaH* mutant of strain 536 (Nagy et al. 2002) indicates that RfaH-dependent biofilm formation and virulence gene expression are mutually exclusive processes and that biofilm formation may not be regarded as a virulence trait per se. This idea is currently reinforced by other studies also showing inverse regulation of virulence and biofilm-promoting factors in bacteria such as *P. aeruginosa*, *Xanthomonas campestris*, and *Bordetella bronchiseptica* (Dow et al. 2003; Goodman et al. 2004; Irie et al. 2004; Kuchma et al. 2005). Recent data consistently indicate that biofilm of uropathogenic *E. coli* must be formed at the right place under appropriate conditions, and that this may also promote virulence under certain growth conditions (Anderson et al. 2003; Justice et al. 2004). These results suggest that biofilm formation definitely plays a role in the persistence of bacteria rather than being directly implicated in the infective mechanism itself. However, biofilms constitute reservoirs of bacteria that are potentially virulent, and the switch from biofilm to virulence and vice-versa might well be controlled by several regulators such as RfaH in *E. coli*.

4.3.5 Influence of Other Global Regulators on *E. coli* Biofilm Formation

Two regulators, H-NS and RpoS, associated with responses to environmental conditions, also play a role in modulating biofilm formation. H-NS is a nucleoid-associated protein that has been shown to regulate a large number of genes in *E. coli* (approximately 5% of the *E. coli* K-12 genome), including numerous cell envelope components such as flagella, type 1 fimbriae, LPS, and colanic acid, most of them linked to environmental stimuli including pH, oxygen, temperature, and osmolarity (Dorman and Bhriain 1992; Sledjeski and Gottesman 1995; Olsen et al.

1998; Soutourina et al. 1999; Hommais et al. 2001; Soutourina and Bertin 2003; Dorman 2004). The pleiotropic nature of the H-NS effect within the cells, as well as the fact that a mutation in the *hns* gene results in a reduction in the growth rate (Barth et al. 1995), is hardly compatible with a clear definition of its role in *E. coli* biofilm formation. Nevertheless, H-NS appears to be necessary for *E. coli* to attach to sand columns when it is grown under oxygen-limited conditions (Landini and Zehnder 2002). H-NS often interferes with the expression of genes that depend on the RpoS sigma factor. This interference occurs both by competing with RpoS for binding to the promoter of these genes and by indirectly repressing *rpoS* translation and stimulating RpoS turnover (Hengge-Aronis 1996, 2002). Whereas *rpoS* expression in *E. coli* seems unchanged between cells grown as a planktonic culture in a chemostat and as a biofilm (Adams and McLean 1999; Schembri et al. 2003b; Beloin et al. 2004), the role of RpoS in biofilm formation remains controversial. Depending on the experimental setup, an *rpoS* mutation has different effects on *E. coli* biofilm development, ranging from a strong negative effect to a positive effect (Adams and McLean 1999; Corona-Izquierdo and Membrillo-Hernandez 2002; Jackson et al. 2002b; Schembri et al. 2003b). These results pinpoint the difficulty of comparing studies performed using different experimental protocols, and consequently definitive conclusions cannot be drawn concerning the role of RpoS in the formation of *E. coli* biofilms.

5 Conclusions

The study of bacterial surface colonization and biofilm formation represents a rapidly expanding field of investigation, and we are only now beginning to fathom the complexity of how bacteria adjust their lifestyle when confronted with surface contact and community growth.

The amenability of *E. coli* to genetic analyses makes it a valuable biofilm experimental model applicable to a wide spectrum of molecular biology approaches, from classical to genome-wide genetic analyses. Indeed, the growth of *E. coli* under different static and dynamic biofilm experimental set-ups, combined with advanced microscopic analyses, have led not only to identification of a large repertoire of adhesins, but also to the uncovering of complex interplays between regulatory networks involved in biofilm lifestyle. Moreover, the progressive switch from the reliable and convenient laboratory workhorse *E. coli* K-12 to wild commensal and pathogenic isolates has revealed unsuspected physiological capacities and unexplored biological resources that can be used by *E. coli* species to operate within a biofilm.

Therefore, in addition to the molecular details of its own biofilm formation, the use of the most extensively studied living organism thus far, *E. coli*, is likely to represent a coherent choice when venturing into the complex area of experimental molecular multispecies microbiology. Indeed, although bacterial flora exist primarily as multicellular biofilm communities, the absence of an appropriate approach has hampered the investigation of these issues in a complex mixed-biofilm context.

We believe that some fundamental questions of the molecular intricacies of commensal-pathogen interactions may also be efficiently addressed in model *E. coli* mixed consortia. Hence, beyond their fundamental, clinical, and ecological relevance, *E. coli* species have probably not yet revealed all of their secrets, and might significantly contribute to bacterial biofilm research in the near future.

Acknowledgements We thank Sandra Da Re, Jaione Valle and Benjamin Le Quéré for helpful suggestions and critical reading of the manuscript. A.R. is supported by the Ministère de l'Education Nationale, de l'Enseignement Supérieur et de la Recherche (MENESR), and by the Fondation pour la Recherche Médicale (FRM); C.B. and J.M.G. are supported by grants from the Institut Pasteur, the CNRS URA 2172, the Network of Excellence EuroPathoGenomics; LSHB-CT-2005-512061 and the Fondation BNP PARIBAS.

References

Aberg A, Shingler V, Balsalobre C (2006) (p)ppGpp regulates type 1 fimbriation of *Escherichia coli* by modulating the expression of the site-specific recombinase FimB. Mol Microbiol 60:1520-1533

Abraham JM, Freitag CS, Clements JR, Eisenstein BI (1985) An invertible element of DNA controls phase variation of type 1 fimbriae of *Escherichia coli*. Proc Natl Acad Sci U S A 82:5724-5727

Adams JL, McLean RJ (1999) Impact of *rpoS* deletion on *Escherichia coli* biofilms. Appl Environ Microbiol 65:4285-4287

Agladze K, Wang X, Romeo T (2005) Spatial periodicity of *Escherichia coli* K-12 biofilm microstructure initiates during a reversible, polar attachment phase of development and requires the polysaccharide adhesin PGA. J Bacteriol 187:8237-8246

Ahmer BM (2004) Cell-to-cell signalling in *Escherichia coli* and *Salmonella enterica*. Mol Microbiol 52:933-945

Aldea M, Hernandez-Chico C, de la Campa AG, Kushner SR, Vicente M (1988) Identification, cloning, and expression of *bolA*, an *fts*Z-dependent morphogene of *Escherichia coli*. J Bacteriol 170:5169-5176

Amabile-Cuevas CF, Chicurel ME (1996) A possible role for plasmids in mediating the cell-cell proximity required for gene flux. J Theor Biol 181:237-243

Amikam D, Benziman M (1989) Cyclic diguanylic acid and cellulose synthesis in *Agrobacterium tumefaciens*. J Bacteriol 171:6649-6655

Amikam D, Galperin MY (2006) PilZ domain is part of the bacterial c-di-GMP binding protein. Bioinformatics 22:3-6

Anderson GG, Palermo JJ, Schilling JD, Roth R, Heuser J, Hultgren SJ (2003) Intracellular bacterial biofilm-like pods in urinary tract infections. Science 301:105-107

Balzer GJ, McLean RJ (2002) The stringent response genes *relA* and *spoT* are important for *Escherichia coli* biofilms under slow-growth conditions. Can J Microbiol 48:675-680

Barnhart MM, Lynem J, Chapman MR (2006) GlcNAc-6P levels modulate the expression of Curli fibers by *Escherichia coli*. J Bacteriol 188:5212-5219

Barrios AF, Zuo R, Ren D, Wood TK (2006) Hha, YbaJ, and OmpA regulate *Escherichia coli* K12 biofilm formation and conjugation plasmids abolish motility. Biotechnol Bioeng 93:188-200

Barth M, Marschall C, Muffler A, Fischer D, Hengge-Aronis R (1995) Role for the histone-like protein H-NS in growth phase-dependent and osmotic regulation of sigma S and many sigma S-dependent genes in *Escherichia coli*. J Bacteriol 177:3455-3464

Beloin C, Ghigo JM (2005) Finding gene-expression patterns in bacterial biofilms. Trends Microbiol 13:16-19

Beloin C, Michaelis K, Lindner K, Landini P, Hacker J, Ghigo JM, Dobrindt U (2006) The transcriptional antiterminator RfaH represses biofilm formation in *Escherichia coli*. J Bacteriol 188:1316-1331

Beloin C, Valle J, Latour-Lambert P, Faure P, Kzreminski M, Balestrino D, Haagensen JA, Molin S, Prensier G, Arbeille B, Ghigo JM (2004) Global impact of mature biofilm lifestyle on *Escherichia coli* K-12 gene expression. Mol Microbiol 51:659-674

Ben Nasr A, Olsen A, Sjobring U, Muller-Esterl W, Bjorck L (1996) Assembly of human contact phase proteins and release of bradykinin at the surface of curli-expressing *Escherichia coli*. Mol Microbiol 20:927-935

Bianco C, Imperlini E, Calogero R, Senatore B, Amoresano A, Carpentieri A, Pucci P, Defez R (2006) Indole-3-acetic acid improves *Escherichia coli*'s defences to stress. Arch Microbiol 185:373-382

Blomfield IC, Kulasekara DH, Eisenstein BI (1997) Integration host factor stimulates both FimB- and FimE-mediated site-specific DNA inversion that controls phase variation of type 1 fimbriae expression in *Escherichia coli*. Mol Microbiol 23:705-717

Blumer C, Kleefeld A, Lehnen D, Heintz M, Dobrindt U, Nagy G, Michaelis K, Emody L, Polen T, Rachel R, Wendisch VF, Unden G (2005) Regulation of type 1 fimbriae synthesis and bio-film formation by the transcriptional regulator LrhA of *Escherichia coli*. Microbiology 151:3287-3298

Bollinger RR, Everett ML, Palestrant D, Love SD, Lin SS, Parker W (2003) Human secretory immunoglobulin A may contribute to biofilm formation in the gut. Immunology 109:580-587

Bollinger RR, Everett ML, Wahl SD, Lee YH, Orndorff PE, Parker W (2006) Secretory IgA and mucin-mediated biofilm formation by environmental strains of *Escherichia coli*: role of type 1 pili. Mol Immunol 43:378-387

Branda SS, Vik S, Friedman L, Kolter R (2005) Biofilms: the matrix revisited. Trends Microbiol 13:20-26

Brombacher E, Baratto A, Dorel C, Landini P (2006) Gene expression regulation by the Curli activator CsgD protein: modulation of cellulose biosynthesis and control of negative determinants for microbial adhesion. J Bacteriol 188:2027-2037

Brombacher E, Dorel C, Zehnder AJ, Landini P (2003) The curli biosynthesis regulator CsgD co-ordinates the expression of both positive and negative determinants for biofilm formation in *Escherichia coli*. Microbiology 149:2847-2857

Bryan A, Roesch P, Davis L, Moritz R, Pellett S, Welch RA (2006) Regulation of type 1 fimbriae by unlinked FimB- and FimE-like recombinases in uropathogenic *Escherichia coli* strain CFT073. Infect Immun 74:1072-1083

Buckles EL, Bahrani-Mougeot FK, Molina A, Lockatell CV, Johnson DE, Drachenberg CB, Burland V, Blattner FR, Donnenberg MS (2004) Identification and characterization of a novel uropathogenic *Escherichia coli*-associated fimbrial gene cluster. Infect Immun 72:3890-3901

Colon-Gonzalez M, Mendez-Ortiz MM, Membrillo-Hernandez J (2004) Anaerobic growth does not support biofilm formation in *Escherichia coli* K-12. Res Microbiol 155:514-521

Cookson AL, Cooley WA, Woodward MJ (2002) The role of type 1 and curli fimbriae of Shiga toxin-producing *Escherichia coli* in adherence to abiotic surfaces. Int J Med Microbiol 292:195-205

Corona-Izquierdo FP, Membrillo-Hernandez J (2002) A mutation in *rpoS* enhances biofilm forma-tion in *Escherichia coli* during exponential phase of growth. FEMS Microbiol Lett 211:105-110

Costerton JW, Cheng KJ, Geesey GG, Ladd TI, Nickel JC, Dasgupta M, Marrie TJ (1987) Bacterial biofilms in nature and disease. Annu Rev Microbiol 41:435-464

Costerton JW, Lewandowski Z, Caldwell DE, Korber DR, Lappin-Scott HM (1995) Microbial biofilms. Annu Rev Microbiol 49:711-745

Cucarella C, Solano C, Valle J, Amorena B, Lasa I, Penades JR (2001) Bap, a *Staphylococcus aureus* surface protein involved in biofilm formation. J Bacteriol 183:2888-2896

Czaja W, Krystynowicz A, Bielecki S, Brown RM Jr (2006) Microbial cellulose - the natural power to heal wounds. Biomaterials 27:145-151

Da Re S, Ghigo JM (2006) A CsgD-independent pathway for cellulose production and biofilm formation in *Escherichia coli*. J Bacteriol 188:3073-3087

Danese PN, Pratt LA, Dove SL, Kolter R (2000a) The outer membrane protein, antigen 43, mediates cell-to-cell interactions within *Escherichia coli* biofilms. Mol Microbiol 37:424-432

Danese PN, Pratt LA, Kolter R (2000b) Exopolysaccharide production is required for development of *Escherichia coli* K-12 biofilm architecture. J Bacteriol 182:3593-3596

De Wulf P, McGuire AM, Liu X, Lin EC (2002) Genome-wide profiling of promoter recognition by the two-component response regulator CpxR-P in *Escherichia coli*. J Biol Chem 277:26652-26661

Di Martino P, Merieau A, Phillips R, Orange N, Hulen C (2002) Isolation of an *Escherichia coil* strain mutant unable to form biofilm on polystyrene and to adhere to human pneumocyte cells: involvement of tryptophanase. Can J Microbiol 48:132-137

Di Martino PD, Fursy R, Bret L, Sundararaju B, Phillips RS (2003) Indole can act as an extracellular signal to regulate biofilm formation of *Escherichia coli* and other indole-producing bacteria. Can J Microbiol 49:443-449

Diderichsen B (1980) *flu*, a metastable gene controlling surface properties of *Escherichia coli*. J Bacteriol 141:858-867

Dionisio F, Matic I, Radman M, Rodrigues OR, Taddei F (2002) Plasmids spread very fast in heterogeneous bacterial communities. Genetics 162:1525-1532

Dobrindt U, Hentschel U, Kaper JB, Hacker J (2002) Genome plasticity in pathogenic and nonpathogenic enterobacteria. Curr Top Microbiol Immunol 264:157-175

Domka J, Lee J, Wood TK (2006) YliH (BssR) and YceP (BssS) regulate *Escherichia coli* K-12 biofilm formation by influencing cell signaling. Appl Environ Microbiol 72:2449-2459

Donlan RM (2002) Biofilms: microbial life on surfaces. Emerg Infect Dis 8:881-890

Dorel C, Lejeune P, Rodrigue A (2006) The Cpx system of *Escherichia coli*, a strategic signaling pathway for confronting adverse conditions and for settling biofilm communities? Res Microbiol 157:306-314

Dorel C, Vidal O, Prigent-Combaret C, Vallet I, Lejeune P (1999) Involvement of the Cpx signal transduction pathway of *E. coli* in biofilm formation. FEMS Microbiol Lett 178:169-175

Dorman CJ (2004) H-NS: a universal regulator for a dynamic genome. Nat Rev Microbiol 2:391-400

Dorman CJ, Bhriain NN (1992) Thermal regulation of *fimA*, the *Escherichia coli* gene coding for the type 1 fimbrial subunit protein. FEMS Microbiol Lett 78:125-130

Dorman CJ, Higgins CF (1987) Fimbrial phase variation in *Escherichia coli*: dependence on integration host factor and homologies with other site-specific recombinases. J Bacteriol 169:3840-3843

Dow JM, Crossman L, Findlay K, He YQ, Feng JX, Tang JL (2003) Biofilm dispersal in *Xanthomonas campestris* is controlled by cell-cell signaling and is required for full virulence to plants. Proc Natl Acad Sci U S A 100:10995-11000

Dudley EG, Abe C, Ghigo JM, Latour-Lambert P, Hormazabal JC, Nataro JP (2006) An IncI1 plasmid contributes to the adherence of the atypical enteroaggregative *Escherichia coli* strain C1096 to cultured cells and abiotic surfaces. Infect Immun 74:2102-2114

Duncan MJ, Mann EL, Cohen MS, Ofek I, Sharon N, Abraham SN (2005) The distinct binding specificities exhibited by enterobacterial type 1 fimbriae are determined by their fimbrial shafts. J Biol Chem 280:37707-37716

Dunne WM Jr (2002) Bacterial adhesion: seen any good biofilms lately? Clin Microbiol Rev 15:155-166

Economou A, Christie PJ, Fernandez RC, Palmer T, Plano GV, Pugsley AP (2006) Secretion by numbers: Protein traffic in prokaryotes. Mol Microbiol 62:308-319

Eisenstein BI, Sweet DS, Vaughn V, Friedman DI (1987) Integration host factor is required for the DNA inversion that controls phase variation in *Escherichia coli*. Proc Natl Acad Sci U S A 84:6506-6510

Ferrieres L, Clarke DJ (2003) The RcsC sensor kinase is required for normal biofilm formation in *Escherichia coli* K-12 and controls the expression of a regulon in response to growth on a solid surface. Mol Microbiol 50:1665-1682

Fletcher M (1988) Attachment of *Pseudomonas fluorescens* to glass and influence of electrolytes on bacterium-substratum separation distance. J Bacteriol 170:2027-2030

Francez-Charlot A, Laugel B, Van Gemert A, Dubarry N, Wiorowski F, Castanie-Cornet MP, Gutierrez C, Cam K (2003) RcsCDB His-Asp phosphorelay system negatively regulates the *flhDC* operon in *Escherichia coli*. Mol Microbiol 49:823-832

Fredericks CE, Shibata S, Aizawa S, Reimann SA, Wolfe AJ (2006) Acetyl phosphate-sensitive regulation of flagellar biogenesis and capsular biosynthesis depends on the Rcs phosphorelay. Mol Microbiol 61:734-747

Freire P, Vieira HL, Furtado AR, de Pedro MA, Arraiano CM (2006) Effect of the morphogene *bolA* on the permeability of the *Escherichia coli* outer membrane. FEMS Microbiol Lett 260:106-111

Frost LS, Ippen-Ihler K, Skurray RA (1994) Analysis of the sequence and gene products of the transfer region of the F sex factor. Microbiol Rev 58:162-210

Gally DL, Leathart J, Blomfield IC (1996) Interaction of FimB, FimE with the fim switch that controls the phase variation of type 1 fimbriae in *Escherichia coli* K-12. Mol Microbiol 21:725-738

Geesey GG (2001) Bacterial behavior at surfaces. Curr Opin Microbiol 4:296-300

Genevaux P, Bauda P, DuBow MS, Oudega B (1999) Identification of Tn10 insertions in the *rfaG*, *rfaP*, and *galU* genes involved in lipopolysaccharide core biosynthesis that affect *Escherichia coli* adhesion. Arch Microbiol 172:1-8

Genevaux P, Muller S, Bauda P (1996) A rapid screening procedure to identify mini-Tn10 insertion mutants of *Escherichia coli* K-12 with altered adhesion properties. FEMS Microbiol Lett 142:27-30

Gerstel U, Park C, Romling U (2003) Complex regulation of *csgD* promoter activity by global regulatory proteins. Mol Microbiol 49:639-654

Ghannoum M, O'Toole GA (eds) (2001) Microbial biofilms. ASM Press, Washington DC

Ghigo JM (2001) Natural conjugative plasmids induce bacterial biofilm development. Nature 412:442-445

Goller C, Wang X, Itoh Y, Romeo T (2006) The cation-responsive protein NhaR of *Escherichia coli* activates *pgaABCD* transcription, required for production of the biofilm adhesin poly-beta-1,6-N-acetyl-D-glucosamine. J Bacteriol 188:8022-8032

Gonzalez Barrios AF, Zuo R, Hashimoto Y, Yang L, Bentley WE, Wood TK (2006) Autoinducer 2 controls biofilm formation in *Escherichia coli* through a novel motility quorum-sensing regulator (MqsR, B3022). J Bacteriol 188:305-316

Goodman AL, Kulasekara B, Rietsch A, Boyd D, Smith RS, Lory S (2004) A signaling network reciprocally regulates genes associated with acute infection and chronic persistence in *Pseudomonas aeruginosa*. Dev Cell 7:745-754

Gotz F (2002) *Staphylococcus* and biofilms. Mol Microbiol 43:1367-1378

Haagmans W, van der Woude M (2000) Phase variation of Ag43 in *Escherichia coli*: dam-dependent methylation abrogates OxyR binding and OxyR-mediated repression of transcription. Mol Microbiol 35:877-887

Hancock V, Klemm P (2006) Global gene expression profiling of asymptomatic bacteriuria *Escherichia coli* during biofilm growth in human urine. Infect Immun 75:966-976

Hanna A, Berg M, Stout V, Razatos A (2003) Role of capsular colanic acid in adhesion of uropathogenic *Escherichia coli*. Appl Environ Microbiol 69:4474-4481

Harris SL, Elliott DA, Blake MC, Must LM, Messenger M, Orndorff PE (1990) Isolation and characterization of mutants with lesions affecting pellicle formation and erythrocyte agglutination by type 1 piliated *Escherichia coli*. J Bacteriol 172:6411-6418

Harshey RM, Toguchi A (1996) Spinning tails: homologies among bacterial flagellar systems. Trends Microbiol 4:226-231

Hasman H, Chakraborty T, Klemm P (1999) Antigen-43-mediated autoaggregation of *Escherichia coli* is blocked by fimbriation. J Bacteriol 181:4834-4841

Hausner M, Wuertz S (1999) High rates of conjugation in bacterial biofilms as determined by quantitative in situ analysis. Appl Environ Microbiol 65:3710-3713

Henderson IR, Meehan M, Owen P (1997a) Antigen 43, a phase-variable bipartite outer membrane protein, determines colony morphology and autoaggregation in *Escherichia coli* K-12. FEMS Microbiol Lett 149:115-120

Henderson IR, Meehan M, Owen P (1997b) A novel regulatory mechanism for a novel phase-variable outer membrane protein of *Escherichia coli*. Adv Exp Med Biol 412:349-355

Henderson IR, Navarro-Garcia F, Desvaux M, Fernandez RC, Ala'Aldeen D (2004) Type V protein secretion pathway: the autotransporter story. Microbiol Mol Biol Rev 68:692-744

Henderson IR, Navarro-Garcia F, Nataro JP (1998) The great escape: structure and function of the autotransporter proteins. Trends Microbiol 6:370-378

Henderson IR, Owen P, Nataro JP (1999) Molecular switches - the ON, OFF of bacterial phase variation. Mol Microbiol 33:919-932

Hengge-Aronis R (1996) Stationary-phase gene regulation. In: Neidhart FC, Curtiss III R, Ingraham JL, Lin ECC, Low KB, Magasanik B, Reznikoff WS, Riley M, Schaechter M, Umbarger HE (eds) *Escherichia coli* and *Salmonella*. Cellular and molecular biology, vol 1. ASM Press, Washington DC, pp 1497-1512

Hengge-Aronis R (2002) Signal transduction and regulatory mechanisms involved in control of the sigma(S) (RpoS) subunit of RNA polymerase. Microbiol Mol Biol Rev 66:373-395

Herzberg M, Kaye IK, Peti W, Wood TK (2006) YdgG (TqsA) controls biofilm formation in *Escherichia coli* K-12 through autoinducer 2 transport. J Bacteriol 188:587-598

Holden NJ, Gally DL (2004) Switches, cross-talk and memory in *Escherichia coli* adherence. J Med Microbiol 53:585-593

Hommais F, Krin E, Laurent-Winter C, Soutourina O, Malpertuy A, Le Caer JP, Danchin A, Bertin P (2001) Large-scale monitoring of pleiotropic regulation of gene expression by the prokaryotic nucleoid-associated protein, H-NS. Mol Microbiol 40:20-36

Huang CT, Xu KD, McFeters GA, Stewart PS (1998) Spatial patterns of alkaline phosphatase expression within bacterial colonies and biofilms in response to phosphate starvation. Appl Environ Microbiol 64:1526-1531

Huang YH, Ferrieres L, Clarke DJ (2006) The role of the Rcs phosphorelay in Enterobacteriaceae. Res Microbiol 157:206-212

Irie Y, Mattoo S, Yuk MH (2004) The Bvg virulence control system regulates biofilm formation in *Bordetella bronchiseptica*. J Bacteriol 186:5692-5698

Itoh Y, Wang X, Hinnebusch BJ, Preston JF 3rd, Romeo T (2005) Depolymerization of beta-1,6-N-acetyl-D-glucosamine disrupts the integrity of diverse bacterial biofilms. J Bacteriol 187:382-387

Jackson DW, Simecka JW, Romeo T (2002a) Catabolite repression of *Escherichia coli* biofilm formation. J Bacteriol 184:3406-3410

Jackson DW, Suzuki K, Oakford L, Simecka JW, Hart ME, Romeo T (2002b) Biofilm formation and dispersal under the influence of the global regulator CsrA of *Escherichia coli*. J Bacteriol 184:290-301

Jenal U, Malone J (2006) Mechanisms of cyclic-di-GMP signaling in bacteria. Annu Rev Genet 40:385-407

Jubelin G, Vianney A, Beloin C, Ghigo JM, Lazzaroni JC, Lejeune P, Dorel C (2005) CpxR/OmpR interplay regulates curli gene expression in response to osmolarity in *Escherichia coli*. J Bacteriol 187:2038-2049

Justice SS, Hung C, Theriot JA, Fletcher DA, Anderson GG, Footer MJ, Hultgren SJ (2004) Differentiation and developmental pathways of uropathogenic *Escherichia coli* in urinary tract pathogenesis. Proc Natl Acad Sci U S A 101:1333-1338

Kaper JB, Nataro JP, Mobley HL (2004) Pathogenic *Escherichia coli*. Nat Rev Microbiol 2:123-140

Kaplan JB, Velliyagounder K, Ragunath C, Rohde H, Mack D, Knobloch JK, Ramasubbu N (2004) Genes involved in the synthesis and degradation of matrix polysaccharide in *Actinobacillus actinomycetemcomitans* and *Actinobacillus pleuropneumoniae* biofilms. J Bacteriol 186:8213-8220

Kikuchi T, Mizunoe Y, Takade A, Naito S, Yoshida S (2005) Curli fibers are required for development of biofilm architecture in *Escherichia coli* K-12 and enhance bacterial adherence to human uroepithelial cells. Microbiol Immunol 49:875-884

Kirillina O, Fetherston JD, Bobrov AG, Abney J, Perry RD (2004) HmsP, a putative phosphodiesterase, and HmsT, a putative diguanylate cyclase, control Hms-dependent biofilm formation in *Yersinia pestis*. Mol Microbiol 54:75-88

Kjaergaard K, Schembri MA, Hasman H, Klemm P (2000a) Antigen 43 from *Escherichia coli* induces inter- and intraspecies cell aggregation and changes in colony morphology of *Pseudomonas fluorescens*. J Bacteriol 182:4789-4796

Kjaergaard K, Schembri MA, Ramos C, Molin S, Klemm P (2000b) Antigen 43 facilitates formation of multispecies biofilms. Environ Microbiol 2:695-702

Klemm P (1986) Two regulatory *fim* genes, *fimB* and *fimE*, control the phase variation of type 1 fimbriae in *Escherichia coli*. EMBO J 5:1389-1393

Klemm P, Vejborg RM, Sherlock O (2006) Self-associating autotransporters, SAATs: functional and structural similarities. Int J Med Microbiol 296:187-195

Ko M, Park C (2000) Two novel flagellar components and H-NS are involved in the motor function of *Escherichia coli*. J Mol Biol 303:371-382

Kuchma SL, Connolly JP, O'Toole GA (2005) A three-component regulatory system regulates biofilm maturation and type III secretion in *Pseudomonas aeruginosa*. J Bacteriol 187:1441-1454

Lacqua A, Wanner O, Colangelo T, Martinotti MG, Landini P (2006) Emergence of biofilm-forming subpopulations upon exposure of *Escherichia coli* to environmental bacteriophages. Appl Environ Microbiol 72:956-959

Landini P, Zehnder AJ (2002) The global regulatory *hns* gene negatively affects adhesion to solid surfaces by anaerobically grown *Escherichia coli* by modulating expression of flagellar genes and lipopolysaccharide production. J Bacteriol 184:1522-1529

Latasa C, Roux A, Toledo-Arana A, Ghigo JM, Gamazo C, Penades JR, Lasa I (2005) BapA, a large secreted protein required for biofilm formation and host colonization of *Salmonella enterica* serovar enteritidis. Mol Microbiol 58:1322-1339

Lebaron P, Bauda P, Lett MC, Duval-Iflah Y, Simonet P, Jacq E, Frank N, Roux B, Baleux B, Faurie G, Hubert JC, Normand P, Prieur D, Schmitt S, Block JC (1997) Recombinant plasmid mobilization between *E. coli* strains in seven sterile microcosms. Can J Microbiol 43:534-540

Lehnen D, Blumer C, Polen T, Wackwitz B, Wendisch VF, Unden G (2002) LrhA as a new transcriptional key regulator of flagella, motility and chemotaxis genes in *Escherichia coli*. Mol Microbiol 45:521-532

Licht TR, Christensen BB, Krogfelt KA, Molin S (1999) Plasmid transfer in the animal intestine and other dynamic bacterial populations: the role of community structure and environment. Microbiology 145:2615-2622

Mack D, Fischer W, Krokotsch A, Leopold K, Hartmann R, Egge H, Laufs R (1996) The intercellular adhesin involved in biofilm accumulation of *Staphylococcus epidermidis* is a linear beta-1, 6-linked glucosaminoglycan: purification and structural analysis. J Bacteriol 178:175-183

Maeda S, Ito M, Ando T, Ishimoto Y, Fujisawa Y, Takahashi H, Matsuda A, Sawamura A, Kato S (2006) Horizontal transfer of nonconjugative plasmids in a colony biofilm of *Escherichia coli*. FEMS Microbiol Lett 255:115-120

Maira-Litran T, Kropec A, Abeygunawardana C, Joyce J, Mark G 3rd, Goldmann DA, Pier GB (2002) Immunochemical properties of the staphylococcal poly-N-acetylglucosamine surface polysaccharide. Infect Immun 70:4433-4440

Majdalani N, Gottesman S (2005) The Rcs phosphorelay: a complex signal transduction system. Annu Rev Microbiol 59:379-405

Mayer R, Ross P, Weinhouse H, Amikam D, Volman G, Ohana P, Calhoon RD, Wong HC, Emerick AW, Benziman M (1991) Polypeptide composition of bacterial cyclic diguanylic acid-dependent cellulose synthase and the occurrence of immunologically crossreacting proteins in higher plants. Proc Natl Acad Sci U S A 88:5472-5476

Molin S, Tolker-Nielsen T (2003) Gene transfer occurs with enhanced efficiency in biofilms and induces enhanced stabilisation of the biofilm structure. Curr Opin Biotechnol 14:255-261

Moreira CG, Carneiro SM, Nataro JP, Trabulsi LR, Elias WP (2003) Role of type I fimbriae in the aggregative adhesion pattern of enteroaggregative *Escherichia coli*. FEMS Microbiol Lett 226:79-85

Nagy G, Dobrindt U, Schneider G, Khan AS, Hacker J, Emody L (2002) Loss of regulatory protein RfaH attenuates virulence of uropathogenic *Escherichia coli*. Infect Immun 70:4406-4413

Newton WA, Snell EE (1964) Catalytic properties of tryptophanase, a multifunctional pyridoxal phosphate enzyme. Proc Natl Acad Sci U S A 51:382-389

Olsen A, Jonsson A, Normark S (1989) Fibronectin binding mediated by a novel class of surface organelles on *Escherichia coli*. Nature 338:652-655

Olsen PB, Schembri MA, Gally DL, Klemm P (1998) Differential temperature modulation by H-NS of the *fimB* and *fimE* recombinase genes which control the orientation of the type 1 fimbrial phase switch. FEMS Microbiol Lett 162:17-23

Oomen CJ, van Ulsen P, van Gelder P, Feijen M, Tommassen J, Gros P (2004) Structure of the translocator domain of a bacterial autotransporter. EMBO J 23:1257-1266

Ophir T, Gutnick D (1994) A role for exopolysaccharides in the protection of microorganisms from desiccation. Appl Environ Microbiol 60:740-745

Orndorff PE, Devapali A, Palestrant S, Wyse A, Everett ML, Bollinger RR, Parker W (2004) Immunoglobulin-mediated agglutination of and biofilm formation by *Escherichia coli* K-12 require the type 1 pilus fiber. Infect Immun 72:1929-1938

Oshima T, Aiba H, Masuda Y, Kanaya S, Sugiura M, Wanner BL, Mori H, Mizuno T (2002) Transcriptome analysis of all two-component regulatory system mutants of *Escherichia coli* K-12. Mol Microbiol 46:281-291

Otto K, Hermansson M (2004) Inactivation of *ompX* causes increased interactions of type 1 fimbriated *Escherichia coli* with abiotic surfaces. J Bacteriol 186:226-234

Otto K, Norbeck J, Larsson T, Karlsson KA, Hermansson M (2001) Adhesion of type 1-fimbriated *Escherichia coli* to abiotic surfaces leads to altered composition of outer membrane proteins. J Bacteriol 183:2445-2453

Otto K, Silhavy TJ (2002) Surface sensing and adhesion of *Escherichia coli* controlled by the Cpx-signaling pathway. Proc Natl Acad Sci U S A 99:2287-2292

Owen P, Meehan M, de Loughry-Doherty H, Henderson I (1996) Phase-variable outer membrane proteins in *Escherichia coli*. FEMS Immunol Med Microbiol 16:63-76

Perna NT, Plunkett G 3rd, Burland V, Mau B, Glasner JD, Rose DJ, Mayhew GF, Evans PS, Gregor J, Kirkpatrick HA, Posfai G, Hackett J, Klink S, Boutin A, Shao Y, Miller L, Grotbeck EJ, Davis NW, Lim A, Dimalanta ET, Potamousis KD, Apodaca J, Anantharaman TS, Lin J, Yen G, Schwartz DC, Welch RA, Blattner FR (2001) Genome sequence of enterohaemorrhagic *Escherichia coli* O157:H7. Nature 409:529-533

Pohlner J, Halter R, Beyreuther K, Meyer TF (1987) Gene structure and extracellular secretion of *Neisseria gonorrhoeae* IgA protease. Nature 325:458-462

Pratt LA, Kolter R (1998) Genetic analysis of *Escherichia coli* biofilm formation: roles of flagella, motility, chemotaxis and type I pili. Mol Microbiol 30:285-293

Pratt LA, Kolter R (1999) Genetic analyses of bacterial biofilm formation. Curr Opin Microbiol 2:598-603

Prigent-Combaret C, Brombacher E, Vidal O, Ambert A, Lejeune P, Landini P, Dorel C (2001) Complex regulatory network controls initial adhesion and biofilm formation in *Escherichia coli* via regulation of the *csgD* gene. J Bacteriol 183:7213-7223

Prigent-Combaret C, Lejeune P (1999) Monitoring gene expression in biofilms. Methods Enzymol 310:56-79

Prigent-Combaret C, Prensier G, Le Thi TT, Vidal O, Lejeune P, Dorel C (2000) Developmental pathway for biofilm formation in curli-producing *Escherichia coli* strains: role of flagella, curli and colanic acid. Environ Microbiol 2:450-464

Prigent-Combaret C, Vidal O, Dorel C, Lejeune P (1999) Abiotic surface sensing and biofilm-dependent regulation of gene expression in *Escherichia coli*. J Bacteriol 181:5993-6002

Probert HM, Gibson GR (2002) Bacterial biofilms in the human gastrointestinal tract. Curr Issues Intest Microbiol 3:23-27

Pruss BM, Wolfe AJ (1994) Regulation of acetyl phosphate synthesis and degradation, and the control of flagellar expression in *Escherichia coli*. Mol Microbiol 12:973-984

Raetz CR (1996) Bacterial lipopolysaccharides: a remarkable family of bioactive macroam-phiphiles. In: Neidhart FC, Curtiss R III, Ingraham JL, Lin ECC, Low KB, Magasanik B, Reznikoff WS, Riley M, Schaechter M, Umbarger HE (eds) *Escherichia coli* and *Salmonella*. Cellular and molecular biology, vol 2. ASM Press, Washington DC, p 69

Raivio TL, Silhavy TJ (2001) Periplasmic stress and ECF sigma factors. Annu Rev Microbiol 55:591-624

Redfield RJ (2002) Is quorum sensing a side effect of diffusion sensing? Trends Microbiol 10:365-370

Reisner A, Haagensen JA, Schembri MA, Zechner EL, Molin S (2003) Development and matura-tion of *Escherichia coli* K-12 biofilms. Mol Microbiol 48:933-946

Reisner A, Höller BM, Molin S, Zechner EL (2006) Synergistic effects in mixed *Escherichia coli* biofilms: conjugative plasmid transfer drives biofilm expansion. J Bacteriol 188:3582-3588

Ren D, Bedzyk LA, Thomas SM, Ye RW, Wood TK (2004a) Gene expression in *Escherichia coli* biofilms. Appl Microbiol Biotechnol 64:515-524

Ren D, Bedzyk LA, Ye RW, Thomas SM, Wood TK (2004b) Differential gene expression shows natural brominated furanones interfere with the autoinducer-2 bacterial signaling system of *Escherichia coli*. Biotechnol Bioeng 88:630-642

Ren D, Bedzyk LA, Ye RW, Thomas SM, Wood TK (2004c) Stationary-phase quorum-sensing signals affect autoinducer-2 and gene expression in *Escherichia coli*. Appl Environ Microbiol 70:2038-2043

Ren D, Sims JJ, Wood TK (2001) Inhibition of biofilm formation and swarming of *Escherichia coli* by (5Z)-4-bromo-5-(bromomethylene)-3-butyl-2(5H)-furanone. Environ Microbiol 3:731-736

Roberts IS (1996) The biochemistry and genetics of capsular polysaccharide production in bacte-ria. Annu Rev Microbiol 50:285-315

Romling U (2002) Molecular biology of cellulose production in bacteria. Res Microbiol 153:205-212

Romling U (2005) Characterization of the rdar morphotype, a multicellular behaviour in Enterobacteriaceae. Cell Mol Life Sci 62:1234-1246

Romling U, Amikam D (2006) Cyclic di-GMP as a second messenger. Curr Opin Microbiol 9:218-228

Romling U, Bokranz W, Rabsch W, Zogaj X, Nimtz M, Tschape H (2003) Occurrence and regula-tion of the multicellular morphotype in *Salmonella* serovars important in human disease. Int J Med Microbiol 293:273-285

Romling U, Gomelsky M, Galperin MY (2005) C-di-GMP: the dawning of a novel bacterial sig-nalling system. Mol Microbiol 57:629-639

Roux A, Beloin C, Ghigo JM (2005) Combined inactivation and expression strategy to study gene function under physiological conditions: application to identification of new *Escherichia coli* adhesins. J Bacteriol 187:1001-1013

Rupp ME, Fey PD, Heilmann C, Gotz F (2001) Characterization of the importance of *Staphylococcus epidermidis* autolysin and polysaccharide intercellular adhesin in the patho-genesis of intravascular catheter-associated infection in a rat model. J Infect Dis 183:1038-1042

Sahu SN, Acharya S, Tuminaro H, Patel I, Dudley K, LeClerc JE, Cebula TA, Mukhopadhyay S (2003) The bacterial adaptive response gene, *barA*, encodes a novel conserved histidine kinase regulatory switch for adaptation and modulation of metabolism in *Escherichia coli*. Mol Cell Biochem 253:167-177

Sailer FC, Meberg BM, Young KD (2003) Beta-lactam induction of colanic acid gene expression in *Escherichia coli*. FEMS Microbiol Lett 226:245-249

Santos JM, Lobo M, Matos AP, De Pedro MA, Arraiano CM (2002) The gene *bolA* regulates *dacA* (PBP5), *dacC* (PBP6) and *ampC* (AmpC), promoting normal morphology in *Escherichia coli*. Mol Microbiol 45:1729-1740

Sauer FG, Mulvey MA, Schilling JD, Martinez JJ, Hultgren SJ (2000) Bacterial pili: molecular mechanisms of pathogenesis. Curr Opin Microbiol 3:65-72

Schembri MA, Dalsgaard D, Klemm P (2004) Capsule shields the function of short bacterial adhesins. J Bacteriol 186:1249-1257

Schembri MA, Hjerrild L, Gjermansen M, Klemm P (2003a) Differential expression of the *Escherichia coli* autoaggregation factor antigen 43. J Bacteriol 185:2236-2242

Schembri MA, Kjaergaard K, Klemm P (2003b) Global gene expression in *Escherichia coli* biofilms. Mol Microbiol 48:253-267

Schembri MA, Klemm P (2001) Coordinate gene regulation by fimbriae-induced signal transduction. EMBO J 20:3074-3081

Schembri MA, Ussery DW, Workman C, Hasman H, Klemm P (2002) DNA microarray analysis of *fim* mutations in *Escherichia coli*. Mol Genet Genomics 267:721-729

Sheikh J, Hicks S, Dall'Agnol M, Phillips AD, Nataro JP (2001) Roles for Fis and YafK in biofilm formation by enteroaggregative *Escherichia coli*. Mol Microbiol 41:983-997

Sherlock O, Dobrindt U, Jensen JB, Munk Vejborg R, Klemm P (2006) Glycosylation of the self-recognizing *Escherichia coli* Ag43 autotransporter protein. J Bacteriol 188:1798-1807

Sherlock O, Schembri MA, Reisner A, Klemm P (2004) Novel roles for the AIDA adhesin from diarrheagenic *Escherichia coli*: cell aggregation and biofilm formation. J Bacteriol 186:8058-8065

Sherlock O, Vejborg RM, Klemm P (2005) The TibA adhesin/invasin from enterotoxigenic *Escherichia coli* is self recognizing and induces bacterial aggregation and biofilm formation. Infect Immun 73:1954-1963

Shin S, Park C (1995) Modulation of flagellar expression in *Escherichia coli* by acetyl phosphate and the osmoregulator OmpR. J Bacteriol 177:4696-4702

Simm R, Fetherston JD, Kader A, Romling U, Perry RD (2005) Phenotypic convergence mediated by GGDEF-domain-containing proteins. J Bacteriol 187:6816-6823

Simm R, Morr M, Kader A, Nimtz M, Romling U (2004) GGDEF, EAL domains inversely regulate cyclic di-GMP levels and transition from sessility to motility. Mol Microbiol 53:1123-1134

Simonsen L (1990) Dynamics of plasmid transfer on surfaces. J Gen Microbiol 136:1001-1007

Sledjeski D, Gottesman S (1995) A small RNA acts as an antisilencer of the H-NS-silenced *rcsA* gene of *Escherichia coli*. Proc Natl Acad Sci U S A 92:2003-2007

Sledjeski DD, Gottesman S (1996) Osmotic shock induction of capsule synthesis in *Escherichia coli* K-12. J Bacteriol 178:1204-1206

Smyth CJ, Marron MB, Twohig JM, Smith SG (1996) Fimbrial adhesins: similarities and variations in structure and biogenesis. FEMS Immunol Med Microbiol 16:127-139

Snyder WB, Davis LJ, Danese PN, Cosma CL, Silhavy TJ (1995) Overproduction of NlpE, a new outer membrane lipoprotein, suppresses the toxicity of periplasmic LacZ by activation of the Cpx signal transduction pathway. J Bacteriol 177:4216-4223

Sohanpal BK, El-Labany S, Lahooti M, Plumbridge JA, Blomfield IC (2004) Integrated regulatory responses of *fimB* to N-acetylneuraminic (sialic) acid and GlcNAc in *Escherichia coli* K-12. Proc Natl Acad Sci U S A 101:16322-16327

Solano C, Garcia B, Valle J, Berasain C, Ghigo JM, Gamazo C, Lasa I (2002) Genetic analysis of *Salmonella enteritidis* biofilm formation: critical role of cellulose. Mol Microbiol 43:793-808

Soutourina O, Kolb A, Krin E, Laurent-Winter C, Rimsky S, Danchin A, Bertin P (1999) Multiple control of flagellum biosynthesis in *Escherichia coli*: role of H-NS protein and the cyclic

AMP-catabolite activator protein complex in transcription of the *flhDC* master operon. J Bacteriol 181:7500-7508

Soutourina OA, Bertin PN (2003) Regulation cascade of flagellar expression in Gram-negative bacteria. FEMS Microbiol Rev 27:505-523

Sperandio V, Torres AG, Kaper JB (2002) Quorum sensing *Escherichia coli* regulators B, C (QseBC): a novel two-component regulatory system involved in the regulation of flagella and motility by quorum sensing in *E. coli*. Mol Microbiol 43:809-821

Starkey M, Gray AK, Chang SI, Parsek M (2004) A sticky business: the extracellular polymeric substance matrix of bacterial biofilms. In: Ghannoum M, O'Toole GA (eds) Microbial biofilms, vol 336. ASM Press, Washington DC, pp 174-191

Stevenson G, Andrianopoulos K, Hobbs M, Reeves PR (1996) Organization of the *Escherichia coli* K-12 gene cluster responsible for production of the extracellular polysaccharide colanic acid. J Bacteriol 178:4885-4893

Sturgill G, Toutain CM, Komperda J, O'Toole GA, Rather PN (2004) Role of CysE in production of an extracellular signaling molecule in *Providencia stuartii* and *Escherichia coli*: loss of CysE enhances biofilm formation in *Escherichia coli*. J Bacteriol 186:7610-7617

Sutherland IW (2001) The biofilm matrix - an immobilized but dynamic microbial environment. Trends Microbiol 9:222-227

Suzuki K, Wang X, Weilbacher T, Pernestig AK, Melefors O, Georgellis D, Babitzke P, Romeo T (2002) Regulatory circuitry of the CsrA/CsrB, BarA/UvrY systems of *Escherichia coli*. J Bacteriol 184:5130-5140

Tenorio E, Saeki T, Fujita K, Kitakawa M, Baba T, Mori H, Isono K (2003) Systematic characterization of *Escherichia coli* genes/ORFs affecting biofilm formation. FEMS Microbiol Lett 225:107-114

Uhlich GA, Cooke PH, Solomon EB (2006) Analyses of the red-dry-rough phenotype of an *Escherichia coli* O157:H7 strain and its role in biofilm formation and resistance to antibacterial agents. Appl Environ Microbiol 72:2564-2572

Ulett GC, Webb RI, Schembri MA (2006) Antigen-43-mediated autoaggregation impairs motility in *Escherichia coli*. Microbiology 152:2101-2110

Valle J, Da Re S, Henry N, Fontaine T, Balestrino D, Latour-Lambert P, Ghigo JM (2006) Broad-spectrum biofilm inhibition by a secreted bacterial polysaccharide. Proc Natl Acad Sci U S A 103:12558-12563

Van Biesen T, Frost LS (1992) Differential levels of fertility inhibition among F-like plasmids are related to the cellular concentration of finO mRNA. Mol Microbiol 6:771-780

Van der Woude MW (2006) Re-examining the role and random nature of phase variation. FEMS Microbiol Lett 254:190-197

Van der Woude MW, Baumler AJ (2004) Phase and antigenic variation in bacteria. Clin Microbiol Rev 17:581-611

Van Houdt R, Aertsen A, Moons P, Vanoirbeek K, Michiels CW (2006) N-acyl-L-homoserine lactone signal interception by *Escherichia coli*. FEMS Microbiol Lett 256:83-89

Van Loosdrecht MC, Lyklema J, Norde W, Zehnder AJ (1990) Influence of interfaces on microbial activity. Microbiol Rev 54:75-87

Vianney A, Jubelin G, Renault S, Dorel C, Lejeune P, Lazzaroni JC (2005) *Escherichia coli tol* and *rcs* genes participate in the complex network affecting curli synthesis. Microbiology 151:2487-2497

Vidal O, Longin R, Prigent-Combaret C, Dorel C, Hooreman M, Lejeune P (1998) Isolation of an *Escherichia coli* K-12 mutant strain able to form biofilms on inert surfaces: involvement of a new *ompR* allele that increases curli expression. J Bacteriol 180:2442-2449

Vieira HL, Freire P, Arraiano CM (2004) Effect of *Escherichia coli* Morphogene *bolA* on Biofilms. Appl Environ Microbiol 70:5682-5684

Voulhoux R, Bos MP, Geurtsen J, Mols M, Tommassen J (2003) Role of a highly conserved bacterial protein in outer membrane protein assembly. Science 299:262-265

Wallecha A, Correnti J, Munster V, van der Woude M (2003) Phase variation of Ag43 is independent of the oxidation state of OxyR. J Bacteriol 185:2203-2209

Wallecha A, Munster V, Correnti J, Chan T, van der Woude M (2002) Dam- and OxyR-dependent phase variation of *agn43*: essential elements and evidence for a new role of DNA methylation. J Bacteriol 184:3338-3347

Wang D, Ding X, Rather PN (2001) Indole can act as an extracellular signal in *Escherichia coli*. J Bacteriol 183:4210-4216

Wang X, Dubey AK, Suzuki K, Baker CS, Babitzke P, Romeo T (2005) CsrA post-transcriptionally represses *pgaABCD*, responsible for synthesis of a biofilm polysaccharide adhesin of *Escherichia coli*. Mol Microbiol 56:1648-1663

Wang X, Preston JF 3rd, Romeo T (2004) The *pgaABCD* locus of *Escherichia coli* promotes the synthesis of a polysaccharide adhesin required for biofilm formation. J Bacteriol 186:2724-2734

Weber H, Pesavento C, Possling A, Tischendorf G, Hengge R (2006) Cyclic-di-GMP-mediated signalling within the sigma network of *Escherichia coli*. Mol Microbiol 62:1014-1034

Wei BL, Brun-Zinkernagel AM, Simecka JW, Pruss BM, Babitzke P, Romeo T (2001) Positive regulation of motility and *flhDC* expression by the RNA-binding protein CsrA of *Escherichia coli*. Mol Microbiol 40:245-256

Welch RA, Burland V, Plunkett G 3rd, Redford P, Roesch P, Rasko D, Buckles EL, Liou SR, Boutin A, Hackett J, Stroud D, Mayhew GF, Rose DJ, Zhou S, Schwartz DC, Perna NT, Mobley HL, Donnenberg MS, Blattner FR (2002) Extensive mosaic structure revealed by the complete genome sequence of uropathogenic *Escherichia coli*. Proc Natl Acad Sci U S A 99:17020-17024

White AP, Gibson DL, Collinson SK, Banser PA, Kay WW (2003) Extracellular polysaccharides associated with thin aggregative fimbriae of *Salmonella enterica* serovar enteritidis. J Bacteriol 185:5398-5407

Whitfield C (2006) Biosynthesis and assembly of capsular polysaccharides in *Escherichia coli*. Annu Rev Biochem 75:39-68

Whitfield C, Roberts IS (1999) Structure, assembly and regulation of expression of capsules in *Escherichia coli*. Mol Microbiol 31:1307-1319

Wolfe AJ, Chang DE, Walker JD, Seitz-Partridge JE, Vidaurri MD, Lange CF, Pruss BM, Henk MC, Larkin JC, Conway T (2003) Evidence that acetyl phosphate functions as a global signal during biofilm development. Mol Microbiol 48:977-988

Wood TK, Gonzalez Barrios AF, Herzberg M, Lee J (2006) Motility influences biofilm architecture in *Escherichia coli*. Appl Microbiol Biotechnol:1-7

Wuertz S, Okabe S, Hausner M (2004) Microbial communities and their interactions in biofilm systems: an overview. Water Sci Technol 49:327-336

Xavier KB, Bassler BL (2003) LuxS quorum sensing: more than just a numbers game. Curr Opin Microbiol 6:191-197

Xavier KB, Bassler BL (2005) Regulation of uptake and processing of the quorum-sensing autoinducer AI-2 in *Escherichia coli*. J Bacteriol 187:238-248

Xie Y, Yao Y, Kolisnychenko V, Teng CH, Kim KS (2006) HbiF regulates type 1 fimbriation independently of FimB, FimE. Infect Immun 74:4039-4047

Yamamoto K, Nagura R, Tanabe H, Fujita N, Ishihama A, Utsumi R (2000) Negative regulation of the bolA1p of *Escherichia coli* K-12 by the transcription factor OmpR for osmolarity response genes. FEMS Microbiol Lett 186:257-262

Yoshioka Y, Ohtsubo H, Ohtsubo E (1987) Repressor gene finO in plasmids R100 and F: constitutive transfer of plasmid F is caused by insertion of IS3 into F finO. J Bacteriol 169:619-623

Zobell CE (1943) The effect of solid surfaces upon bacterial activity. J Bacteriol 46:39-56

Zogaj X, Bokranz W, Nimtz M, Romling U (2003) Production of cellulose and curli fimbriae by members of the family Enterobacteriaceae isolated from the human gastrointestinal tract. Infect Immun 71:4151-4158

Zogaj X, Nimtz M, Rohde M, Bokranz W, Romling U (2001) The multicellular morphotypes of *Salmonella typhimurium* and *Escherichia coli* produce cellulose as the second component of the extracellular matrix. Mol Microbiol 39:1452-1463

Index

Current Topics in Microbiology and Immunology

Volumes published since 1989

Vol. 295: **Sullivan, David J.; Krishna Sanjeew (Eds.):** Malaria: Drugs, Disease and Post-genomic Biology. 2005. 40 figs., XI, 446 pp. ISBN 3-540-25363-7

Vol. 296: **Oldstone, Michael B. A. (Ed.):** Molecular Mimicry: Infection Induced Autoimmune Disease. 2005. 28 figs., VIII, 167 pp. ISBN 3-540-25597-4

Vol. 297: **Langhorne, Jean (Ed.):** Immunology and Immunopathogenesis of Malaria. 2005. 8 figs., XII, 236 pp. ISBN 3-540-25718-7

Vol. 298: **Vivier, Eric; Colonna, Marco (Eds.):** Immunobiology of Natural Killer Cell Receptors. 2005. 27 figs., VIII, 286 pp. ISBN 3-540-26083-8

Vol. 299: **Domingo, Esteban (Ed.):** Quasispecies: Concept and Implications. 2006. 44 figs., XII, 401 pp. ISBN 3-540-26395-0

Vol. 300: **Wiertz, Emmanuel J.H.J.; Kikkert, Marjolein (Eds.):** Dislocation and Degradation of Proteins from the Endoplasmic Reticulum. 2006. 19 figs., VIII, 168 pp. ISBN 3-540-28006-5

Vol. 301: **Doerfler, Walter; Böhm, Petra (Eds.):** DNA Methylation: Basic Mechanisms. 2006. 24 figs., VIII, 324 pp. ISBN 3-540-29114-8

Vol. 302: **Robert N. Eisenman (Ed.):** The Myc/Max/Mad Transcription Factor Network. 2006. 28 figs., XII, 278 pp. ISBN 3-540-23968-5

Vol. 303: **Thomas E. Lane (Ed.):** Chemokines and Viral Infection. 2006. 14 figs. XII, 154 pp. ISBN 3-540-29207-1

Vol. 304: **Stanley A. Plotkin (Ed.):** Mass Vaccination: Global Aspects – Progress and Obstacles. 2006. 40 figs. X, 270 pp. ISBN 3-540-29382-5

Vol. 305: **Radbruch, Andreas; Lipsky, Peter E. (Eds.):** Current Concepts in Autoimmunity. 2006. 29 figs. IIX, 276 pp. ISBN 3-540-29713-8

Vol. 306: **William M. Shafer (Ed.):** Antimicrobial Peptides and Human Disease. 2006. 12 figs. XII, 262 pp. ISBN 3-540-29915-7

Vol. 307: **John L. Casey (Ed.):** Hepatitis Delta Virus. 2006. 22 figs. XII, 228 pp. ISBN 3-540-29801-0

Vol. 308: **Honjo, Tasuku; Melchers, Fritz (Eds.):** Gut-Associated Lymphoid Tissues. 2006. 24 figs. XII, 204 pp. ISBN 3-540-30656-0

Vol. 309: **Polly Roy (Ed.):** Reoviruses: Entry, Assembly and Morphogenesis. 2006. 43 figs. XX, 261 pp. ISBN 3-540-30772-9

Vol. 310: **Doerfler, Walter; Böhm, Petra (Eds.):** DNA Methylation: Development, Genetic Disease and Cancer. 2006. 25 figs. X, 284 pp. ISBN 3-540-31180-7

Vol. 311: **Pulendran, Bali; Ahmed, Rafi (Eds.):** From Innate Immunity to Immunological Memory. 2006. 13 figs. X, 177 pp. ISBN 3-540-32635-9

Vol. 312: **Boshoff, Chris; Weiss, Robin A. (Eds.):** Kaposi Sarcoma Herpesvirus: New Perspectives. 2006. 29 figs. XVI, 330 pp. ISBN 3-540-34343-1

Vol. 313: **Pandolfi, Pier P.; Vogt, Peter K. (Eds.):** Acute Promyelocytic Leukemia. 2007. 16 figs. VIII, 273 pp. ISBN 3-540-34592-2

Vol. 314: **Moody, Branch D. (Ed.):** T Cell Activation by CD1 and Lipid Antigens, 2007, 25 figs. VIII, 348 pp. ISBN 978-3-540-69510-3

Vol. 315: **Childs, James, E.; Mackenzie, John S.; Richt, Jürgen A. (Eds.):** Wildlife and Emerging Zoonotic Diseases: The Biology, Circumstances and Consequences of Cross-Species Transmission. 2007. 49 figs. VII, 524 pp. ISBN 978-3-540-70961-9

Vol. 316: **Pitha, Paula M. (Ed.):** Interferon: The 50th Anniversary. 2007. VII, 391 pp. ISBN 978-3-540-71328-9

Vol. 317: **Dessain, Scott K. (Ed.):** Human Antibody Therapeutics for Viral Disease. 2007. XI, 202 pp. ISBN 978-3-540-72144-4

Vol. 318: **Rodriguez, Moses (Ed.):** Advances in Multiple Sclerosis and Experimental Demyelinating Diseases. 2008. XIV, 376. ISBN 978-3-540-73679-9

Vol. 319: **Manser, Tim (Ed.):** Specialization and Complementation of Humoral Immune Responses to Infection. 2008. XII, 174. ISBN 978-3-540-73899-2

Vol. 320: **Paddison, Patrick J.; Vogt, Peter K. (Eds.):** RNA Interference. 2008. VIII, 273. ISBN 978-3-540-75156-4

Vol. 321: **B. Beutler (Ed.):** Immunology, Phenotype First: How Mutations Have Established New Principles and Pathways in Immunology. 2008. ISBN 978-3-540-75202-8

Vol. 322: **Romeo, Tony (Ed.):** Bacterial Biofilms. 2008. , . ISBN 978-3-540-75417-6

Printed in the United States
148027LV00001B/56/P

9 783540 754176

[Processor is 2500 k
[MOBO Asus P8268 -pro   Combo

Memory — 4x4 gB   Mushkin
Power supply — Corsair entusist 650 8ot
Heat sink - no
HDD— to buy — 80
SSD want ✓ (intel, crucial?)
Windows 7 - ✓
Video card — 460 Fermi EVGA
case — to buy   →134) 160
DUD — ✓ (Asus)

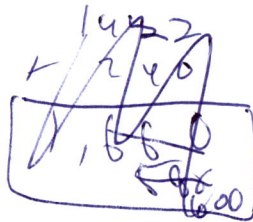

142
✓ 2 40   1506 + 170 HD
1660
460
1600   1576